Cell Phone Investigations

Search Warrants, Cell Sites and Evidence Recovery

Aaron Edens

POLICE PUBLISHING

Terre Haute, Indiana

Dedicated to the law enforcement community

POLICE PUBLISHING books may be purchased for educational or sales promotional use. For information please write: POLICE PUBLISHING, 661 Poplar Street, Terre Haute, IN 47807.

Designed by POLICE PUBLISHING
Manufactured in the United States of America

The Library of Congress Cataloging-in-Publication Data
Edens, Aaron
Cell Phone Investigations: searches, Cell Sites and Data/ by Aaron Edens.
p. cm.
Includes Index.
ISBN: 978-1-63180-006-1 (PB)
Control Number: 2014955886

Table of Contents

Forward

"If I have seen a little further it is by standing on the shoulders of giants."

Isaac Newton, 1676

The cellular phone forensics field is advanced by a small group of visionary and tenacious hackers, technicians, programmers, and code writers, some who are law enforcement and some who are not.

Evidentiary examinations of cell phones are a complex and often frustrating process. However, the rewards can frequently justify the time and costs associated with the process. This book will provide an introduction to the basics of cell phone forensic examination, but is in no way a complete or comprehensive treatise. Cell phone forensics is a dynamic field and the available tools do not always keep up with technology, which is one of the constant challenges to those involved in forensic examinations.

The National Institute of Standards and Technology (NIST) in their 2007 report entitled *Guidelines of Cell Phone Forensics* states: "Mobile phone forensics is the science of recovering digital evidence from a mobile phone under forensically sound conditions using accepted methods."

Often cell phone forensics requires an examiner to use techniques that have not been evaluated as an "accepted method." The examination technique involves using non-forensic methods such as the manufacturer's content management software, highly technical programming or repair tools, and third party software designed to modify the operating systems of phones.

Law enforcement has resisted the concept that a cell phone carries the same evidentiary weight as any other item of evidence or contraband. Perhaps it is the ubiquity that causes veteran investigators to treat a cell phone differently than a firearm or package of narcotics. Trends such as this only change course when a significant case is lost or an adverse legal precedent is established.

From physical evidence contained on the handset itself, to the digital evidence contained within the device, there is a wealth of evidence and intelligence to be recovered from even the most primitive cell phones. Combined with the information retained by the cell service providers, it is possible to glean critical insights into criminal activities and conspiracies that eluded earlier generations of law enforcement officers.

Writing a book is a challenge, particularly when the field involves the topics of law, technology, and forensic principals. My personal goal is to arm the reader with knowledge of the tools available in field of cell phone forensics as well as avoid the pitfalls you may encounter in this new and ever-evolving field. I hope that by reading this book you will gain insight into these various disciplines and be able to apply some of the investigative techniques into the noble profession of law enforcement.

I would like to express my gratitude to those who comprise the tip of the spear in the field of cell phone forensics. I would especially like to thank those who freely make their research, tools, techniques, and programs available to the law enforcement community. One of the greatest aspects of being a police officer is the community that exist nationwide. No other job or career sees the outpouring of support from departments and agencies across the nation when one of our brothers or sisters falls.

Stay Safe,

Aaron Edens

Chapter 1: Search Warrants

On June 25, 2014, the Supreme Court of the United States announced its decision in two cases, People v. Wurrie and Riley v. California. The result of those decisions requires a warrant, or other valid warrantless search exception, before a mobile device may be searched. This decision presents unique challenges and opportunities for both veteran and entry level law enforcement officers.

The first challenge is writing the search warrant affidavit itself. For many law enforcement agencies, seeking a search warrant is the exclusive dominion of seasoned detectives, narcotics officers, and other investigators. However, the Supreme Court decision is likely to push the responsibility for seeking and executing search warrants for mobile devices down to the patrol level.

Nearly everyone has a cell phone. A 2013 study by the Pew Research Center's Internet and American Life Project found that cell phone ownership among adults in the United States is 91 percent. Consequently, there is a statistical probability a criminal suspect will own a cell phone or other mobile device. There is also an increasing likelihood a cell phone or other device will be integral to criminal investigations.

Cell phone investigations are a common tool in conspiracy crimes investigations where communication between co-conspirators is an operational necessity. However, using a cell phone or other mobile device as a communication mechanism is not limited to these types of crimes. Certainly there are crimes where a mobile device is unlikely to be used, but those crimes are few and far between. Common criminal investigations involving the use of mobile devices include:

- Pedophiles using their mobile devices to communicate with and recruit victims and to store images, videos, and other mementos of their heinous crimes.
- Intoxicated drivers who communicate their location, plans, and intoxication level using social media, digital photos and videos displaying themselves in their intoxicated state as well as phone calls and text messages from their mobile devices.
- Burglars targeting specific locations and communicating about the presence of alarm systems, dogs, street lighting conditions, access points, and egress routes.
- Organized retail theft 'booster rings' targeting specific stores and possessing digital 'shopping lists' of preferred items to steal and fence.
- Car thieves looking for specific makes and models of vehicles and describing the best way to steal them as well as communications regarding police pursuit policies in the area.

Since most people use one mobile device in almost every major crime, proper search warrant preparation and procedure are necessary. For law enforcement agencies it isn't sufficient to wait for a detective to write the search warrant. Not only will this burden an investigator with completing search warrant affidavits for which he or she has no direct knowledge, but it will rob the other officer of an incredibly valuable learning experience. Search warrant preparation is one of the key tools in law enforcement investigations. Once an officer has experienced the process a few times it becomes progressively easier and faster to complete one. Officers who might otherwise delay or abandon a case because a search warrant is required to advance the investigation have increased confidence in their ability to seek, and be granted, a search warrant.

Law enforcement officers sometimes struggle with the proper nomenclature and technology for mobile devices. They rely on search warrant templates or 'boilerplate' handed down from other investigators. Sometimes these 'go-bys' have not been updated to reflect the changes in technology. In some cases, they still refer to antiquated technology associated with rotary dial telephones and numeric pagers. To address the unique needs of a modern mobile device search warrant, it is helpful to start by addressing the correlation between the crime and the device.

Some crimes inherently require using mobile communication devices. For example, it is almost impossible to be a successful narcotics dealer without using a mobile phone. Some crimes, such as those involving gangs, are inherently conspiratorial, frequently using cell phones or tablets to coordinate the group's activities. Articulating these facts is essential in an affidavit to support searching any seized device. When you consider applying for a search warrant, think about the device's nexus to the crime. Is it probable the suspect(s) used the device?

Before the commission of the crime?

As noted above, crimes such as narcotics violations and gang activity require communication and coordination before the act is carried out. Another example is robbery. While street level robbers may commit the crime with little or no pre-planning, many perform surveillance on the target person or location. They will coordinate with lookouts and getaway drivers to ensure there is no immediate police presence and easy escape routes from the scene. Articulating this knowledge is based on the prior training and experience of the officer and may be documented in a search warrant example:

> *Based on your Affiant's prior training and experience and the experience and training of other veteran law enforcement officers, your Affiant is aware [robbery/narcotics/weapons trafficking] is an inherently conspiratorial crime. The nature of the conspiracy requires participating members to communicate in order to coordinate their planning prior to the commission of the crime as well as during the actual commission of the crime. Such communications are commonly made using mobile devices such as tablet computers, mobile phones, and Wi-Fi capable portable gaming consoles.*

During the commission of the crime?

Similarly, narcotics, robbery, and other crimes require the suspects to actively communicate while the crime is being committed:

> *Based on your Affiant's prior training and experience and the experience and training of other veteran law enforcement officers, your Affiant is aware [robbery/burglary] is commonly a conspiratorial crime involving the use of others to assist during the commission of the crime. This assistance comes in the form of other suspects, both known and unknown, who will monitor police radio traffic and alert the perpetrators of the impending arrival of law enforcement, lookouts who will maintain visual surveillance on the approaches to the crime scene to alert their associates of the presence of law enforcement, and getaway drivers who will assist the perpetrators with their escape. Communications between co-conspirators is essential to the successful commission of the crime and the subsequent escape from the area of the crime. Such communications are commonly made using mobile devices such as tablet computers, mobile phones, and Wi-Fi capable gaming consoles.*

After the commission of the crime?

Once a crime has been committed, the suspect is also likely to use his phone. Law enforcement officers are aware those who commit crimes will take steps to conceal their activities, making the subsequent law enforcement investigation more difficult. This is true of almost every crime, including those not traditionally associated with conspiratorial activities. For example, many domestic violence investigators will tell you suspects frequently contact the victims after the crime to apologize, thereby demonstrating consciousness of guilt. Others will take steps to create alibis for their whereabouts during the crime, conspire with others to conceal or destroy evidence, or make plans to flee the jurisdiction where the crime occurred:

> *Based on your Affiant's prior training and experience and the experience and training of other veteran law enforcement officers, your Affiant is aware suspects who commit the crime of [homicide/assault/domestic violence] will often attempt to conceal or destroy physical evidence of the crime, conspire with others to create alibis, boast or brag about the commission of the crime, and/or attempt to flee the jurisdiction where the crime occurred. Communications between the perpetrator and those who knowingly or unknowingly assist them is essential to the concealment of the crime and possible flight from the area of the crime. Such communications are commonly made using mobile devices such as tablet computers, mobile phones, and Wi-Fi capable gaming consoles.*

Common Search Warrant and Affidavit Errors

Another challenge is adapting existing search warrants to make sure all available evidence is being collected while preventing additional issues due to antiquated language or omissions in the search warrant affidavit.

Once probable cause that a mobile device was used before, during, or after the commission of a crime is established, the next step is to describe with reasonable particularity the items to be searched for and seized. However, a review of current law enforcement search warrant affidavits revealed common errors and deficits in the standard language used to search a cell phone or other mobile device. Law enforcement investigators should review their search warrants and consider the following common errors and omissions:

Failing to request the corresponding call detail records

Law enforcement has traditionally treated searching mobile devices and investigating the records that cellular service providers maintain as two separate disciplines. In fact, both avenues of investigation should be explored simultaneously. The Supreme Court ruling provides an opportunity to blend the investigation of the mobile device and the corresponding records from the cell phone company together. If an officer has probable cause to search a mobile device's contents for evidence he also has probable cause to search service providers' records for corresponding evidence. In many cases, a search warrant was required to obtain cell site location information (CSLI), GPS data, other historical location data, and the stored content the provider maintained, such as incoming or archived voicemail messages and email messages. By adding a few lines to a search warrant affidavit and the face sheet of the warrant, law enforcement officers can access sources of data that were previously overlooked. The following language was adopted from a search warrant by Santa Clara County Deputy District Attorney Mike Galli.

The items listed may be contained in or at the cellular service provider. [INSERT PROVIDER] has been determined to have provided service to the listed phone number [INSERT PHONE NUMBER] associated with the seized device:

a. Subscriber information, including by way of example and not limitation:
    a. Subscriber name
    b. Subscriber address
    c. Identifying information such as date of birth, driver license number, and/or social security number
    d. Subscriber contact information including electronic mail addresses, contact phone numbers also referred to as 'can be reached numbers'
b. Billing and credit information, including by way of example and not limitation:
    a. Method and source of payment information including credit card numbers, electronic funds transfers, and locations of cash payments.
    b. Credit information including any credit report run by the provider prior to authorizing service
c. Service information, including by way of example and not limitation:
    a. Purchase and activation location
    b. Types of service subscribed to
    c. Additional phone numbers associated with the same account
    d. Make, model, and serial numbers of the phone(s) associated with the account
d. Call detail records
    a. Incoming/outgoing phone calls, SMS/MMS text message, including the content thereof, data events date, time, and duration of same
    b. Cell site location information including beginning and ending cell sites for any recorded events, per call measurement data (PCMD) and/or timing advance (TA) information alternately known Real Time Data or Round Trip Data.

Using overly broad language such as "Any and all"

Law enforcement officers have used the phrase "any and all" in writing search warrants for years. Unfortunately, several United States courts find the phrase fails to meet the reasonable particularity requirement of the Fourth Amendment of the Constitution. Depending on the local jurisdiction's requirements, some law enforcement agencies have begun using the alternate phrase "including by way of example and not limitation."

Failing to support the items to be searched in the affidavit

Every item to be seized pursuant to the search warrant must be supported in the affidavit. However, some search warrant templates include a long list of items to be seized from the mobile device without supporting language. There are two competing schools of thought with regards to the appropriate wording in a search warrant. The first, commonly used by federal law enforcement agencies, is a 40-60 page affidavit describing in painstaking detail all of the locations to be searched including: contact information stored in the electronic phone book, phone call logs, short message service (SMS) and multimedia message service (MMS) text/media messages, and calendar events. The problem with this type of search warrant is if officers miss anything by not specifying it in the warrant, such as the file system or application data, they risk losing it during a suppression hearing. The second school of thought is to specify evidence of the crime under investigation wherever it may be located on the mobile device. The challenge with this type of warrant is it could be viewed as too broad. The solution is a warrant that addresses the potential negatives from either type; specific enough to search for and locate the relevant evidence and while not requiring the time and often unnecessary obsession with detail. DDA Galli has created an ideal balance between brevity and meeting the reasonable particularity requirement. DDA Galli has made his template available to the law enforcement community via this download link http://policetechnical.com/model-cell-phone-search-warrant/. While specific to California and referencing California statute, DDA Galli's warrant also includes language to obtain information from the cellular service provider.

Not listing external storage media, including Micro SD cards

Micro SD cards are external storage media inserted into a mobile device to increase the storage capacity. Micro SD cards are an important source of information during mobile device forensic examinations, as they have large storage capacity and can retain deleted data, such as images and videos, for a considerable length of time. However, most law enforcement search warrants do not specify Micro SD cards as items to be searched. While there is no specific case law addressing this issue, an argument could be made the Micro SD card is a separate container within the mobile device and, if not specifically addressed in the search warrant, would fall outside the purview of the search warrant. One can look at current law enforcement practice to find the similarities. A good search warrant for a residence is not limited to a search at 123 Main Street. Most search warrants include language such as, "any outbuildings, garages, sheds, basements, and safes." Even though these buildings and containers are on the property to be searched, they are viewed as separate containers and judicial authorization is specifically sought to include them in the search. Similarly, Micro SD cards and other external storage media should be treated in the same manner.

Failing to authorize password bypass

When a person engages handset security on his mobile device, he expresses a higher expectation of privacy. While there is no specific case law requiring judicial authorization for bypassing a handset security lock, the actual mechanics of the process may make proactively seeking it a convenience. Apple and Google are able to either circumvent handset security features and provide data from the device or remotely reset the security features. These methods require specific judicial authorization to compel the companies to provide assistance. There are also technical methods for retrieving or bypassing the handset security codes from many of the popular models and operating systems. However, these methods are invasive and can, at times, be destructive to the evidence. When seeking a search warrant for an Apple or Google device, it may be prudent to include the appropriate language in the original warrant to prevent repeated trips to the magistrate.

Failing to specify deleted data

A user who deletes data from his mobile device considers the material to be gone. For many models of mobile devices and operating systems, modern forensic techniques are able to recover data even after it has been deleted. While no known precedent exists regarding specific judicial authorization to recover deleted data, it may be wise to proactively address the issue in the initial search warrant. This limits potential suppression challenges by criminal defense attorneys.

Failing to list the possible need for assistance from other agencies

Law enforcement agencies may not have the requisite equipment or expertise to examine a mobile device. It is common to rely on larger municipal, state, and/or federal law enforcement agencies to assist them with the recovery of evidence from a device. While not a statutory requirement, some law enforcement agencies require specific judicial authorization to assist with an investigation. This is particularly true of federal law enforcement agencies who assist their municipal, state, county, or tribal partners. For law enforcement officers who routinely send their mobile devices to another agency for examination, it is good practice to build judicial authorization for this practice into the search warrant.

Failing to investigate applications installed on the device

Increasingly, suspects are using alternate communications methodologies from their mobile devices. Most frequently these are applications and services installed on a device that use Wi-Fi for voice, messaging, and email outside of the services the cellular company provides. Additionally, there are covert applications used to store images, videos, text and multimedia messages, contacts, files, and other data that may be concealed as an innocent appearing application during a manual or forensic examination. Knowing what applications were downloaded by the suspect can help identify any other communications channels for follow-up investigations. Both Google and Apple are able to search for applications downloaded and installed on devices associated with their customers' accounts, device serial numbers, and/or phone numbers.

Failing to search associated cloud storage and backup files

There are cloud storage applications that allow a user to store and transfer files. Depending on the type of case under investigation, cloud storage services such as Dropbox and Google Drive may contain relevant evidence that is not recovered during a routine forensic examination of the mobile device. Additionally, Apple allows customers to back up their data to the iCloud storage system including contact lists, photos, application data, Internet browsing tabs and bookmarks, calendar events, and electronic mail. There may be evidence located in the backup files that was deleted or inaccessible on the mobile device. Some cellular service providers have additional backup functionality that can be accessed with a search warrant. AT&T's Mobile Locate service periodically backs up contacts, photos, and videos from its customer's mobile devices. If this information is not specified in a search warrant, the company will not provide it or even notify law enforcement of its existence.

Failing to search the operating system provider

The two major operating system providers in use today, Apple and Google, both store information from devices in associated cloud, storage, or backup files. The data that both of these operating system providers retain can be recovered, with an appropriate search warrant, even if the device is handset locked or protected by other security features such as encryption. The available data from Apple consists of associated mail accounts, contacts, calendar entries, notes, reminders, and documents created using Apple's Pages, Numbers, and Keynote applications

Operating System Warrants: Google

The amount of recoverable data from Apple is relatively minimal when compared with Google. In fact, the amount of data available from Google is more than some current forensic hardware and software programs recover from the mobile device. In order to create a search warrant affidavit that captures all relevant data, it is helpful to examine exactly what kind of data Google retains.

The following pages demonstrate the information available from Google pursuant to a search warrant. As long as law enforcement officers are now required to seek a search warrant to examine a mobile device, there is no reason to leave data "on the table" by the operating system provider. This is particularly true when the breadth and depth of data collected by Google is fully understood.

One of the first arguments encountered at this stage is "how do I know the suspect has a Google account and how do I find it?" The first half of the argument is relatively simple to answer. If the phone uses the Android operating system there is likely an associate Google account. When a customer takes his brand new Android phone out of the box and turns it on one of the first things he is prompted to do is to associate an email account for recovery purposes. The system prompts the user to associate an existing account or create a new one in case the user forgets his password. When this account is created it automatically begins collecting data even if the user does not primarily use Google's service such as Gmail.

Google is able to search its records by the device's phone number and the unique serial numbers: the international mobile equipment identifier (IMEI) and the mobile equipment identifier (MEID.) Both of these numbers can be located by removing the battery cover and the battery and examining the manufacturer's sticker underneath.

To see what information Google collects click on the icon located in the upper right corner of the main search page and select the **Account** link.

Under the **Personal info** tab at the top of the screen are the email accounts and phone numbers linked to the Google account. In this example there are two email accounts, a primary email and a recovery email. This is an opportunity to discover any previously unknown email accounts and follow-up with additional legal demands. There are also phone numbers linked to this account. In this example there are two phone numbers: the mobile number for the device and a Google Voice number.

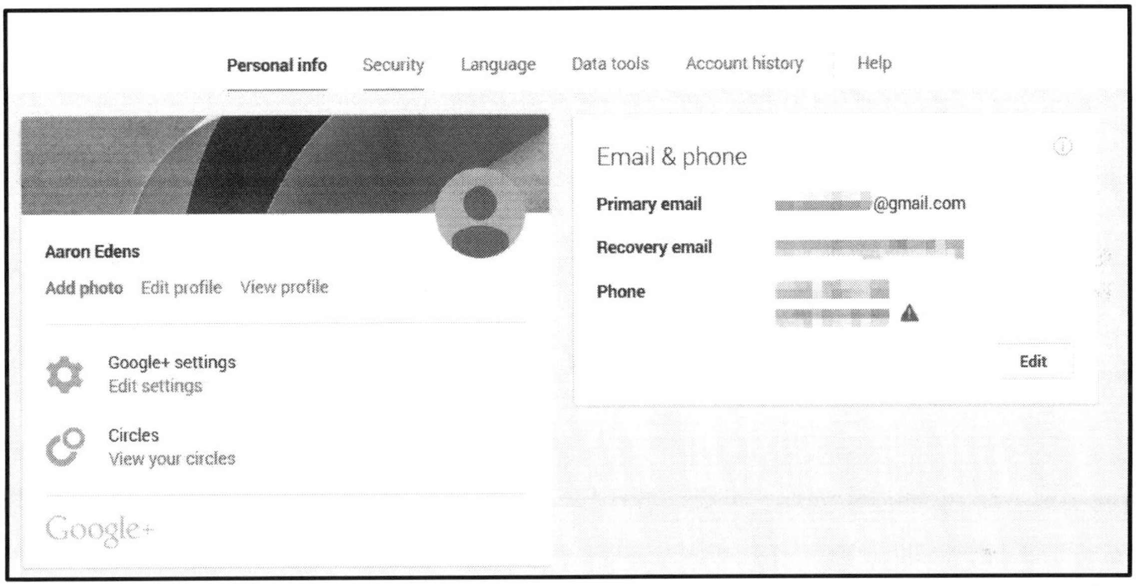

The **Account history** tab at the top of the screen and has additional information of significant investigative value. This data is available elsewhere, but this interface is a very user friendly for exploring some of the data Google collects and retains.

The first option is the search history. If you have ever started typing a search query into the Google search bar you have no doubt encountered the auto fill option. These responses are based on previous searches and are stored by Google to assist the user. However, this data can provide evidence and intelligence. Obtaining the search history from Google was previously limited to cases involving known or suspected uses of the search engine in a criminal investigation. Depending on the type of investigation, evidence can be found in the search history. For example, a marijuana cultivator may use the Google search engine to obtain hydroponic growing equipment or to research a potential grow location. It is up to the investigator to associate the search history with the criminal investigation. This is not a difficult task as the Google search engine is the default on Android phones and Google accounts for approximately 66-68 percent of search queries worldwide.

Selecting the **Manage history** link will demonstrate the types of search query data available from Google. There is a user friendly graph of search activity by hour and day. There is also a calendar function. The calendar function shows the specific search history by day. This data can be critical in date/time specific crimes such as homicide. However, the search data as a whole can also be relevant in conspiracy crimes investigations like narcotics, gangs, and organized crime.

The Manage history link also has an available feature to narrow the search results by category such as web search, image search, maps, etc. Clicking on any of the hyperlinks will take you to the search results.

The YouTube search history is self-explanatory. It shows the search history and links to the viewed video. This data is also available from other locations, but may be of investigative interest depending on the type of criminal case.

Perhaps one of the most intriguing and useful data sets collected by Google and obtainable with a search warrant is the **Location history**. Most Android users are unaware there is passively collected historical GPS capability built into the operating system. From the interface above or it can be accessed at the website https://maps.google.com/locationhistory/b/0.

The historical GPS location data is not comprehensive as the data is captured when a GPS application is activated on the device. However, it can provide valuable leads and location based intelligence. Note there can be large areas and time periods where GPS location data is not recorded, such as the movements of the device from West to East. Alternately, the regularly spaced data points highlighted with the red arrow are recordings taken approximately every minute while a GPS navigation application was active. The table below the illustration gives a visual representation of the time periods when the most GPS data was captured.

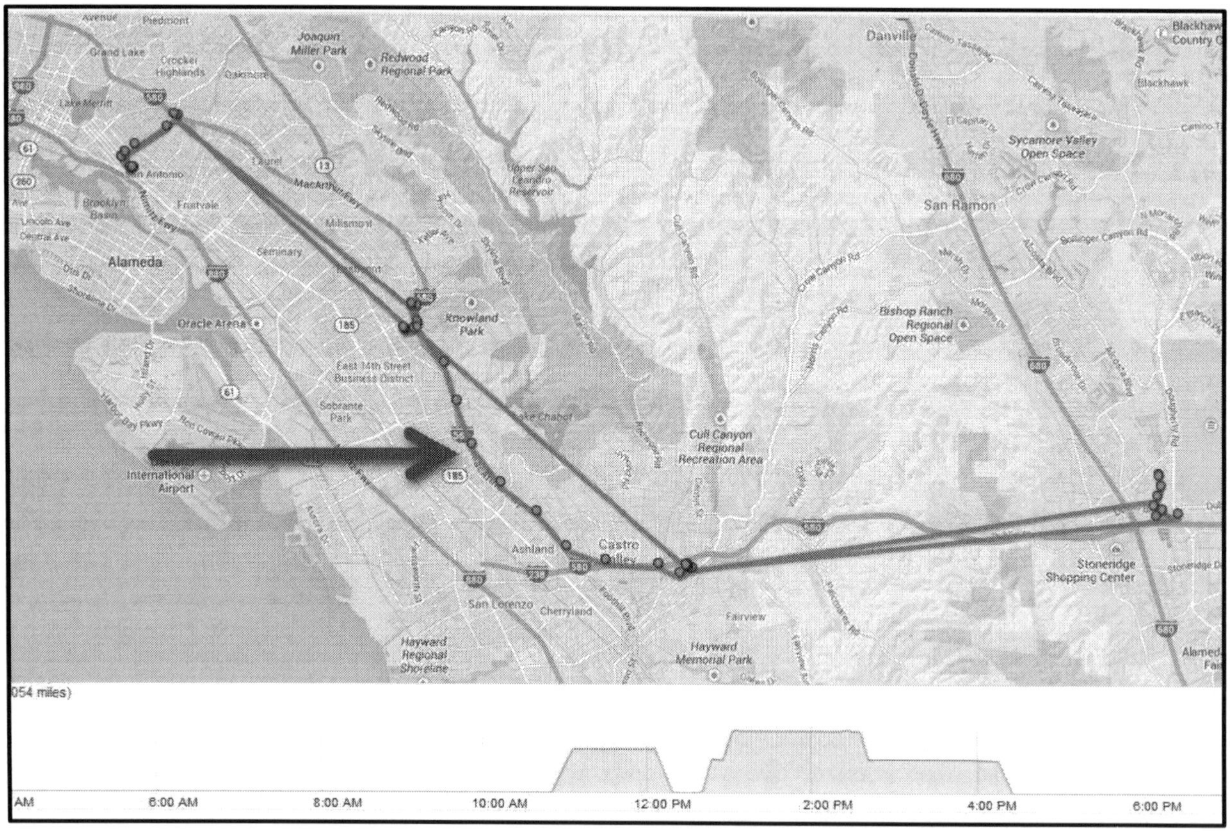

On the location history map page is a calendar feature. Individual days may be selected by clicking on a specific data on the calendar. Selecting **Show timestamps** will display all of the times a GPS location was recorded on the selected date. Selecting a time will show the corresponding hit on the interactive map.

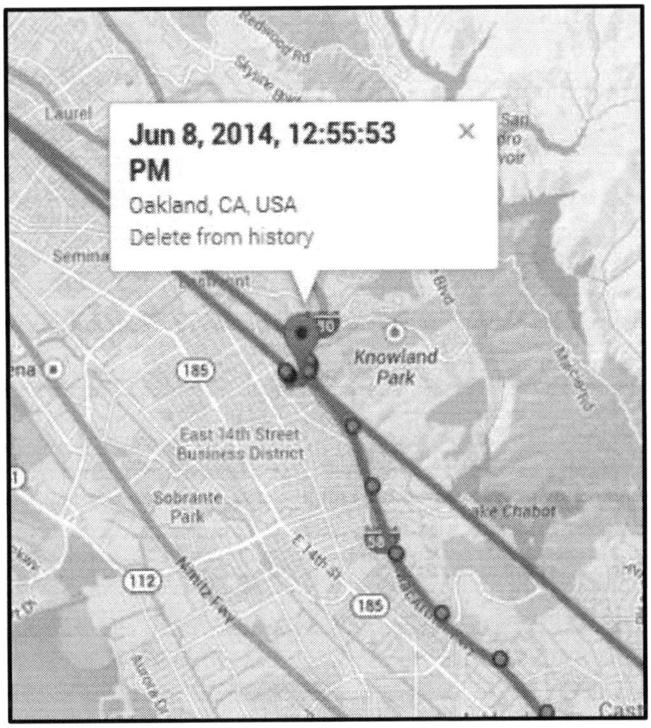

There is also a drop down feature that will show the aggregated location data over a 1,2,3,5,7,14, and 30 day time period.

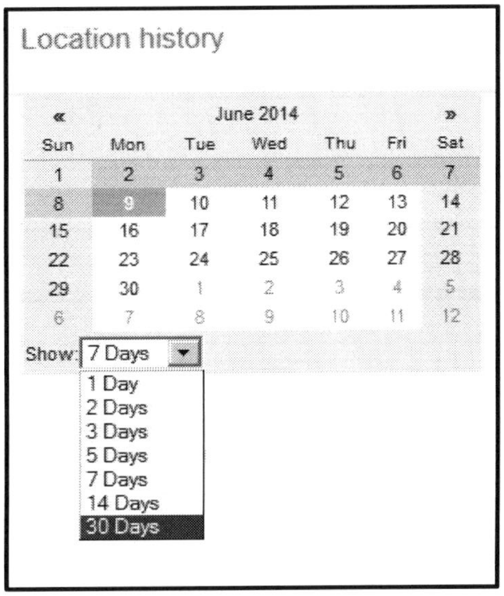

The collection and recording of location based data by Google appears to extend for at least 10 months. In June 2014, I was able to view the movements of the mobile device as far back as August, 2013. Selecting the left and right scroll arrows to access data points beyond the maximum 30 days permitted in the drop down box.

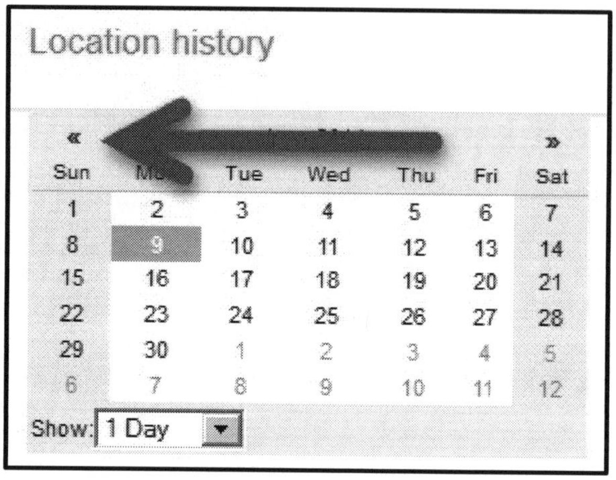

Data Points

Google is less than transparent about when location data is collected. Its privacy policy states: "When you use a location-enabled Google service, we may collect and process information about your actual location." The inference is this location information is collected when the user activates a Google service such as Google maps. However, experimentation and research shows this data is passively collected when there are events on the phone. This appears to include browsing the Internet and email pushes to the mobile device, but the frequency this data is collected is not entirely known. For example, the data points in the illustration below were obtained when there was no activity on the device — no phone calls, no browsing, no application usage, or mapping functions.

How the data is collected

Google states: "Location Services uses GPS, Wi-Fi hotspots, and cellular network towers to determine your location. Location Services uses Wi-Fi only if it's enabled on your device. Because Wi-Fi provides better accuracy than cellular towers, we recommend that Wi-Fi you enable Wi-Fi." Experimentation has shown this to be the case. Disabling the GPS decreases the accuracy of historical location data, but it does not eliminate it. It appears the use of cell site location information and Wi-Fi access points give the non-GPS enabled location data nearly the same accuracy.

Aberrant Data

Likely every law enforcement investigator who has dealt with GPS location data has been confounded by the appearance of aberrant data — specifically, GPS plots that seem illogical in the context of the larger data set. For example, the screen capture below shows GPS activity on the left part of the screen and then two outliers in the highlighted area on the right of the image. The device was never in those locations and, based on the limited understanding of Google's historical GPS data collection, the appearance of those data points is unexplainable. Equally confounding is the fact that those data points later disappeared from the map.

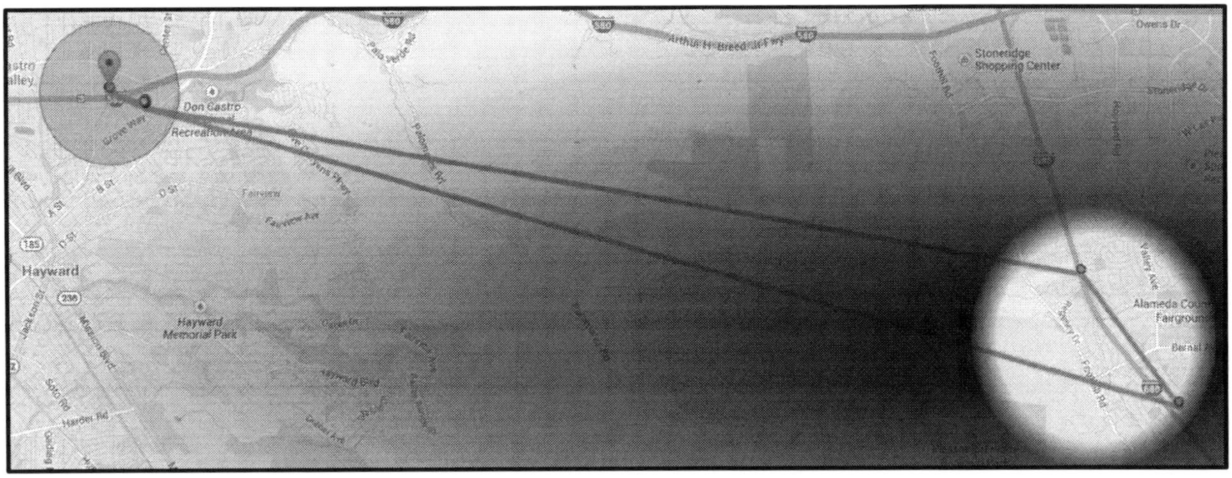

To further understand the available data from Google it is helpful to move beyond the user friendly **Account history** tab into the **Data tools**. From the tab select **View account data**.

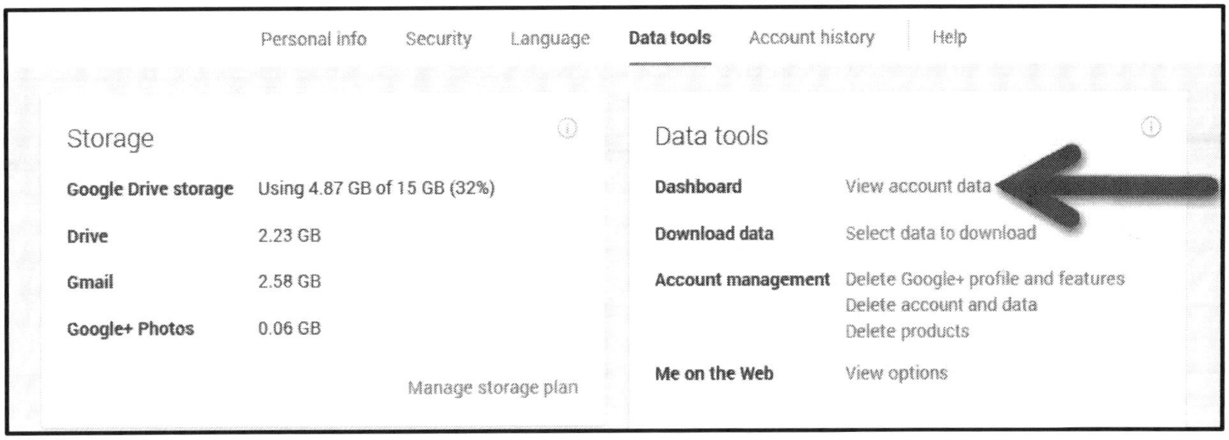

To fully appreciate the types and amount of data available from Google, select the **Expand all** tab. This dramatically increases the display of information available to an investigator.

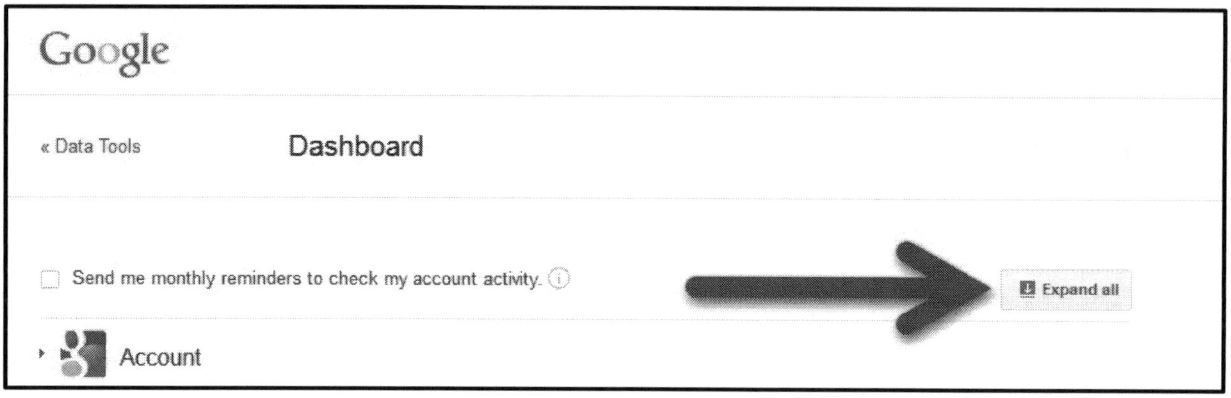

Each category has specific data sets contained within the drop down menu. The facts and circumstances of each investigation are different, but there are some common areas of interest. The first is the **Account** field. Initially this information appears to be fairly innocuous, particularly when the view is not expanded. However, selecting the **See the full list** link brings up recent log-in activity. This is not general activity, but it indicates when the user logged into a Google account from his mobile device, laptop, or desktop computer. Geographic information about where the device used to log in was located is included.

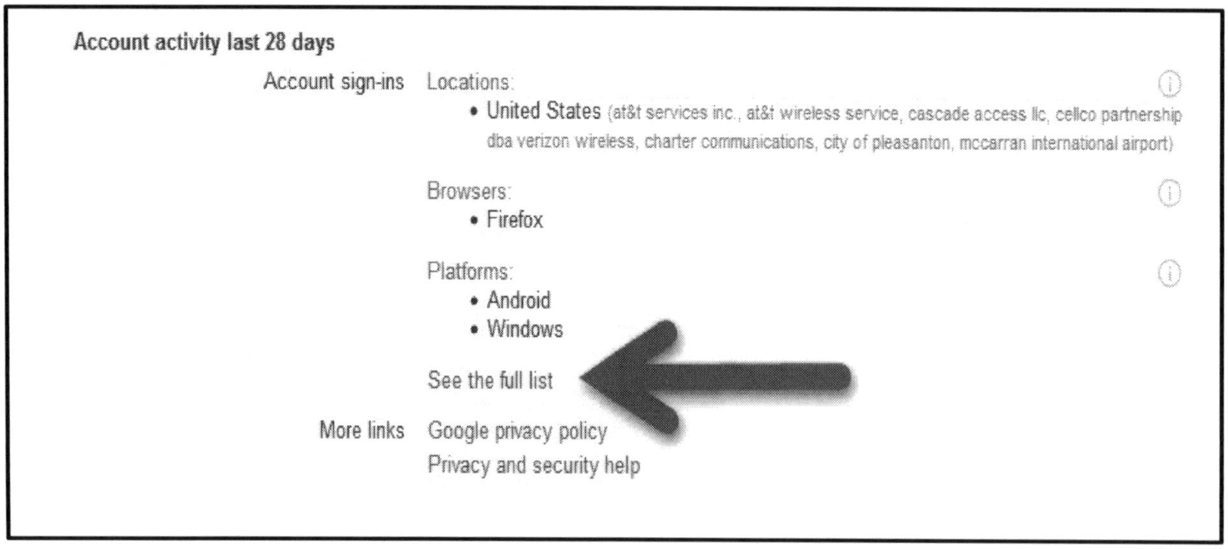

Selecting any of the dates and events brings up a map showing the general geographic area and the Internet protocol (IP) address of the connection. The IP address can be resolved to a service provider, which, in turn, can provide the location of the connection such as a Wi-Fi hot spot.

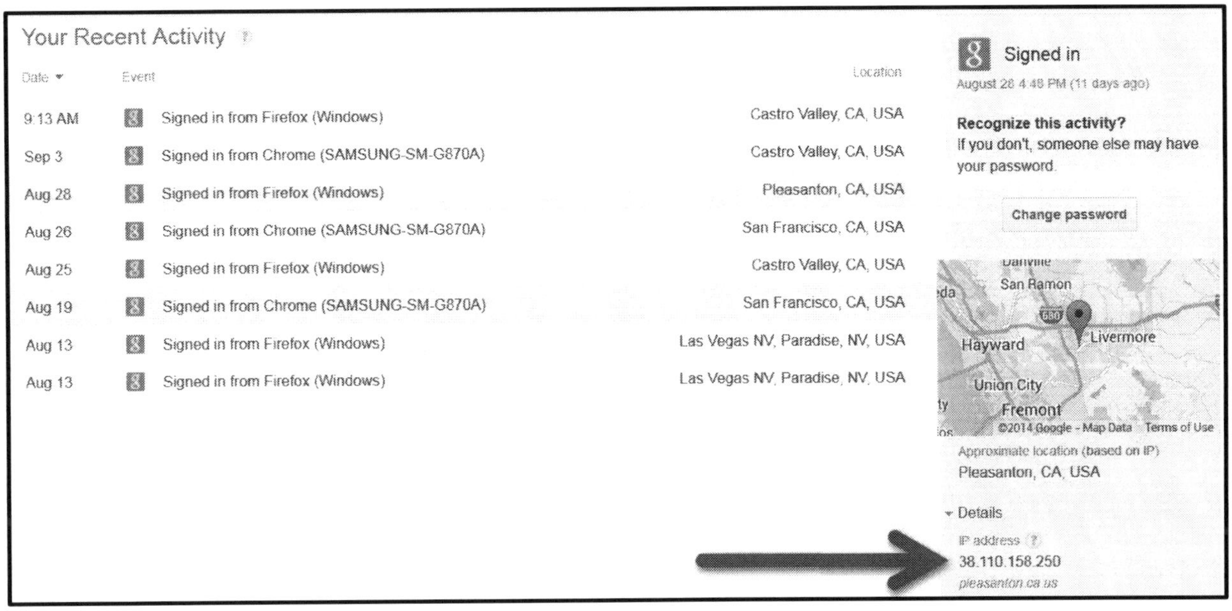

The **Android** section displays all of the devices that were linked to Google's services and displays the IMEI or MEID and file backup dates and times.

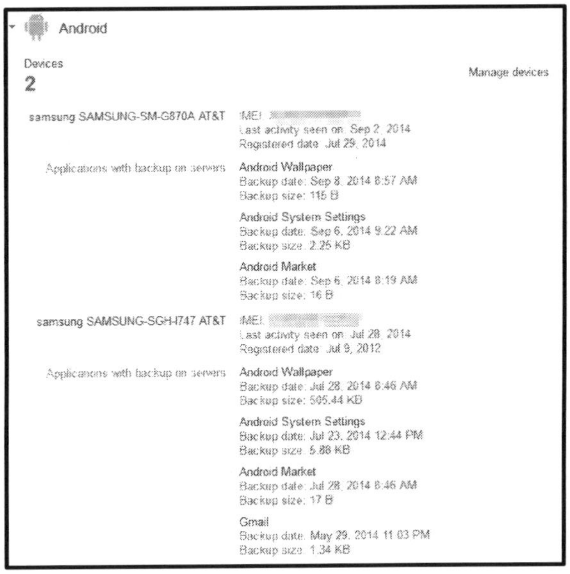

Calendar, unsurprisingly, shows the Google calendar linked to the account. The calendar events may be relevant in certain types of investigations. Additionally, in the case of linked calendars, it may be beneficial to see who has access to the calendar. This would likely be someone closely linked either personally or professionally.

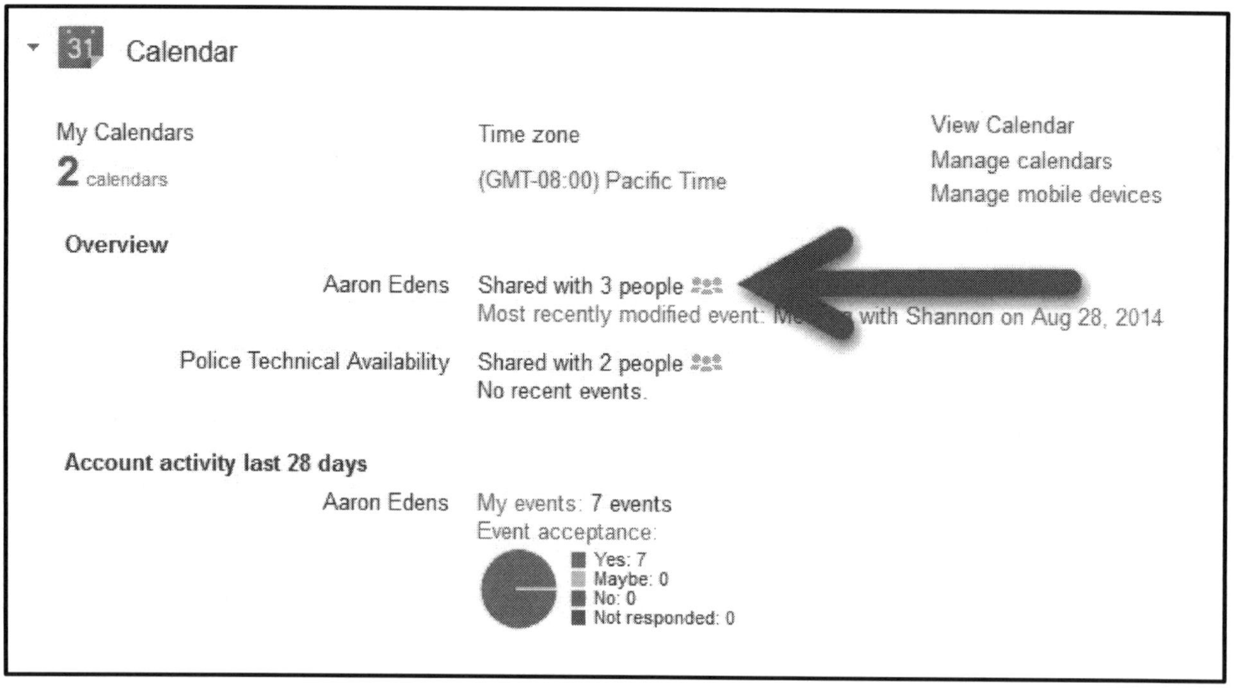

The **Contacts** section has a complete listing of all contacts the user entered or imported from his Google Mail. The data includes anything such as names, phone numbers, pictures, and email addresses. This may be an alternative source of contact information in the event a law enforcement officer is confronted with a locked Android device and no way around the passcode. By serving Google with a search warrant an investigator can replicate the contacts section on the Android device without actually accessing the device itself.

The **Docs** section shows links to the user's Google docs and any downloaded or received documents using this service. Google Docs is increasing in popularity as a replacement to Microsoft's products as Google's are available for free.

Many street level criminals are unlikely to make use of Google's **Finance** services, but the data may be useful in organized crime and money laundering investigations. The **Gmail** section contains information regarding any associated Gmail accounts. It is troublesome that many criminal investigators will obtain call detail records and other data from the cellular service provider and neglect to obtain email conversations. The same probable cause for searching a mobile device is likely applicable to email content, as well. In fact, email content may have more investigative value than text and multi-media messages as the content is not limited to a finite number of characters or by certain file types. An often overlooked feature in Gmail is that deleted emails are stored in the Trash file and are not actually deleted unless the user empties that file. Therefore, it is possible to recover 'deleted' emails from Google with a search warrant. Note that once the Trash file is emptied the email files are not recoverable from Google but may still be recoverable from the device or an associated computer.

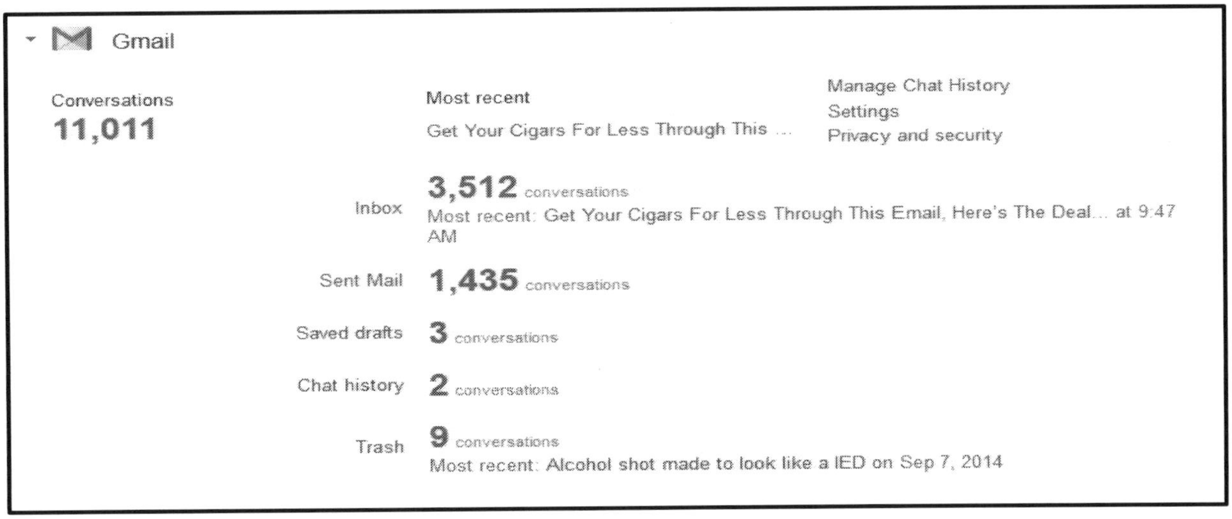

Google Photos stores images shared with the Google+ social network and any pictures taken with a mobile device that are uploaded to Google Drive cloud storage. Depending on the user, the images in Google Photos may be the same as the images taken with the associated mobile device. Investigators who are familiar with EXIF data, also called metadata, should be aware the GPS coordinates of where the picture was taken are not encoded in the image. This feature is turned off by default on the Google Pictures settings and the user must turn it on to activate.

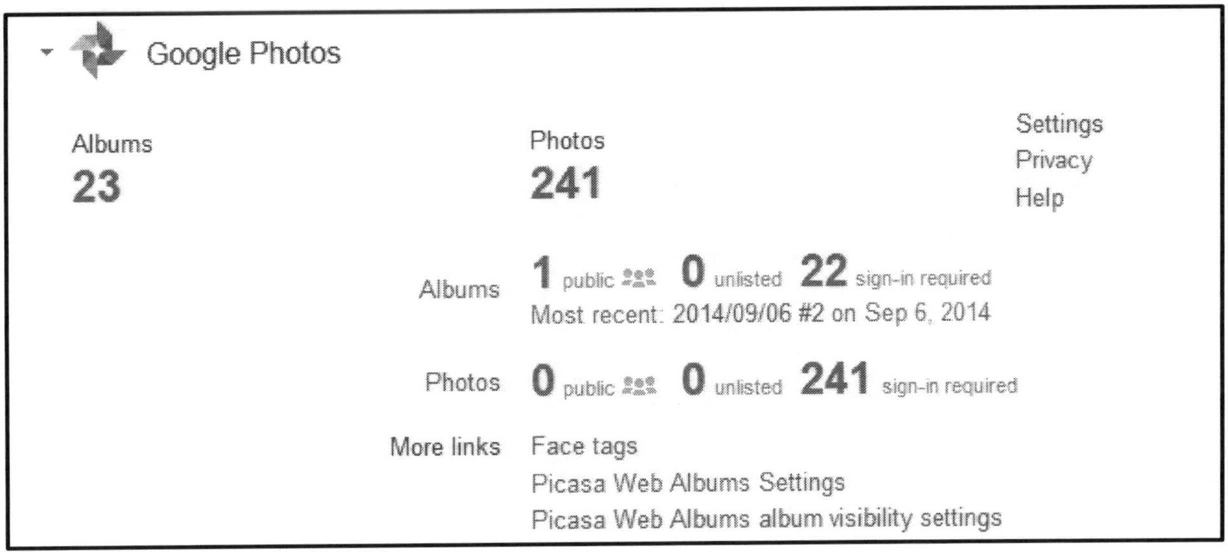

A commonly overlooked area of mobile device investigations are the applications installed. Records of purchased and downloaded applications are available from Google's **Play Store** via a search warrant. This provides an abundance of leads regarding a suspect's purchasing history through apps like Amazon.com, financial institution information indicated by a banking application, cloud storage accounts, and social media sites.

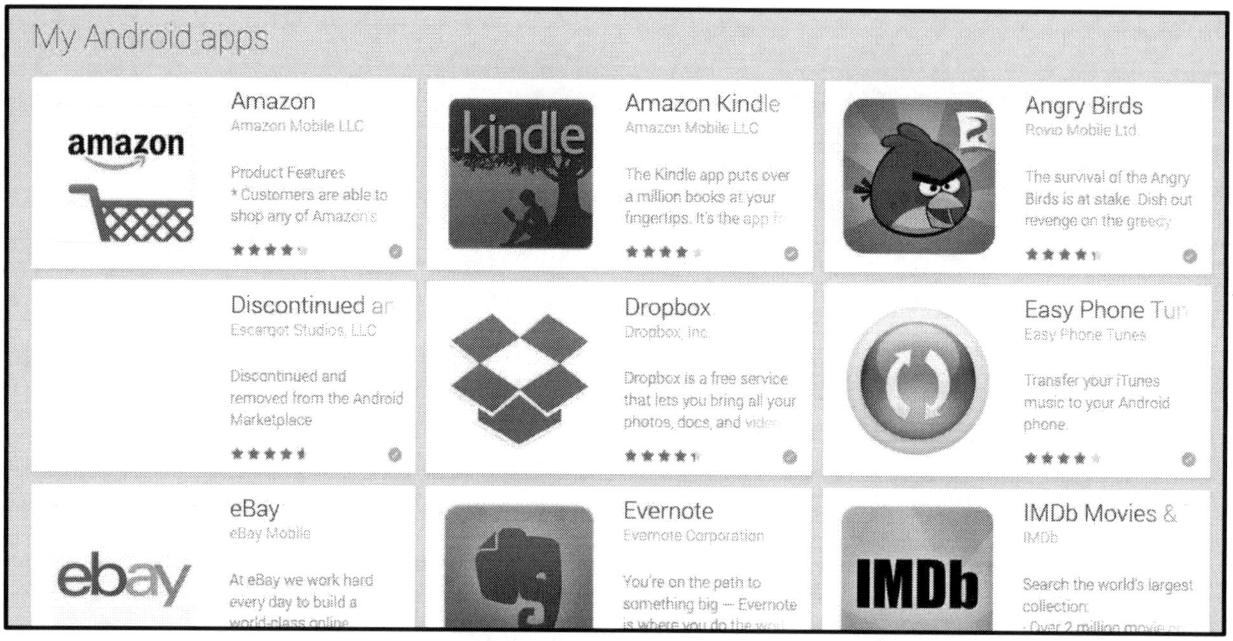

According to two Nielson surveys the average smartphone user has 41 applications installed on his device. The average user also spends 30 hours, 15 minutes using apps per month. Compare this with the use of the cellular phone function of a device at 7 hours 35 minutes per month.

The Nielson statistics are revealing and have a significant imperative for law enforcement investigators. If you are not investigating the applications and are only focusing on call detail records you are missing a substantial amount of evidence and intelligence. The use of communications applications on smartphones is roughly half that of the cellular use. Neglecting communications applications means you may be missing as much as 50 percent of the available records and data.

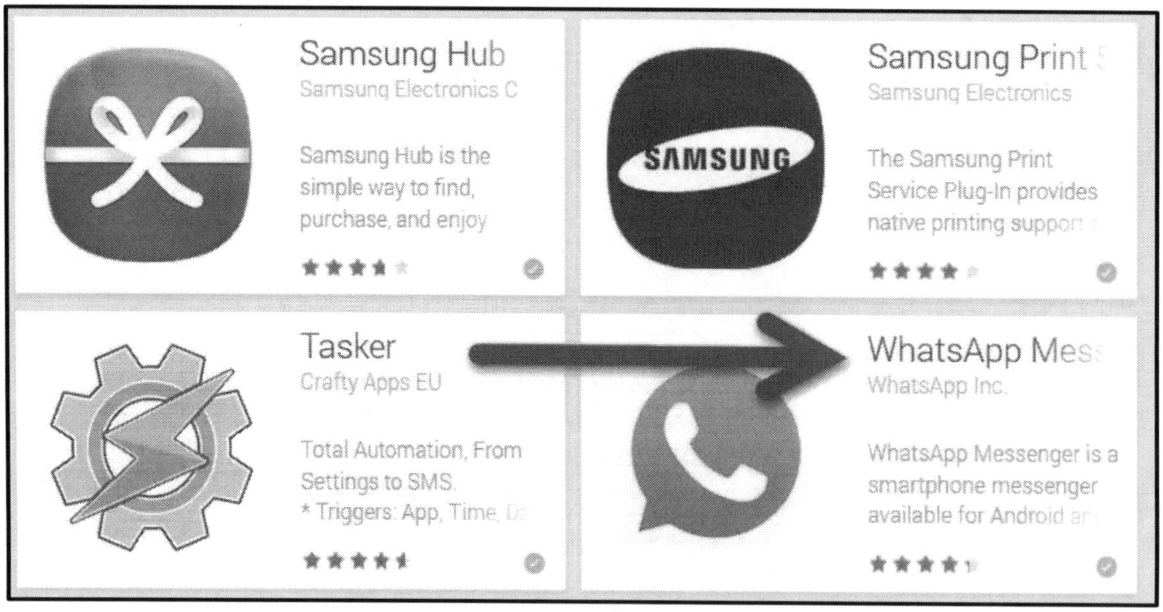

If the suspect in a criminal investigation is using Google's **Voice** service then there is a substantial amount of information available from the company pursuant to a search warrant. Investigators may lament the minimal call detail record information they obtain from Google. While minimalist, the wealth of information from the other features, specifically text messaging and voicemail, more than compensate. This data is retained indefinitely in Voice until the user deletes it. This translates into potentially hundreds of text messages and voicemail messages available pursuant to a search warrant.

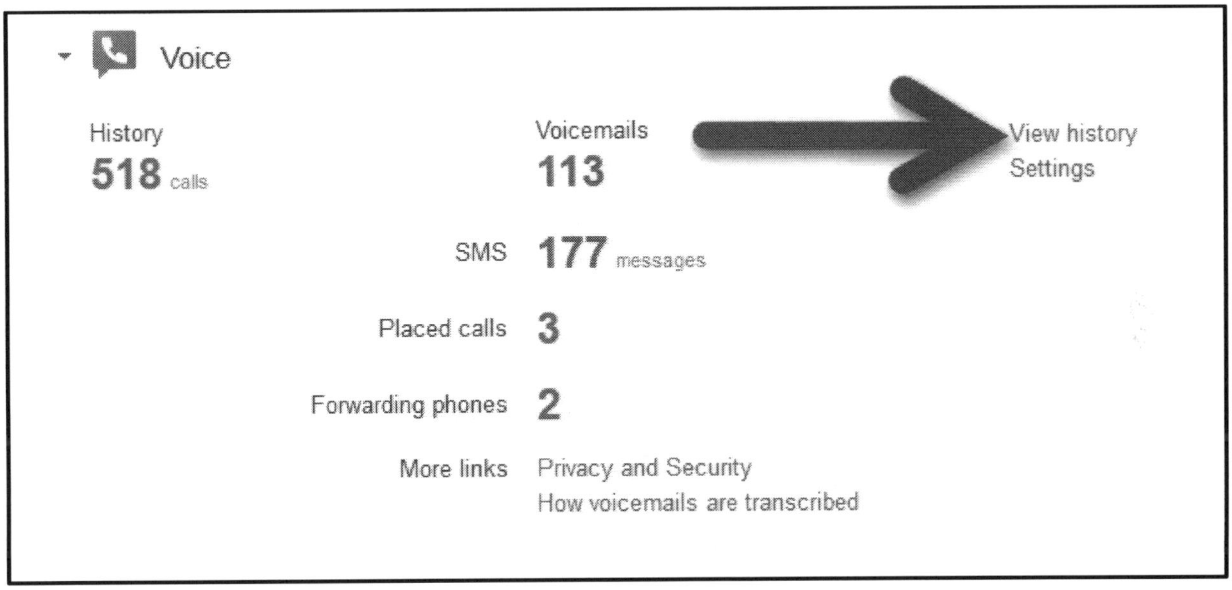

Google Voice makes an attempt at translating voicemail messages. Depending on the clarity of the speaker it is possible to quickly preview voicemail messages of investigative interest and bypass those that do not appear relevant.

In order to purchase applications for an Android device, a subject must make use of Google's **Wallet/Checkout** service. While most users are familiar with this in conjunction with application purchases, Google's service also works for other Internet purchases. It is possible to view the user's entire purchase history made with the Wallet/Checkout service. There is also information about linked credit cards, which may be of assistance in organized crime, money laundering, and fugitive investigations.

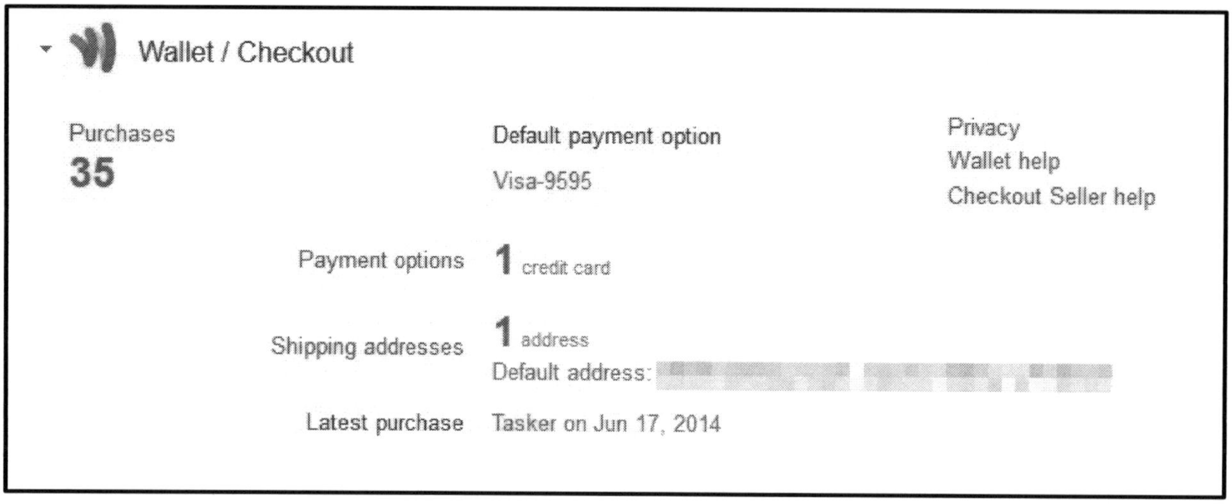

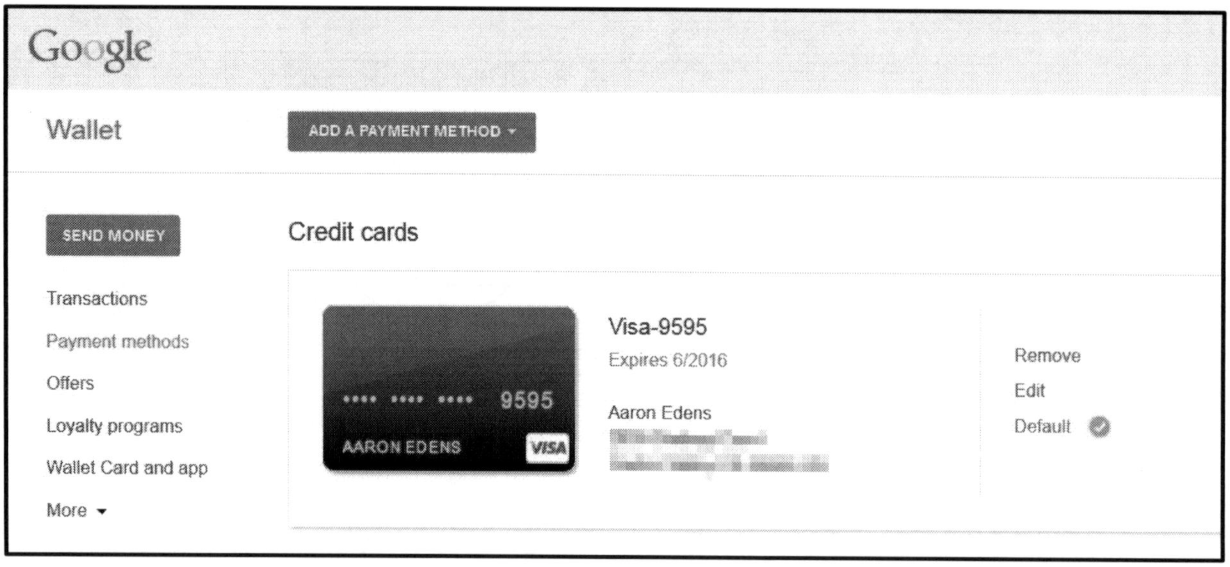

Each investigation is different and the information requirements are based on what the investigator needs. The item to be seized from Google pursuant to a search warrant must be supported in the probable cause affidavit. Some suggested language may include:

> *Based on your Affiant's prior training and experience, I am aware users of Android operating system phones, such as the one list is Exhibit 1A and which this affidavit seeks a search warrant for, commonly have an associated account with Google. When a user purchases and activates a mobile device one of the initial prompts during the set-up phase is to associate a Google Gmail account with the device. The purposes of this account are to facilitate a password reset in the event the consumer forgets their passcode or pattern unlock. If the consumer does not have an existing Gmail account, the operating system prompts the user to create a new account. Whether the Gmail account is new or existing the association of the account with the device allows Google to collect and store information relevant to this criminal investigation. This information includes, by way of example and not limitation:*

Account Information-User name, primary email address, secondary email addresses, connected applications and sites, and account activity for the previous 28 days, including account sign in locations, browser information, platform information, and Internet protocol (IP) addresses.

Google maintains information about its customers including primary email addresses, secondary email addresses for account password recovery, applications, websites, and services that are allowed to access their Google accounts or use them as a password logins. The company also records activity such as geographic areas where the users logged in, what type of Internet browser and device they were using, and the Internet protocol (IP) address they logged in from. The IP address is roughly analogous to a telephone number the Internet service provider assigned to a computer. The IP can be resolved back to a physical address such as a residence or business with Wi-Fi access or residential cable Internet. I believe this information will assist in the investigation by identifying previously unknown email accounts and location history information tending to show the movements of the suspect, his mobile device, and/or computers.

Android Information-Device make, model, and International Mobile Equipment Identifier (IMEI) or Mobile Equipment Identifier (MEID) of all associated devices linked to the Google accounts of the target device.

Google stores information about mobile devices associated with the user's Google account. This includes the make, model, and unique serial numbers of all linked devices. I believe this information will identify any previously unknown cell phones or other mobile devices associated with the suspect's account and/or known device(s).

Calendar-All calendars, including shared calendars and the identities of those with whom they are shared, calendar entries, notes, alerts, invites, and invitees.

Google offers a calendar feature that allows users to schedule events. This calendar function is the default option in the Android operating system and remains so unless the user adds a third party application. Calendar events may include dates, times, notes and descriptions, others invited to the event, and invitations to events. This information will identify relevant dates and appointments germane to an investigation as well as identify previously unknown co-conspirators and/or witnesses.

Contacts-All contacts stored by Google including name, all contact phone numbers, emails, social network links, and images.

When users link their Android devices to their Google accounts the names, addresses, phone numbers, email addresses, notes, and pictures associated with them are transferred to the phone and vice versa. This process is continuously updated so when a contact is added, deleted, or modified using either the Google account or the mobile device the other is simultaneously updated. This information is pertinent to the investigation as it will assist with identifying previously unknown co-conspirators and/or witnesses.

Docs (Documents): All Google documents including by way of example and not limitation, Docs (a web-based word processing application), Sheets (a web-based spreadsheet program), and Slides (a

web-based presentation program). Documents will include all files whether created, shared, or downloaded.

Google offers their users access to free, web-based alternatives to existing word processing, spreadsheet, and presentation software. These documents are stored in the user's account and are accessible from any device or platform as long as the user knows the password. These documents can include those the user creates, modifies, edits or shares. This information may contain notes, files, and spreadsheets containing information relevant to this investigation including recordation of sales, communications with unknown co-conspirators and/or witnesses, and other information concerning the ongoing investigation.

Finance: All records of securities, funds, and portfolios associated with the target Google account and/or target device.

Google allows users to create custom portfolios of stocks, bonds, and mutual funds. These portfolios are updated based on the market conditions and contain the investment type and the dollar value of the investment. I believe this data may contain information about income, investments, and previously undiscovered assets derived as proceeds from the ongoing criminal enterprise.

Gmail: All email messages, including by way of example and not limitation, such as inbox messages whether read or unread, sent mail, saved drafts, chat histories, and emails in the trash folder. Such messages will include all information such as the date, time, internet protocol (IP) address routing information, sender, receiver, subject line, any other parties sent the same electronic mail through the 'cc' (carbon copy) or the 'bcc' (blind carbon copy), the message content or body, and all attached files.

As noted previously, when user of an Android device first activates the device they are prompted to associate the device with a Google mail, commonly referred to as Gmail, account. The purpose of this account is to facilitate password recovery in the event the user forgets his password or pattern lock. If the user does not have an existing Gmail account he is prompted to create one. The Gmail account may be used to send and receive email messages and chat histories. These messages include incoming mail, sent mail, and draft messages. Messages deleted from Gmail are not actually deleted. They are moved to a folder labelled Trash and are stored there until the user empties the Trash file.

Additionally, users can send and receive files as attachments. These files may include documents, videos, and other media files. I believe these messages would reveal motivations, plans and intentions, associates, and other co-conspirators.

Google Photos: All images, graphic files, video files, and other media files stored in the Google Photos service.

Google users have the option to store, upload, and share digital images, graphic files, video files, and other media files. These images may be downloaded from the Internet, sent from other users, or uploaded from a user's mobile device. In many cases, an Android user may configure his device to automatically upload pictures taken with a mobile device to his Google account. I believe a review of these images would provide evidence depicting the suspect, his associates, and others performing incriminating acts. I also believe these image files may assist investigators with determining geographic locations such as residences, businesses, and other places relevant to the ongoing criminal investigation.

Location History: All location data whether derived from Global Positioning System (GPS) data, cell site/cell tower triangulation/trilateration, precision measurement information such as timing advance or per call measurement data, and Wi-Fi location. Such data shall include the GPS coordinates and the dates and times of all location recordings from the period ____ to ____.

Google collects and retains location data from Android enabled mobile devices. The company uses this information for location based advertising and location based search results. Per Google, this information is derived from Global Position System (GPS) data, cell site/cell tower information, and Wi-Fi access points. While the specific parameters of when this data is collected, it appears that Google collects this data whenever one of its services is activated and/or whenever there is an event on the mobile device such as a phone call, text messages, Internet access, or email access. I believe this data will show the movements of the suspect's mobile device and assist investigators with establishing patterns of movement, identifying residences, work locations, and other areas that may contain further evidence relevant to the ongoing criminal investigation.

Play Store: All applications downloaded, installed, and/or purchased by the associated account and/or device.

Google operates an online marketplace whereby Google and other third party vendors offer for sale applications such as games, productivity tools, and social media portals. Many of these applications can communicate outside a mobile device's service by using Wi-Fi Internet. These various applications facilitate communication using voice over Internet protocol (VOIP) technology, short message system (SMS) text messages, multi-media message system (MMS) text messages, audio transmission of recorded messages, and recorded or live video messages. As these services operate independently of the cellular service network there is no corresponding information regarding communications from the provider. Identifying communications applications purchased, downloaded, and/or installed on the mobile device would assist investigators by determining what application provider should be served with additional search warrants. Furthermore, identifying the user's applications would assist investigators with determining banking and other financial institution information and social media sites used. Identifying the purchased or installed applications would assist locating those with potentially criminal implications such as applications that appear to the observer to be a calculator or other innocuous appearing program, but, in actuality, are used to conceal pictures, videos, and other files. These concealment applications are commonly missed during manual and forensic examinations of mobile devices, as existing technologies are not designed to detect and locate them and the information they conceal.

Search History: All search history and queries, including by way of example and not limitation, such as World Wide Web (web), images, news, shopping, ads, videos, maps, travel, and finance.

Google retains a user's search history whether it is done from a mobile device or from a traditional computer. This history includes the searched for terms, the date and time of the search, and the user selected results. Furthermore, these searches are differentiated by the specific type of search a user performed into categories. These categories include a general web search, and specialty searches where the results are focused in a particular group such as images, news, videos, and shopping. I believe a review of the suspect's search history would reveal information relevant to the ongoing criminal investigation by revealing what information the suspect sought and when he sought it.

Voice-All call detail records, connection records, short message system (SMS) or multi-media message system (MMS) messages, and voicemail messages sent by or from the Google Voice account associated with the target account/device.

Google offers users access to a free voice over Internet protocol (VOIP) communications system called Google Voice or simply Voice. This system is layered on top of any existing cellular service. Users are provided with a phone number they select from a pool of available numbers. These numbers can be from whatever area code and prefix they desire and have no correlation with the user's actual location when the number is selected. Google allows users to access this system to make and receive phone calls and text messages. The service also has a voicemail feature where incoming phone calls are permitted to leave a message that is subsequently transcribed by Google and delivered by electronic mail and/or text message. Google maintains call detail records similar to those of a traditional cellular or wireline telephone company. Additionally, they also store the text message content of sent and received text messages as well as any saved voicemail messages and the associated transcriptions.

Wallet/Checkout-All information contained in the associated Google Wallet account including transactions, purchases, money transfers, payment methods, including the full credit card number and/or bank account numbers used for the transactions, and address book.

Google operates a financial services division that allows users to make online purchases through Google and other vendors as well as send and receive money from other users. Applications that are purchased and installed on a mobile device are handled by Google's Wallet/Checkout service. The purchase and installation of applications on a mobile device requires the use of the Google Wallet service. Therefore, any applications installed on the suspect's mobile phone have a transaction record in Google Wallet. Google stores information regarding the transactions including the date and time of the purchase. Additionally, it has method of payment information such as associated credit card numbers used to facilitate the purchase. Other data includes the billing address of any linked credit card and any addresses where purchased products were shipped to. I believe identifying the method of payment information would assist investigators with identifying any previously unknown financial institutions and that these financial institutions may have additional relevant information pertinent to this investigation.

A thorough understanding of the data Google Services retains reveals an unprecedented amount of information relevant to criminal investigations. Failing to include Google as a place to be searched in investigations involving Android devices may lead to critical evidence and intelligence being 'left at the table.'

Chapter 2: Phone Records

"There is no constitutionally protected reasonable expectation of privacy in the numbers dialed into a telephone system."[1]

It is important to look at a cell phone and its phone call records as a whole. Too often law enforcement officers will only view the contents of the handset and neglect to extend their search to records the cellular service provider maintains. This occurs because law enforcement officers are unaware of the record-obtaining process or they believe the provider doesn't keep records because it is a prepaid service.

Let's look at information obtained from a cell phone handset and records from a provider. The cell phone handset contains an electronic phone book and call logs, which record a finite amount of incoming, outgoing, and missed phone calls. These "call logs" are limited to a number the manufacturer sets. On some basic devices the number of entries stored can be as few as 20. Examining these call logs will only reveal the phone user's most recent call activity. The electronic phone book will list the user-entered name, phone number, and other contact information for an entity. These entries are important to the investigator because they detail how phone's user refers to the person. For example, a gang member will list his contacts by their monikers or street names, not by their full given names. This is helpful for an investigator to establish a relationship between the handset's user and other potential gang members. The user entered information may also contain data that is unknown to the investigator.

However, failure to dig deeper into transactional phone calls that are not visible in call logs will often cause investigators to miss leads, co-conspirators, relational information, and calling patterns. For instance, it is helpful for an investigator to establish who a dealer contacts after an informant places a phone call to order narcotics. Almost all drug dealers have to obtain their supply from a higher level dealer. Looking at the call logs on a cell phone handset will frequently fail to disclose this information. A detailed examination of the phone records is necessary in order to identify the source of supply for a lower-level narcotics dealer.

There are a number of legal processes available to obtain information from cellular service providers. These include, Administrative Subpoenas, National Security Letters (available to federal law enforcement officers conducting terrorism investigations pursuant to the Patriot Act), subpoenas, search warrants, and court orders. Certain jurisdictions require law enforcement officers to draft a search warrant, based upon probable cause, to obtain these records. However, the Stored Communications Act Of 1986 allows law enforcement officers to get these records using a lesser legal method than the probable cause search warrants require. The federal government enacted the Electronic Communications Privacy Act (ECPA) and incorporated it into the United States Code, specifically Title 18 sections 2701-2712. Although it is a federal regulation, it applies to state and local law enforcement agencies, as well.

One of the most significant sections of the act is that is establishes a requirement for obtaining records from communications providers that does not mention probable cause. This has confused prosecuting agencies that assist state and local law enforcement. Some state and county prosecuting attorneys find the less than probable cause standard alarming and continue to require officers to obtain records from communications companies with a search warrant. Yet, this is not the case.

When examining ECPA, it appears to apply to electronic mail as "remote computing services." However, 18 USC 2711 references section 2510 for other definitions to which ECPA applies. It is from these definitions that ECPA applies to telecommunications providers such as phone companies.

---

[1] *Smith v. Maryland* (1979) 442 US 735

§ 2703. Required disclosure of customer communications or records

(a) **Contents of Wire or Electronic Communications in Electronic Storage.** – A governmental entity may require the disclosure by a provider of electronic communication service of the contents of a wire or electronic communication, that is in electronic storage in an electronic communications system for one hundred and eighty days or less, only pursuant to a warrant issued using the procedures described in the Federal Rules of Criminal Procedure by a court with jurisdiction over the offense under investigation or equivalent State warrant. A governmental entity may require the disclosure by a provider of electronic communications services of the contents of a wire or electronic communication that has been in electronic storage in an electronic communications system for more than one hundred and eighty days by the means available under subsection (b) of this section.

(b) Contents of Wire or Electronic Communications in a Remote Computing Service. –

(1) A governmental entity may require a provider of remote computing service to disclose the contents of any wire or electronic communication to which this paragraph is made applicable by paragraph (2) of this subsection—

(A) without required notice to the subscriber or customer, if the governmental entity obtains a warrant issued using the procedures described in the Federal Rules of Criminal Procedure by a court with jurisdiction over the offense under investigation or equivalent State warrant; or

(B) with prior notice from the governmental entity to the subscriber or customer if the governmental entity—

(i) uses an administrative subpoena authorized by a Federal or State statute or a Federal or State grand jury or trial subpoena; or

(ii) obtains a court order for such disclosure under subsection (d) of this section; except that delayed notice may be given pursuant to section 2705 of this title.

(2) Paragraph (1) is applicable with respect to any wire or electronic communication that is held or maintained on that service—

(A) on behalf of, and received by means of electronic transmission from (or created by means of computer processing of communications received by means of electronic transmission from), a subscriber or customer of such remote computing service; and

(B) solely for the purpose of providing storage or computer processing services to such subscriber or customer, if the provider is not authorized to access the contents of any such communications for purposes of providing any services other than storage or computer processing.

(c) Records Concerning Electronic Communication Service or Remote Computing Service. –

(1) A governmental entity may require a provider of electronic communication service or remote computing service to disclose a record or other information pertaining to a subscriber to or customer of such service (not including the contents of communications) only when the governmental entity—

(A) obtains a warrant issued using the procedures described in the Federal Rules of Criminal Procedure by a court with jurisdiction over the offense under investigation or equivalent State warrant;

(B) obtains a court order for such disclosure under subsection (d) of this section;

(C) has the consent of the subscriber or customer to such disclosure;

(D) submits a formal written request relevant to a law enforcement investigation concerning telemarketing fraud for the name, address, and place of business of a subscriber or customer of such provider, which subscriber or customer is engaged in telemarketing (as such term is defined in section 2325 of this title); or

(E) seeks information under paragraph (2).

(2) a provider of electronic communication service or remote computing service shall disclose to a governmental entity the—

(A) name;

(B) address;

(C) local and long distance telephone connection records, or records of session times and durations;

(D) length of service (including start date) and types of service utilized;

(E) telephone or instrument number or other subscriber number or identity, including any temporarily assigned network address; and

(F) means and source of payment for such service (including any credit card or bank account number),

of a subscriber to or customer of such service when the governmental entity uses an administrative subpoena authorized by a Federal or State statute or a Federal or State grand jury or trial subpoena or any means available under paragraph (1).

(3) A governmental entity receiving records or information under this subsection is not required to provide notice to a subscriber or customer.

(d) Requirements for Court Order. – A court order for disclosure under subsection (b) or (c) may be issued by any court that is a court of competent jurisdiction and shall issue only if the governmental entity offers specific and articulable facts showing that there are reasonable grounds to believe that the contents of a wire or electronic communication, or the records or other information sought, are relevant and material to an ongoing criminal investigation. In the case of a State governmental authority, such a court order shall not issue if prohibited by the law of such State. A court issuing an order pursuant to this section, on a motion made promptly by the service provider, may quash or modify such order, if the information or records requested are unusually voluminous in nature or compliance with such order otherwise would cause an undue burden on such provider.

(e) No Cause of Action Against a Provider Disclosing Information Under This Chapter. – No cause of action shall lie in any court against any provider of wire or electronic communication service, its officers, employees, agents, or other specified persons for providing information, facilities, or assistance in accordance with the terms of a court order, warrant, subpoena, statutory authorization, or certification under this chapter.

(f) Requirement to Preserve Evidence. –

(1) In general. – A provider of wire or electronic communication services or a remote computing service, upon the request of a governmental entity, shall take all necessary steps to preserve records and other evidence in its possession pending the issuance of a court order or other process.

(2) Period of retention. – Records referred to in paragraph (1) shall be retained for a period of 90 days, which shall be extended for an additional 90-day period upon a renewed request by the governmental entity.

(g) Presence of Officer Not Required. – Notwithstanding section 3105 of this title, the presence of an officer shall not be required for service or execution of a search warrant issued in accordance with this chapter requiring disclosure by a provider of electronic communications service or remote computing service of the contents of communications or records or other information pertaining to a subscriber to or customer of such service.

Examine some of the specifics of this law, specifically section (d). The order for production of records "may be issued by any court that is a court of competent jurisdiction." Therefore, it is not limited to federal courts or federal law enforcement. Any law enforcement may file an application to the court for an order under this section as long as it is not prohibited by state law.[2]

> *Shall issue only if the governmental entity offers specific and articulable facts showing that there are reasonable grounds to believe that the contents of a wire or electronic communication, or the records or other information sought, are relevant and material to an ongoing criminal investigation.*

To understand the provisions of this code it is important to know the difference between contents and records. Contents, actual files stored in the account, are defined by 18 USC Section 2510 (8) as follows: "contents, when used with respect to any wire, oral, or electronic communication, includes any information concerning the substance, purport, or meaning of that communication." For example, stored e-mails and voicemails are "contents."

"Records" includes material that is not content such as basic subscriber and transactional information. One court clarified the distinction of a record as "something stored or archived. The term information is synonymous with data."[3] The courts have held that records include transactional records, such as intranet account logs that track account usage and cell site/ tower data for calls.[4]

Basic subscriber and session information are permitted in this section. This includes the name, address, local and long-distance telephone connection records, the length of service with start date, services utilized, telephone number, and source of payment (including any credit card or bank account number).

While it is difficult to equate what "reasonable grounds to believe" are in relation to probable cause and suspicion it likely hovers somewhere below the reasonable suspicion threshold. A law enforcement officer is still required to provide "specific and articulable facts" to a judge. Furthermore, the law enforcement officer must also show the information sought is "relevant and material to an ongoing criminal investigation." According to the United States Department of Justice: "[What is required is] a short factual summary of the investigation and the role that the records will serve in advancing the investigation."[5] The "specific and articulable facts" standard is derived from the U.S. Supreme Court decision *Terry v. Ohio,* which established the framework for detentions and pat searches when the subject is not under arrest. In the decision the Court stated, "in justifying the particular intrusion the police officer must be able to point to specific and articulable facts which, taken together with rational inferences from those facts, reasonably warrant the intrusion."

There are less requirements for this section are much less than those to obtain a search warrant. This gives law enforcement officers the ability to advance their investigations in cases when they have yet to reach the probable cause threshold.

---

[2] *Brown* v. *Waddell* (4C 1995) 50 F3 285, 290

[3] United States, 509 F. Supp. 2d 76, 80 (D. Mass. 2007).

[4] Application of United States, 509 F. Supp. 76, 80 (D. Mass. 2007)

[5] *Searching and Seizing Computers and Obtaining Electronic Evidence in Criminal Investigations*, Computer Crime and Intellectual Property Section, Criminal Division, US Department of Justice (July 2002)

Voicemail, e-mail, text messages, and multimedia messages are known as "stored communications" because the senders know that his messages are recorded and stored for playback and forwarding. As stored communications are not considered as private as real-time phone conversations, a wiretap order is not required to obtain copies. Law enforcement officers may gain stored communications via a standard search warrant. However, if the stored communications are older than 180 days they may be procured with a court order.

Section (d)

Section (d) is the heart of the statute as it applies to law enforcement legal demands associated with records that cell service providers maintain. As noted in the first sentence, an order may be issued by any court that is a court of competent jurisdiction. This is an extremely useful clause in the law because it allows any court to issue an order pursuant to this federal statute. Section (d) is one of the areas that state, county, and local magistrates have difficulty adjusting to. They are used to interpreting and ruling on state law and, yet, here is a federal statute that allows them to issue an order under federal law.

The order may be issued by a court of competent jurisdiction once the law enforcement officer demonstrates "specific and articulable facts showing there are reasonable grounds to believe ... are relevant and material to an ongoing criminal investigation." The phrase "specific and articulable facts" requires the investigator to show the reviewing magistrate there is something that indicates the records sought are germane to the investigation. In other words, an order issued under this section is not a fishing expedition to acquire records from cell service providers for target phone numbers, especially when there is little information or suspicion about their involvement in criminal activity.

However, the phrase "relevant and material to an ongoing criminal investigation" does give the law enforcement community the opportunity to expand its investigations, particularly into conspiratorial crimes, by the use of this section. For examples, narcotics trafficking is an inherently conspiratorial crime. A dealer must enlist the aid of others in order to accomplish his intended goal. A street level dealer must obtain the narcotics they sell from a mid-level dealer. The mid-level dealer must obtain his supply from an upper level dealer. An upper level dealer must obtain his drugs from a smuggler or drug trafficking organization. But along the way there are additional people involved in aiding and abetting the crime.

The street level dealer may be assisted by lookouts, associates who are in charge of running the money or drugs back and forth from a secure location, transporters who deliver the drugs to their clients, and others who may be used as security to protect the dealer and collect debts. Similarly, the mid-level dealer communicates with his supplier and other subordinate street level dealers as well as others who may be involved in packaging, storage, and drug delivery.

The upper level dealer also must communicate with a subordinate network of dealers and a similar logistics network to include those involved with the integration of money derived from the sales of drugs into the legitimate financial system. Each step of the ladder and every relationship with an associate requires communication in order to facilitate the crime. Communication may take many forms but is commonly done via electronic means such as cell phones. An investigator may articulate this relationship and the requirement for communication between members of the conspiracy, using it to satisfy the reasonable grounds standard found in 18 USC 2703(d); they do this by showing the records are relevant and material to the ongoing criminal investigation. In order to use this section to address conspiratorial crimes it is helpful to understand the definitions to the words "relevant" and "material." The dictionary definitions are:

- Relevant — bearing upon or connected with the matter in hand; pertinent

- Material — a group of ideas, facts, data, etc., that may provide the basis for or be incorporated into some integrated work

To satisfy the two-prong requirement under 18 USC 2703(d), an officer must show a connection between the crime under investigation and the fact that the inherent conspiratorial nature of the crime is part of a larger integrated work. Investigators can apply for court orders when there are other unknown factors involved in the commission of the crime.

Over the course of six months a medium-sized local law enforcement agency arrested gang members during criminal investigations. Not all of the arrestees were members of the same gang. In fact, some were from rival sets or cliques that were feuding with others. Pursuant to the arrest, any gang member's cell phone was searched and the results were copied onto an Excel spreadsheet. Investigators began to notice there were common contacts stored in the suspects' electronic phones books; even among those who were feuding rivals. The questions arose. Who were those common contacts and why were members of different gangs contacting them?

Every law enforcement officer knows gang membership and associated crimes are inherently conspiratorial. To even be defined as a gang most states require certain criteria to be classified as such. These requirements vary by jurisdiction, but they usually need elements such as three or more people, a common name or identifying sign, and a pattern of criminal activity. Gangs also have certain needs that must be met in order to operate. They frequently commit crimes such as drug trafficking and must obtain narcotics from their sources of supply. Gang members arm themselves and do so by either stealing guns or purchasing them from others. Gang members also have a logistics infrastructure with sources for transportation and housing. By combining the knowledge of gangs and the likely communication with those who assist them, investigators in this case were able to write court orders pursuant to 18 USC 2703(d) for more than 50 common contacts. This transformed the investigation into a proactive intelligence-led one that resulted in the identification of drug suppliers and weapons traffickers. It also allowed investigators to target individuals based on multiple gang members' information, which lead to stash houses, hideouts, and other previously unidentified gang members or associates.

The provisions of 18 USC 2703(d) permit investigators assigned to conspiracy crimes, such as narcotics and gangs, to advance their inquiries when traditional enforcement techniques are ineffective. During a narcotics investigation, the standard method of investigation is to arrest a dealer and then try and "flip" him as an informant. However, in many cases the investigation stalls when the suspect refuses to cooperate. Examining the dealer's call detail records can help identify his suppliers as well as other unidentified co-conspirators.

Due to the lower threshold of "reasonable grounds to believe" there is little justification for not obtaining the call detail records of suspects involved in conspiracy crimes. Unfortunately, law enforcement agencies fail to obtain these records or only look to them as a potential investigative lead when other methods have failed. If you are investigating conspiracy crimes you should be compiling the suspect's call detail records into a consolidated database to identify the hierarchy of the organization, common contacts, and other co-conspirators.

Narcotics and gangs are two obvious examples of utilizing 18 USC 2703(d) to obtain records from cellular service providers to advance the investigation. However, this section can be used on just about any crime as long as the "relevant and material" requirements are met. Proper articulation makes it easier to apply for these court orders in just about any criminal investigation. Law enforcement officers are hesitant to use this section for crimes that are not conspiratorial. Without specific information they believe they are unable to obtain a court order for the cell service provider, particularly when the crime only involves one person. A law enforcement officer to obtain an order under 18 USC 2703(d) as long as he has proper documentation.

Without additional information it might be difficult to document sufficient probable cause to obtain a search warrant for the cell service provider's records. The language below may help when applying for an 18 USC 2703(d) court order:

> *Those who operate a motor vehicle while intoxicated will have often communicated their plans and actions to friends and associates. These communications may include the suspects' intent to become intoxicated, their locations and destinations, and their companions who are or will be present with them. These communications are commonly done via phone call and/or text message. The identification of the friends and associations will provide investigators with the opportunity to identify potential witnesses who can testify as to the substance and nature of those communications, as well as, to possibly recover digital evidence showing the suspect's level of intoxication. Additionally, suspects' call detail records maintained by the cellular service provider for the suspect's phone routinely captures cell site location data (CSLI) during the normal course of business. A review of the CSLI in this case may help identify locations where suspects were when they sent or received messages or phone calls leading investigators to locations that may be able to provide additional witness testimony as to the suspect's level of impairment.*

How about the assault and battery case?

> *A review of the call detail records maintained by the cellular service provider for the suspect's phone may reveal additional information relevant to the investigation. Suspects who commit assault and battery may have announced their intention to the commit the crimes to their friends and associates, documented the crimes using their phone's camera or video functions, and sent those images to others. They might have communicated about the crime to others to brag, and/or conspired with others to provide transportation away from the crime scene as well as provide alibis for the suspect's location during the commission of the crime. A review of the suspect's phone records that the cellular service provider maintains will grant investigators information relevant and material to the ongoing criminal investigation including any communication the suspect had with others before, during, or after the commission of the crime.*

You can see that it is possible to apply the provision of 18 USC 2703(d) across a broad spectrum of law enforcement investigations, not just conspiracy crimes.

Warrantless Searches: Legal Challenges to 18 USC 2703(d)

Naturally, a law that allows the government to obtain information without a search warrant would draw the ire of civil rights and defense organizations. Frequently referred to as "warrantless searches" to give the impression there is no judicial review, these groups agitate against the use of 18 USC 2703(d) to obtain records from cellular service providers. They also point to a decision in the Sixth Circuit Court of Appeals known as *United States v. Warshak*.

Warshak is an interesting decision because it is part of case involving a well-known herbal supplement. You probably remember the commercials with "Bob," who used the natural male enhancement supplement Enzyte. Presumably "Bob" was smiling because the product worked. However, everything about the product and the company was a scam. The product didn't work, the medical reviews were false, and the company enrolled customers in automatic shipments even when they didn't want them. Eventually, the Internal Revenue Service (IRS) began an investigation into the company and its owners.

As part of the investigation, the IRS used 18 USC 2703(d) to obtain email records involving the defendants. During the appeals process, the Sixth Circuit Court of Appeals found the use of 18 USC 2703(d) to be unconstitutional. While frequently listed as the definitive ruling to the constitutionality of the law there are certain issues with the case that prevents it from being widely applied to transactional data involving phone records.

In the Warshak case, the IRS used 18 USC 2703(d) to obtain more than 27,000 emails sent between key personnel in the company. These email messages included the content, which is the primary source of contention. Even though 18 USC 2703(d) authorizes a provider to disclose "content," it is interpreted to authorize the disclosure of the transactional information about a communication maintained by a provider instead of the actual message content itself. As noted by the court: "We find the government *did* violate Warshak's Fourth Amendment rights by compelling his Internet Service Provider ("ISP") to turn over the contents of his email."

What the court did not rule on was what the standard use of 18 USC 2703(d) is: recovering information about the transaction of the email or other electronic communication and not the message itself. Consider the analogy of a letter. The sender and the receiver of a letter have an expectation of privacy in the contents. However, there is no expectation of privacy for the outside of the envelope. Everyone who handles the envelope can see what is printed on the outside and the letter's sender knows this. It is impossible for the letter to be delivered without that information on the outside of the envelope and the sender also understands this. Even the Six Circuit Court used the letter analogy in their decision stating: "If we accept that an email is analogous to a letter or a phone call, it is manifest that agents of the government cannot compel a commercial ISP to turn over the contents of an email without triggering the Fourth Amendment." Furthermore, the courts did not address the use of 18 USC 2703(d) orders to obtain information about the "envelope" the emails are contained in.

As described, a person cannot send a letter expecting the writing on the outside to be private. The same principal applies to phone calls. Do people believe the digits dialed on their phone are somehow private and their telecommunications provider does not record them in order to reconcile their bill? Curiously, though, the Sixth Circuit neglected to mention the case of *Smith v. Maryland,* in which the Supreme Court found there was no expectation of privacy in the dialed digits of a phone call because "petitioner voluntarily conveyed numerical information to the telephone company."[6]

The *Smith v. Maryland* case would have appeared to have some relevance to this decision in that it specifically addresses information voluntarily handed over to third parties. The Sixth Circuit did discuss the applicability of *United States v. Miller* 425 U.S. 435 (1976) in which the Supreme Court held that a bank depositor does not have a reasonable expectation of privacy in the contents of bank records, checks, and deposit slips. The Court's holding in Miller was based on the fact that bank documents, "including financial statements and deposit slips, contain only information voluntarily conveyed to the banks and exposed to their employees in the ordinary course of business." I thought the courts would have assigned a higher degree of privacy to someone's financial information than messages sent over the Internet.

But even I would have to acknowledge nothing in *Smith v. Maryland* or the application of 18 USC 2703(d) authorizes the interception of message content without a warrant. I promote the practice of using a search warrant to obtain the content of any electronic communication – phone call, text message, or email. However, it is false to ascribe an expectation of privacy in the transactional information contained in cell service providers' records during the normal course of their business.

The Warshak decision muddies the waters a bit because it deals with email messages and message content that are specifically addressed in the statute. What this case does not address is the applicability

[6] *Smith v. Maryland*, 442 U.S. 735 (1979

to call detail records obtained from a cellular service provider. However, this does not stop certain individuals and groups from pointing to 18 USC 2703(d) and repeating the words "unconstitutional" as uttered by the Sixth Circuit Court of Appeals. Read the decision and statute. Discuss both with your prosecutors and see if they believe, as I do, the "unconstitutional" aspect of this ruling applies to the content of emails and not the transactional records of phone calls.

Cell Site Location Information and 18 USC 2703(d)

Using cell site location information pursuant to 18 USC 2703(d) is objectionable to a number of legal groups including the American Civil Liberties Union and the Electronic Frontier Foundation. A major point of contention is the inclusion of information regarding historical cell site information, frequently referred to as Cell Site Location Information (CSLI), contained within the call detail records that are provided to law enforcement pursuant to a court order. There are those in the legal community who believe CSLI should only be obtained via a search warrant based upon probable cause.

Their belief was reinforced with the *United States Supreme Court United States v. Jones* 10-1259 decided 1/23/2012. In this case, Jones was a suspected narcotics trafficker. As part of the joint FBI-Metropolitan Police task force investigation, officers used a variety of investigative techniques, including surveillance cameras, a pen register, a wiretap, and a GPS tracker installed on Jones' vehicle. That's when the problems started. The tracker was not installed until the 11th day after the search warrant was signed and the installation occurred in Maryland, not in the District of Columbia where the warrant was signed. The tracker was used constantly over the next 28 days and "established the vehicle's location within 50 to 100 feet.... It relayed more than 2,000 pages of data over the 4-week period."

The Court found that the tracker's update every 10 seconds, 24 hours a day, along with the tracking of his vehicle into a garage adjoining his residence, violated Jones' rights under the Fourth Amendment. The ACLU and the EFF quickly used this decision and have attempted to apply it to CSLI. As the EFF's Civil Liberties Director Jennifer Granick said, "This same logic applies in cases of cell phone tracking, and we hope that this decision will be followed by courts that are currently grappling with the question of whether the government must obtain a warrant before using your cell phone as a tracking device." It's not immediately clear if they are deliberately misstating the differences or if they truly do not understand the differences between GPS and CSLI.

Global Positioning System Data versus Cell Site Location Information

The differences between GPS tracking and CSLI are numerous and significant. First, cell service providers retain CSLI as records kept in the normal course of business. Law enforcement is not asking the providers to do anything they are not already doing and there is no intrusion into the cell phone owner's personal property to attach or install anything to collect the data.

Second, CSLI data is historical in nature. Absent a dialed number recorder/trap trace device, referred to as a pen register or a wiretap, it is impossible for law enforcement to obtain real time "live" CSLI from the provider. Even if it were conceivable to continuously track a cell phone using CSLI it would be inadvisable to do so.

A GPS tracker has a large battery with the sole function to power the tracker and relay the data via a cellular network. A mobile device's battery supports multiple functions including phone calls, Internet, and applications, all of which may operate simultaneously. The continuous monitoring of CSLI would cause a significant drain on the device's battery and potentially alert the user due to the significantly decreased battery life.

Third, CSLI information is not continuously captured as was the case with the GPS tracker in the Jones decision. CSLI is only recorded when the user makes or receives a phone call using the provider's network. There are some providers that record CSLI for events other than phone calls, such as text messages and mobile data events, but they are the exception not the rule.

Finally, standalone CSLI is nowhere near as accurate as a GPS device. In published documents, websites, and court case filings, some attorneys propagate the argument that CSLI is extremely accurate and causes an unconstitutional violation of the Fourth Amendment as it can pinpoint a mobile phone inside a house or other structure that would otherwise require probable cause to search. As we will discuss in subsequent chapters, a single cell site can cover anywhere from a city block to five miles or greater in suburban and rural environments. While CSLI can give an investigator general information about a particular cell device's location it is does not provide the same accuracy as GPS.

Preservation Letters

There is another provision in 18 USC 2703 that can greatly assist law enforcement officers in their investigations; paragraph (f). 18 USC 2703(f) let investigators submit a request to service providers for records associated with a target phone number or email address while a search warrant, court order, or subpoena is sought. Much of the information a cell service provider maintains is perishable. Very few companies retain text messages once they are delivered to the subscriber's device. Providers may retain Cell tower or cell site data only for as long as it takes the hard drive of the computer servicing the tower to fill up. Some providers have drastically reduced the time they retain call detail records "to save space on the system" to as little as 90 days. All of these situations can be remedied with the submission of a preservation letter pursuant to 18 USC 2703(f).

With these sections law enforcement officers can submit a formal request to providers asking them to retain the records while legal process is sought. The submission does not need to be done with anything more than a fax on departmental letterhead and does not need to a judge or supervisor's signature. The investigating officer, or his designee, can sign and submit the form.

Upon receipt, the provider is requested to preserve the specified records for a period of 90 days to allow an investigator time to obtain legal authority to seize and search those records. While this is easily done with an 18 USC 2703(d) court order, some jurisdictions are still unwilling to permit their officers to use them and require a search warrant based upon probable cause. As the probable cause standard can take time to develop, submitting a preservation letter will help ensure the records sought still exist by the time the warrant is signed.

18 USC 2703(f) also includes a provision granting law enforcement officers permission to request a single extension of another 90 days if they approach the end of their first 90 days. Here's an important tip: Don't wait until day 89 to submit your second preservation letter. The cell service provider's subpoena compliance department may get hundreds, or even thousands, of fax and email submissions daily. If you send you preservation letter in at 1700 on day 89 it may not get processed for several days. My recommendation is if you find yourself at day 80 no closer to getting the search warrant signed you should submit your extension request.

Here is another recommendation, particularly to law enforcement officers working crimes against person: *submit a preservation letter as soon as you identify a telephone number in the investigation*. My reasons are twofold. Too often in felony crimes like homicide, rape, or robbery someone who was identified early in the investigation as a witness is later determined to be a suspect. These investigations can be lengthy, as failing to submit a preservation letter early in an investigation has resulted in the loss of critical information. My second reason is the simplicity of the process. Submitting a preservation letter only takes a letterhead, fax, and an officer's signature. There is minimal time associated it and no cost to the investigating agency. Therefore, there is no real reason not to send a preservation letter any time a phone number is identified in the investigation. Make it a habit to send these letters out because I have seen too many cases where the lead investigator did not send a preservation letter and regretted it later.

An important distinction brought up in the Warshak case is preservation letters are retroactive and not proactive. There is nothing in the statute that permits collecting data or records. You cannot use a preservation letter to request providers store text messages or other content as it is delivered to their subscriber. Not only does the Sixth Circuit Court of Appeals say so, but the Department of Justice's Compute Crimes Division feels the same way. In the Electronic Search and Seizure Manual, they say: "Agents who send § 2703(f) letters to network service providers should be aware of two limitations. First, § 2703(f) letters should not be used prospectively to order providers to preserve records not yet created. If agents want providers to record information about future electronic communications, they should comply with the electronic surveillance statutes discussed." In other words, if you want prospective or real time monitoring of content you need an electronic surveillance or wiretap order to gather this type of information.

There are two schools of thought on what to put into a preservation letter and I subscribe to them both. The first school sends a preservation letter consisting of about two lines that say, "Pursuant to an ongoing criminal investigation conducted by the Haywood Police Department I am requesting you preserve all records associated with cellular phone number XXX-XXX-XXXX for a period of 90 days."

In my experience this works for the vast majority of cases. However, there is a second school of thought that lists every conceivable piece of information they believe a cellular service provider maintains. These preservation letters are several pages and extremely detailed. There is a use for these types of records, but they shouldn't be used for every case.

Providers will preserve all of the information they have available using the first example of a preservation letter. My concern with the excruciatingly detailed preservation letter is that if you miss something the provider could later say: "You didn't include it in your preservation letter, so we didn't save it." Cell service providers retain a significant amount of information most law enforcement officers don't know about and if you don't know about it you can't include it in the preservation letter. This brings me to where I believe a detailed preservation letter is crucial.

There are two types of data cell service providers retain that are not routinely preserved using the simple preservation request. The first is cell site data because large volume of data must be preserved and demands for cell site data are handled by departments outside of the provider's routine subpoena compliance division.

If you need cell site data preserved for a future cell site dump (to be discussed in detail in a later chapter) you must specify the towers you want saved. You can ask the provider to protect this information in a simple request, but they are not going to do it. In order for the perseveration letter to work for cell site data you will need to list all the towers out. This requires the identification of the cell sites that cover the crime scene.

Other information that must be specifically requested in a preservation letter is, Per Call Measurement Data (PCMD). PCMD's measurement data can reveal the distance of a cellular phone handset from a cell site. This information is extremely perishable and certain providers only retain it for seven to 14 days. After that it is gone forever. Other than these two specific examples I have not encountered a case where the simple preservation letter did not capture the required information.

One of the downsides to 18 USC 2703(f) is there are no specified penalties for failing to comply with a preservation letter. I have no doubt cellular service providers will make a good faith effort to comply with the preservation letter they receive. However, I encountered one particular pre-paid provider that has, on several occasions, said they did not receive the preservation letter and I was unable to obtain the information I needed for the investigation. Unfortunately, there is no remedy and complaints to various regulatory agencies have failed to gain the same traction as those for non-compliance with search warrants, court orders, or subpoenas. Lack of judicial involvement is one of the factors contributing to disinterest in enforcing compliance with preservation requests issued pursuant to 18 USC 2703(f).

Search Warrants

Despite the ease of obtaining records from cellular service providers using 18 USC 2703(d) jurisdictions still require their investigators to use subpoenas or search warrants. There are some advantages to using a search warrant over a court order. Specifically, a search warrant can be used to obtain content from the provider such as text messages, voicemails, and email.

There are distinct challenges when using a search warrant to obtain records from the service provider. Most states have reasonable particularity requirements that mandate an officer describe in detail the places to be searched and the items to be seized. The challenge is making sure the investigator asks for, and receives, all of the information he needs. Thus begins the impossible quest – the search for the perfect warrant template. Unfortunately, it doesn't exist. The information sought from service providers change based on facts and circumstances of the case and information cell service providers retain. Too often I have seen law enforcement officers use templates, go-bys, or other "boiler plate" language obtained from co-workers or other agencies. Some of these templates are dated and do not reflect the providers' current technology or capabilities. In numerous cases these antique templates ask for information the investigator really doesn't need or fails to capture all of the information he requires to advance his case.

There are also different styles of search warrant language, particularly in the affidavit. Some agencies and jurisdictions write their search warrant affidavits in the third person, past tense language, whereas others write theirs in the first person. There is no right or wrong way to craft an affidavit as every jurisdiction is different. The examples below are taken from search warrants authored and served by federal, local, and state law enforcement agencies in California.

Basic Search Warrant

> SUBSCRIBER, BILLING, CREDIT, TOLL RECORDS, TEXT MESSAGES (including content), MULTIMEDIA MESSAGES (including content) and CALL DETAIL INFORMATION (CELL SITE and ADDRESS LOCATIONS and GPS location information if enabled) for the monthly billing periods that covers September 1, 2012 through November 9, 2012 for 30 days after this court order is issued for the telephone number listed below. Please deliver the requested data in an "Excel" or delimited "CSV" (non-pdf) format preferably via email to [insert email]

The language here is relatively generic and covers all of the basic information retained by the major cell service providers. Here is some additional clarification regarding the details of the information sought in the warrant.

Subscriber: This example does not specify what exactly subscriber information is or what the cellular service provider is required to turn over. Usually, the employee in the provider's subpoena compliance office is going to be able to see the basic information available to other customer service employees, including subscriber name, billing address, other contact phone numbers, length of service, and services subscribed to.

Billing: information may include the responsible party on the account for those situations where the customers using the service are not paying the bill. In theory, this category should also include means and source of payment information. However, the company does not routinely provide this data and it should be specified if you really need to know how or where the customer pays his bills.

Credit: Post-paid providers (those who bill their customers after the month of service as opposed to prepaid providers who bill in advance for service) run a credit check on their customers before agreeing to service. Boiler plate search warrants ask for this information, but I have yet to see a cellular service provider release the information pursuant to a search warrant, court order, or subpoena. When I've asked, the response has been providers do not actually run a complete credit report listing of all of the customer's credit accounts but actually run a quick credit report score and base their decision on that number.

Toll Records: Toll records is antiquated language found in many search warrants, court orders, and subpoenas. The reference to toll records also applies to landline telephones and does not apply to cellular devices. In the old days, a local landline call did not incur a charge and was not usually captured by the telephone company's billing system. However, calls across town or over longer distances did incur a toll charge. Hence, the term toll records. It doesn't really need to exist in a search warrant and can be captured using the industry standard term "call detail records."

Text Messages: You can ask for them, but that doesn't mean you are going to receive them. Most major cellular service providers do not store text message content once the message has been delivered. It is important to note that if the target device is not communicating with the network, like when it is powered off, the provider does store those text messages, including the content. But once the phone is turned on the messages are delivered and no longer available. You will have to be content with viewing the transactional data of the text message that is captured by the provider.

MMS Messages: Multimedia messages or MMS contain media such as pictures, video, or audio. They may or may not include additional text added by the sender. Similar to text messages, most major service providers do not store multimedia messages content once they have been delivered to the subscriber's phone. The notable exceptions are Sprint and AT&T. Sprint does store MMS content for a limited time, but the time frame in which they store the messages varies. So if the information may be important to your case a preservation letter pursuant to 18 USC 2703(f) is highly recommended. AT&T stores MMS content that has been saved by the subscriber in an online account and the information can be recovered using a search warrant.

Call Detail Information: Call detail information is transactional data maintained by the cellular service provider that documents calling number, the called number, date and time of the call, duration, special features (such as three-way calling) used, and cell site information. This data is different than the information the customer sees. For example, a customer may be able to view the called number and the duration, but he will not see cell site data.

GPS Information: Good luck with this one. I don't know of any provider who stores historical GPS information or why they would. There is no profit motive for cellular service providers to retain this data. However, a search warrant can be used to proactively monitor GPS in real-time such as through the use of Sprint's L-site service or similar "ping on demand" features built into most pen register orders and some search warrants.

Delivery Format: You can ask for the information in whatever format you want. That doesn't necessarily mean you are going to get it. Smaller regional providers still respond to search warrant and court orders with Adobe pdf files or worse, scanned copies of paper documents.

Additional Instructions

Most search warrants also contain additional instructions to the provider. These additional orders vary by the fact and circumstances in the case. For example:

> *Additional Orders – It is also ordered as follows:*
> *GPS Locations: PROVIDER is directed to provide real time cell tower information on request, as well as, to query the location of the phone if it is enabled with a Global Positioning System (GPS) interface.*

This is a proactive GPS "pinging" order and directs the provider to provide law enforcement with the cell site information and/or GPS coordinates on demand. If you anticipate using someone else to do the location request to the provider I would strongly recommend changing the language above to include the request may be made by the affiant or his designee. The case agent and search warrant affiant can be out in the field leading the surveillance or investigation team and they need someone back at the office or the station to query the provider for the suspect's GPS location. Some providers will only give the information to the author of the search warrant which can be problematic if he or she is out in the field at the time.

> *Release of Information: PROVIDER, Inc. shall furnish affiant with all information gathered pursuant to this order in a reasonable time (five business days) while this order is in effect.*

Depending on the jurisdiction, law enforcement officers have between five and ten days to serve the search warrant. This does not necessarily mean the provider will return the results within that time frame. It is common for some providers to return search warrant results 30 to 45 days after the receipt of the warrant. We'll discuss remedies for this later in this chapter, but you should know that just because you give the company a date to provide the information doesn't mean they will.

> *Non-disclosure: Having determined that there is probable cause to believe that the disclosure of the existence or execution of this order would impede a criminal investigation, it is ordered that PROVIDER, Inc., its officers, employees and agents not disclose to any person any information regarding the existence or execution of this order, or any information obtained pursuant to this order, unless ordered to do so by this Court, as such disclosure would impede the investigation.*

Numerous jurisdictions allow their law enforcement and prosecuting agencies to seal the search warrant affidavit. Court documents are public records and anyone can go to the court clerk and ask to review the search warrants on file. Obviously if the existence of the search warrant were known to the suspect he would immediately ditch the phone and destroy or conceal other evidence. Some cell service providers *will* notify their customers upon receipt of a search warrant unless they are specifically directed not to. If your investigation is still on-going, non-disclosure language should be included in the search warrant.

Stored Content

As discussed above, some providers such as AT&T and Sprint store media content as a service to their customers. If you wish to obtain this information via a search warrant it is important to specify the service and file types. Asking for all stored content may be overly broad and fail to meet the reasonable particularity requirement. I also like to put the provider on notice that I know what I'm talking about. If I'm sending a search warrant to AT&T I make sure to specify: "Photos and videos stored in AT&T's Locker Service, text, calls, and voicemail data retained in AT&T's Message Service, and stored contact information in AT&T's Address Book Service." If you are unsure what data a particular cell service provider stores check out the web page for the company and look for additional services, data, and backup plans.

Specialized Search Warrant Information

You may want some additional information from the cell service provider. It is critically important to remember that every item listed in a search warrant must be supported in the affidavit. Search warrants might contain a laundry list of data to be recovered, but a closer examination of the affidavit often reveals there is no supporting documentation of probable cause. Don't just cut and paste the language from another search warrant because it looks good and it appears to cover everything you could possibly want from the provider. Each item requires its own justification based upon probable cause for inclusion in the warrant.

Here are some examples of search warrant language for some of the more common specialized searches:

Account Comments: Anytime the provider has contact with the customer or modifies the customer's account, a notation will be made by a service representative on the account. The account comments can be revealing, particularly after the account holder has been arrested. It is not uncommon to see customer service representatives' comment indicating other subjects, called the provider to discontinue service, or to attempt to remotely delete the contents of the device after the suspect has been arrested.

Physical Address of Cell Sites and RF Coverage Map: Most cell service providers will include a key to deciphering cell site's location in the call detail records. However, sometimes they do not include the list of cell sites or the information is incomplete. You may choose to include this language in your search warrants to make sure you get the cell site key along with the call detail records. I've never had a problem contacting providers and asking them to provide information without further legal process.

Radio frequency coverage maps are the results of a cell service provider's survey of the coverage area and relative signal strength of a cell site. None of the major providers routinely send this information with their response so you may need to re-contact them to obtain it. Be forewarned. Cell service providers do not like to give these up, even with a search warrant. They consider the data to be proprietary and fear if it is introduced into evidence and becomes part of the court record their competitors will gain insight into their business. I doubt this is the case, but that is the argument most companies put up as a reason for not complying.

If you really need a coverage map for a particular cell site try to limit the amount of data you are requesting so they do not try and quash the warrant as overly broad. Also, make sure your prosecutor and judge are willing to back you up on this demand because companies will either try to delay fulfilling the demand and hope you go away or they will deny the maps' existence.

I can't help you with the first situation because it all depends on how much you want to fight to get the information. If the provider tells you the coverage map does not exist it is lying to you. Whenever a cell site is installed the provider evaluates the resulting coverage by conducting what is known as a drive survey. Multiple engineers perform a drive survey in a minivan festooned with antennas on the roof. The engineers will drive a grid pattern around the cell site while continuously monitoring the signal strength of the cell site. The results of the drive survey are incorporated into the RF coverage map. If a subpoena compliance clerk tells you the information does not exist ask to speak to a manager or supervisor. This data is not routinely accessible by the clerks and is obtained from another department within the cell service provider.

Additional cellular telephone numbers and subscriber information that dial the same number as XXX-XXX-XXXX.

I've seen this phrase creep into many search warrants and I don't really think the affiant knows what it is he is asking for and doesn't realize he hasn't received it. This type of search is known as a calls-to-destination search and is more thoroughly discussed in the chapter on Specialized Searches and Techniques.

Essentially, this language tells the provider to search its database for any other numbers your target is in contact with and provide its subscriber and call detail record information. It is not likely a cell service provider is going to comply with this order and, in most cases, I don't blame them. For the company to comply it would have to review all of the phone numbers contained in the call detail records of the target, identify its customers who call that number, and provide you with the call detail records and subscriber information. I think it could make a good case that the workload would be unreasonable and the legal demand overly broad. If you really want to acquire this information, obtain the targets call detail records and then follow the steps outlined in Call Detail Record Tool and Techniques.

> *All of the above records whether possessed by cellular service provider [target of warrant] or any other cellular service provider.*

At first glance this sentence does not make a lot of sense. How can cell service provider A give you information or records concerning cell service provider B? This language in a search warrant is designed to assist an investigator with two situations: a subscriber who travels to an area where the primary cell service provider does not have coverage and relies on a partner corporation, and the increasing prevalence of Mobile Virtual Network Operators (MVNOs). A MVNO is a cell phone company that does not own any of the actual infrastructures involved in the transmission of cellular calls.

A MVNO enters into a cooperative agreement with another larger cell service provider to handle the routing and transmission of the calls. The MVNO obtains the customers through retail points of sale and handles billing and collection but does not own the cell sites, switches, or fiber optic cables the data are transmitted across.

You would normally have to address the search warrant to two providers, the MVNO and the partner corporation who services the transmission equipment, to get information from a MVNO. The MVNO will be able to provide subscriber and billing information and the partner corporation will provide the call detail records and cell site information. However, some MVNO's have partner agreements with multiple cell service providers and there is no way of knowing which provider handled the particular transaction. This language remedies that situation by compelling the production of records from the partner corporation with resubmitting another search warrant.

> *All stored communications or files, including voice mail, email, digital images, buddy lists, and any other files associated with user accounts identified as: [insert identifying information such account(s) xxxxxx, mobile numbers (xxx) xxx-xxxx, or email account bigdog123@gmail.com.]*

Cell service providers now offer similar services to an Internet Service Provider (ISP) and maintain the same type of records such as text messaging, email, and file storage for the transfer of data including digital pictures. To access stored voicemail messages the provider may have to reset the user's password, which might alert him to the investigation. Some investigators borrow the language found in search warrants for cell Internet service providers and attempt to apply it to cellular companies that also provide data services. As with other specialized searches this information is not always available to a subpoena compliance clerk and they may need to obtain the records from other divisions within the company. You may consider using the following language to obtain Internet records:

- All connection logs and records of user activity for each such account including:

- Connection dates and times.
- Disconnect dates and times.
- Method of connection (e.g., telnet, ftp, http)
- Data transfer volume.
- User name associated with the connections.
- Telephone caller identification records.
- Any other connection information, such as the Internet Protocol (IP) address of the device, source of the connection, and information related to the connection from cellular service provider.

NOTE: The above is a standard request made to an ISP to track connection information. Remember with the type of cellular service offered today the user can send a message from the phone or from the associated account via a computer or other access device. However, the IP address information from most cell service providers is not as uniquely identifying as the IP address information from an Internet or email provider. Cell service providers have a limited number of IP addresses and they route all data through them. So, conceivably, thousands of phone numbers may be associated with a single IP address.

PUK for SIM card #: A Subscriber Identity Module (SIM) is a smart card inside of a GSM cell phone that encrypts voice and data transmissions as well as stores data about the specific user so the user can be authenticated to the network. The SIM also stores data such as personal phone settings specific to the user and phone numbers. SIM cards can be password protected by the user. Even with this protection SIM cards may still be unlocked with a personal unlock key (PUK) that is available from the service provider. Note that after ten wrong PUK codes, the SIM card locks forever.

Additional specialized search and search warrant language will be covered in additional chapters. I also compiled a book titled, *150 Search Warrants, Court Orders, and Affidavits* that have examples of commonly used telecommunications search warrants among others.

Smartphone Manufacturers May Have Additional Information

Smartphone manufacturers also provide a suite of services for their products, which can be independent of those cell service providers give. Research In Motion (RIM), the manufacturer of Blackberry phones maintains a significant amount of information regarding services provided through its company. For example, the cellular service provider upholds records relating to the call details, text messages, data events, and billing information. RIM keeps information regarding Blackberry Internet service email, their proprietary Blackberry Messenger service, their unique PIN messaging service, and some web browsing history.

Likewise, Apple maintains information regarding services of iPhone users. These records are also independent of those the cell service provider preserves and can include information regarding the purchase and activation of the device, application information, and data regarding any associated iTunes accounts.

More information regarding manufacturer specific data and services can be found in the Appendix of this manual.

Extensions for Business Records

In recognition that some legal demands take substantial time to fulfil and some data takes a business entity longer than the statutory requirement for compliance, several states have established a procedure for extending the compliance date. Known as extensions for business records the court may reasonably lengthen the time required for corporations to produce records upon showing that good cause exists for and the extension would not be detrimental to the investigation. If a cell service provider is having difficulty complying as a result of the volume of the information, not because it is "backed up" or "really busy," it is a good idea to seek an extension of the time to comply with the legal demand.

Non-Compliance

An investigator should start requesting call detail records and subscriber information from the cellular services provider *as early in the investigation as possible*. Depending on the provider, it may take several weeks or even months to get a response back. Early legal demands will ensure that you receive them in time to add value to the investigation. If you delay too long in demanding the records then your case may stall while awaiting the results. But what happens when the cellular service provider takes weeks or months to respond?

Consider the "routine" traffic stop performed by law enforcement officers across the county every single day. When you, as a sworn law enforcement officer of the executive branch of your city or state, make a traffic stop and tell the driver to stay in the car, are you asking them or telling them? If the driver does get out of the car and refuses your directions to get back in, are they subject to detention or arrest? If the driver continues to refuse, are they likely to end up face down on the pavement with their hands behind their back? Yet, when we send a search warrant, court order, or subpoena to a cellular service provider law enforcement officers are content to politely wait and allow their investigations to stall because the company has not responded.

Calls to the subpoena compliance departments for most major cellular service providers are met with responses such as "We take all requests in the order they were received" or "We receive hundreds of requests a day" or "We are backed up 30 days." I have not seen a search warrant, court order, or subpoena yet that says the words "request" or "please." These documents are orders from a magistrate or judge commanding the company to provide the records. They are not requests. The failure of a multi-billion dollar telecommunications company to adequately staff their subpoena compliance departments is not my problem. Just like the non-compliant driver on the traffic stop there must be consequences for failing to comply with the orders of an executive or judicial branch officer.

When drafting a subpoena, search warrant, or court order it is standard to include a date by which the records are to be returned to the requesting law enforcement officer. Unfortunately, most telecommunications service providers view their requirements to provide these records in a timely manner as an unnecessary cost and an adverse effect on their bottom line. Their subpoena compliance offices are frequently outsourced or staffed with moderately paid employees who are responsible for thousands of legal requests daily. Each incoming request is processed in the order it was received, often times regardless of the severity of the crime under investigation. While I am sympathetic to the individual employees, I have zero tolerance for the corporations that foster an environment that hinders any law enforcement officer's investigation.

Some law enforcement officers ask, "Well, what can I do about it? I work for Small Town Police Department and they are a giant company located in another state." If a suspect in a crime that occurred in your jurisdiction was out of state would that stop your investigation? Would you throw your hands up in the air and say there's nothing that can be done?

Officers and judges may believe that because the company is located or headquartered in another state they lack standing to compel them to comply. Not true. If the company is doing business in your jurisdiction by providing service or calls to and from the area, then most states have statutes that give the courts the ability to compel them to produce records.

All telecommunications providers are heavily regulated public companies. Every state in the Union has some sort of regulatory agency which governs the conduct of those corporations in the state, even though they may be headquartered outside the state. For example, California Penal Code Section 1524.2(b)(1) states "When properly served with a search warrant issued by the California court, a foreign corporation subject to this section shall provide to the applicant, all records sought pursuant to that warrant within five business days of receipt, including those records maintained or located outside this state."

There are a couple of methods of redress for prolonged response times to law enforcement legal process. While I do not advocate immediately jumping down the throat of the first subpoena compliance clerk who answers the phone, I do recommend the following steps:

1. **Be clear with your urgency**: If the case is significant, tell the provider that up front. Include it on your fax cover sheet and ask for expedited processing. Follow up the request with a phone call and ask to speak to a supervisor. Explain your situation, offer some understanding of their position, and explain some non-sensitive details of the case that may assist in the prompt return of your records. Some providers are now offering shorter response times for a fee. If the case warrants it, spend the money. It's not yours; it's your agencies' and the taxpayers'.
2. **Follow up call:** Remember to call the day the records are due and ask the estimated time of arrival. Ask to speak to a supervisor. Don't be obnoxious and don't threaten to arrest them (unless you have the backing of your law enforcement agency, your prosecutor, and the magistrate who signed the order. If you have all three and you end up arresting one of these people, please call me so I can buy you a beer.)
3. **Bring in the enforcers:** Most states have a Public Utilities Commission that is responsible for regulating one or more intra or inter-state business entities, including cell service providers. If your investigation is stalled because of the company's failure to comply, take the time out of your day to call the regulating agency and file a formal complaint with them. If the situation is emergent or the case is significant ask a representative from the commission to contact the service provider's legal department. I have done this before with varying degrees of success. At times the representative from the commission was less than enthusiastic with helping me. But the majority of the time I receive a shockingly rapid response from the cellular service provider and the records I need.

   In California the state PUC has telecommunications specialists who represent the state's consumers in dispute with providers. These same specialists can respond to complaints from law enforcement agencies regarding non-compliance with legal process. They require some sort of written documentation that should include a copy of the order, proof of service (you did save the fax confirmation, right?), and a summary of the issue including dates, times, and names of company representatives who have been contacted regarding the issue. I have *no love* for cell service providers and I hope more law enforcement officers will complain to their regulatory agencies when the company fails to respond in a timely manner. That being said, use this knowledge judiciously and not five minutes after the order was signed and delivered. Save it for the obnoxious "we'll get to it when we get to it" providers.
4. **Report it to the Feds**: The Federal Communications Commission (FCC) regulates telecommunications companies. Nothing seems to get their attention more than a complaint filed with the FCC. In fact, I have received several nasty responses from the legal department of one large cellular service provider in response to complaints I filed with the FCC for their failure to provide information in a timely bases and respond to exigent circumstances requests. Before taking this step, make sure you have copies of fax confirmations and all details of communication with the company including dates, times, and the names of the persons you spoke with.

5. **Order to show cause:** If the cellular service provider is not responding in a timely fashion, one of the tactics you may consider would be to return to the judge who signed the order. Ask the judge about his or her willingness to do either of the following:
    a. Enforce a contempt order against the provider
    b. Ask for an Order to Show Cause compelling the provider to appear before the court and explain why they are failing to comply. Admittedly, most judges are hesitant to do either because of the perceived jurisdictional issues.

Remember, if you are to take the time out of your day to file a complaint against incompetent cell phone companies, make sure you complain to those who can actually solve your problem.

Paying for Records in a Criminal Investigation

Not only do some cellular service providers have the temerity resist a lawfully served judicial order in the specified time, they next seek to present law enforcement with a bill for their "services." The question arises "Does law enforcement have to pay for records in a criminal investigation?" To answer the question, I rely heavily on the published articles of California Deputy Attorney General Robert Morgester. While partially specific to California law enforcement officers, DAG Morgester illustrates some key characteristics applicable to any agency presented with a cell service provider's bill. In his articles, he points out 18 USC 2706 "requires the law enforcement agency obtaining the contents of communications, records, or other information to pay the person or entity assembling or providing such information a fee or reimbursement of reasonably necessary costs, which were directly incurred in searching for, assembling, reproducing or otherwise providing such information."

However, in the case of California law enforcement agencies, the State Legislature has limited the recovery of costs to civil cases and has applied them to records obtained using a subpoena, not via a search warrant or other legal demand. So even if a law enforcement agency were obligated to pay for the records sought, what is the definition of reasonable costs? DAG Morgester looks to California Evidence Code section 1563, subdivision (b)(1), which states:

Reasonable cost, as used in this section, shall mean ten cents ($.10) per page for standard reproduction of documents of a size of 8 1/2 by 14 inches or less, and actual costs for reproduction of oversize documents or documents the reproduction of which require special processing, necessarily done in responding to the subpoena together with additional reasonable clerical costs incurred in locating and making the records available. The clerical costs may be billed at the rate of ten dollars ($10.00) per hour per person, computed on the basis of two dollars and fifty cents ($2.50) per quarter hour or fraction thereof, and actual costs, if any, charged to the witness by a third person for retrieval and return of records held by such a third person.

As DAG Morgester also points out, California case law stipulates a custodian of record is not allowed to charge for computer costs for retrieving data. Additionally, a corporation is not permitted to pass along charges made by one of its own divisions.

Cell service providers handle bills in two fashions. The majority of them will send records followed by a bill. However, some will attempt to withhold records pending payment. The first situation is easily dealt with by informing the provider that your agency is not paying and the company has a right to challenge it. If the provider is withholding the records pending payment then the investigator may petition the court for an Order to Show Cause citing excessive fees, compelling the provider to produce the records. As an important footnote, the prevailing party is entitled to expenses and attorney's fees. A sample billing dispute letter is included as an appendix to this book.

California statutory and case law aside, examine the provisions regarding cost recovery in 18 USC 2706.

(a) Payment – Except as otherwise provided in subsection (c), a governmental entity obtaining the contents of communications, records, or other information under section 2702, 2703, or 2704 of this title shall pay to the person or entity assembling or providing such information a fee for reimbursement for such costs as are reasonably necessary and which have been directly incurred in searching for, assembling, reproducing, or otherwise providing such information. Such reimbursable costs shall include any costs due to necessary disruption of normal operations of any electronic communication service or remote computing service in which such information may be stored.

I recognize the "shall pay" language as compelling a law enforcement agency to reimburse for records. However, my point of contention is the "costs as are reasonably necessary" language. For example, why is my agency expected to pay $50.00 for records? What are the costs associated with it? A search of salary survey information reveals a subpoena compliance specialist working in Parsippany, NJ (home of T-Mobile's subpoena compliance division) shows the average annual salary is $54,000.[7] Presumably, a 40-hour work week and two weeks off for vacation translates into 2,000 work hours at pre-tax $27 an hour. If the expense is personnel related that means my agency would be reimbursing them for a little less than two hours of work. I find it difficult to believe downloading a spreadsheet, printing a screen, and sending a pdf file takes two hours of work. As all of the documents are returned electronically there are no printing costs, so it appears their costs are limited to personnel. To make sure I routinely ask a cellular service provider to produce accounting of the fees related to fulfilling my court order or search warrant. To date I have yet to receive a response from any company.

(b) Amount – The amount of the fee provided by subsection (a) shall be as mutually agreed by the governmental entity and the person or entity providing the information, or, in the absence of agreement, shall be as determined by the court which issued the order for production of such information (or the court before which a criminal prosecution relating to such information would be brought, if no court order was issued for production of the information).

The cell service provider and the law enforcement agency have to agree to the fees. I don't agree to pay for any records from any cell service provider pursuant to 18 USC 2703(d) or a state search warrant. Some law enforcement agencies have passed down the same search warrant, subpoena, or court order template for years and that template may include boilerplate language agreeing to pay for records. Take that language out otherwise you've agreed to the charges. If they want to sue your department for the $50.00 they have to do so in the court that issued the order. Maybe then they will provide a list of their expenses.

(c) Exception – The requirement of subsection (a) of this section does not apply with respect to records or other information maintained by a communications common carrier that relate to telephone toll records and telephone listings obtained under section 2703 of this title. The court may, however, order a payment as described in subsection (a) if the court determines the information required is unusually voluminous in nature or otherwise caused an undue burden on the provider.

---

[7] http://www.indeed.com/salary/q-Subpoena-Compliance-Specialist-I-Parsippany,-NJ.html accessed 4/15/2013

This section specifically states subsection (a) shall not apply to records or other information maintained by the provider that related to telephone toll records and telephone listing obtained under 18 USC 2703(d). However, even using a conservative definition of the term "toll records" it appears there is little justification for a cell service provider to attempt cost recovery for most of the records provided pursuant to 18 USC 2703(d).

To clarify, I am an advocate for paying for specialized searches and services performed outside the normal course of business. There is no statutory exception for cell tower dumps, GPS pings, pen registers, or wiretaps.

Law enforcement officers often use template or "boiler plate" search warrant or court order language taken from other investigators. It is common to find language hidden in these older documents that commits the requesting law enforcement agency to reimburse reasonable costs. As an example, the District Attorney's Office for my employer's county distributes standardized templates for Court Orders and Search Warrants. Buried on the first page is language that states: "Affiant's employing law enforcement agency shall compensate and/or reimburse ABC Cellular Service Provider for all reasonable charges and expenses incurred in complying with this order."

If you continue to include this language or similar statements regarding your intention to give these companies money no matter what is in your legal demands, you may have committed your agency to paying for those records regardless; this is pursuant to 18 USC 2706(b).

Chapter 3: Tools for Examining Records

Not much has changed since the days when pagers were used to deal narcotics. The technology is different, but the process is the same. A user needs narcotics so he calls or sends a text message to his supplier. The supplier replies with a location where they will meet for the transaction. Searching a suspect's phone can provide some leads. But most phones do not have sufficient storage capacity to store enough incoming and outgoing calls to establish comprehensive patterns. By examining the larger set of information contained in your suspect's cell phone records you can establish patterns of behavior. When those patterns are combined with other law enforcement techniques such as surveillance you can gain a better understanding of your suspect's actions.

Inevitably, call detail record analysis will lead to additional subpoenas, court orders, or search warrants. However, the end result can be an intricate diagram a criminal organization's structure. Creating a diagram can be a time-and-resource-intensive process, but the end product is worth the effort.

In this chapter we are going to tackle a lot of territory. We will move into some hands-on work, examining, extrapolating, and diagramming information from provider records, but not before we understand providers and their processing systems. In situations where you are requesting these records, time is of the essence. Because most providers move at a snail's pace, your window of opportunity to retrieve records will leave you with just enough time for one reques – make it good. Lastly, you get what you ask, so know exactly what's available.

Understanding How Providers Work

Before we discuss how to handle and examine provider records we need to spend some time discussing the types of providers because they track and bill their customers differently. The more you know about their systems, the easier it will be for you to move quickly and build stronger investigative leads when time is scarce.

Post Paid and Prepaid Providers

There are two main types of cell service providers: postpaid and prepaid. With postpaid providers, most telephone companies bill their customers for the preceding month of services. Because the phone company extends credit to the customer, they are more likely to run some sort of credit check and require identification prior to initiating service.

In the prepaid provider market, a consumer must purchase an amount of credits before the phone is used and when the credits run out, the consumer buys more. There are two sub-types of prepaid providers, but they work almost identically. National chain stores and neighborhood retail stores may offer "off-the-shelf" phones that require the periodic purchase of credits in order to maintain service. These credits can come in the form of cards with a set number of minutes on them. Credits are replenished electronically via the Internet or by submitting credit card information over a toll-free number. The other type of prepaid service allows a walk-in consumer to pay with cash at a national service provider's retail or franchise store.

Account Information Retained by Providers

Regardless of the provider, most cell service companies retain three types of data related to a specific account: subscriber information, financial information, and call detail records (CDR). In the next sections, we will cover these categories in detail.

Subscriber Information

Subscriber information refers to the information retained by the cellular service provider and can include:

- **Name**: This is the name provided by the subscriber. In the case of prepaid cell phone providers there will be no information or the information is provided by the subscriber and is not verified.
- **Address**: As with the name provided by the subscriber, prepaid providers do not check the validity of the address so the subscriber can give whatever information they want, or none at all.
- **Identifying Information**: Such as SSN or CDL. For postpaid services, the provider requires a credit check in order to provide services. These credit checks need identifying information in the records the cell service provider maintains.
- **Other Contact Numbers** (known in the industry as "can be reached" numbers): These are additional contact numbers for the subscriber other than the phone number the service provider assigns. They are usually the home or work telephone numbers of the customer and are limited to postpaid providers.
- **Other Phone Numbers on the Same Account**: This can be an important source of leads in conspiracy investigations. You may see multiple phone numbers on the same account, paid for by the same consumer.
- **Email Addresses**: This information is found with postpaid providers but can offer additional leads in locating targets as the Internet service company can give the location where the e-mail account is accessed.
- **Make, Model, and Serial Numbers of the Phone**: This includes the make, model, and unique serial numbers of a cell phone as well as the accompanying subscriber identification module (SIM), if equipped. Investigators discount this information as irrelevant. However, I believe it is crucial as it establishes dominion and control over a prepaid subscriber's cell phone.
- **Customer Service Rep Comments and Notes**: When a subscriber calls his provider a customer service representative will record and document the discussion. These notes can include billing disputes, changes of addresses, and additional services requests.

Financial Information

When it comes to investigating prepaid service accounts, examining the payment provider records is often neglected. Even if the device was pulled off the shelf at a large retail chain it must still be activated at a point of sale to operate. The service provider retains this information, just like all of the data listed on the future payments. Therefore, valuable clues can be uncovered when investigating both the locations and method of payment outlined on the records. For instance, the locations that stock and sell prepaid phone and calling cards are also the targets of robberies and organized retail theft crimes. So often they have sophisticated inventory tracking systems and electronic video surveillance. These retailers usually cooperate with law enforcement investigations. Even if the prepaid provider operates a network of retail locations where its customers can pay cash for their service, its retail location can provide clues to your subject's geographic location.

Postpaid providers, which grant service to higher end phones, usually require a preliminary credit check to verify the identity of the customer at the point of sale. Cell service providers might retain the salesperson's identity during this transaction. This creates an additional investigative avenue but also points to a critical operative in a larger criminal conspiracy investigation. Once, during the investigation of a large narcotics distribution network, I noted that one employee had activated all of the cell phones at a particular store. That employee was an associate of the narcotics trafficker and was responsible for providing the suspects with access to an almost unlimited supply of secure communications.

Cell Phones in Prisons

Cell phones inside correctional institutions are a serious problem. They are being used to circumvent monitored communications inside the institution and provide avenues for inmates to continue ongoing criminal conspiracies "on the streets," arrange escapes, facilitate narcotics transactions, and direct witness intimidation. Unfortunately, they are often smuggled inside the institution by correctional staff members. I was surprised to find that not all of the phones seized by officers with the California Department of Corrections and Rehabilitation (CDCR) were low-end prepaid devices. Recently, investigators with CDCR presented me with two high-end phones seized from high-security inmates, an Apple iPhone 3Gs and a Motorola Android.

These phones were activated and serviced by a postpaid provider. Even if they were not high-end smartphones but were prepaid devices, they had to be purchased and activated at a retail point of sale. Ongoing service for the devices has to be provided with either cash payments at a retail center or using prepaid cards to add additional minutes to the account. Both of those methods of payment give way to investigative leads.

Call Detail Records (CDR)

Call detail records are much more thorough than the monthly billing records a subscriber sees. Call detail records (CDR) are the transactional records for every phone call made on an account. These records can also include sent and received text or multimedia messages, Internet usage, and user initiated GPS events. CDR contains fields for the targeted phone number, direction of the call (incoming, outgoing), special features used during the call (call waiting, three-way calling), number dialed for outgoing calls, and number called for incoming calls. The CDR also usually list cell tower information such as tower used at the beginning of the call and the one used at the termination of the call.

The following is the items that may be in the CDR:

- **Number Called** – This is the phone number dialed out from the target phone or the incoming phone number if it is available.
- **Direction of Call** (incoming or outgoing) – Some providers' call detail records will reflect an incoming call that went to voicemail as an outgoing phone call from the target number. Pay attention to the Special Features column of the call detail records.
- **Number Billed** – It is possible to bill a third party number for a phone call, although many providers no longer allow this practice.

- **Duration of Calls** – This is the hours, minutes, and seconds of the call. There is no standard in the industry for how this information is listed in the call detail records. Some providers round up to the nearest minute while others reflect the call duration in the total number of seconds. Also, there is no routine for how a completed call is recorded. Providers may reflect the duration from when the call was answered, but others start recording as soon as the number is dialed or the send button is pressed.
- **Caller ID** (sometimes) – Several companies capture the phone number of the incoming caller whether they have caller ID blocking enabled or not, particularly if the call is between two customers using the same provider. Unfortunately, there is no standard and providers do not always capture the incoming caller's number.
- **Special Features Used** (such as three-way calling) – If a particular call utilized a special feature such as three-way calling or call forwarding, it will be noted. Examining the use of special features can be particularly important for investigations involving inmates as they will often utilize an outside confederate to initiate a three-way call or to forward their call to another number.
- **Beginning and Ending Cell Tower Information** – Nearly all cell service providers capture at least the cell tower and sector used to initiate a phone call and many capture the terminating cell tower and sector, as well.

Smartphone Manufacturers May Differ

Smartphone manufacturers also provide a suite of services for their products. These services can be independent of those provided by the cell service provider. Research In Motion maintains a significant amount of information regarding services provided through their company. For example, the cell service provider maintains records relating to the call details, text messages, data events, and billing information. RIM maintains information regarding Blackberry Internet service email, their proprietary Blackberry Messenger service, their unique PIN messaging service, and some web browsing history.

Likewise Apple maintains information regarding iPhone users' services. These records are also independent from those cell service providers maintain and can have information regarding the purchase and activation of the device, application information, and data regarding any associated iTunes accounts.

Information on manufacturer specific data and services can be found in the Appendix of this book.

Dealing with Burner Phones

"A separate phone used for business purposes, usually for drug deals."

urbandictionary.com

"A cell phone meant to be used for only a short term and then thrown away in criminal circumstances; made popular by the HBO television show The Wire."

Wikipedia.com

On more than one occasion I have seen veteran law enforcement investigators throw their hands up in the air in surrender when encountering a prepaid or "burner" phone. They, like the crooks, believe, because the phone was pulled off the hanger at a convenience store, that there are no investigative leads. This is simply not true. There is a wealth of information contained in the records associated with a burner phone and these avenues should not be discarded merely because the phone is serviced by a prepaid provider.

The first bit of good news is the crooks think they are anonymous so they do not use some of the more complicated tricks to conceal their activities. Criminal suspects also suffer from stupidity and laziness. This is reflected in the prepaid providers' subscriber information. I have written a lot of court orders and search warrants to prepaid cellular service providers. I reviewed several hundred results from one of the largest prepaid providers that do not require any sort of identification in order to open an account. My reviews showed that about 25 percent of the subscribers gave their actual name or an easily recognizable variation, such as inverting first and last names. Additionally, many times the suspect used an established alias or moniker that made them tentatively identifiable. In my retrospective review I found another 25 percent of the subscribers gave an actual address or easily recognizable variation thereof. In summary a review of hundreds of prepaid subscriber records revealed only 50 percent of the records contained completely false information for name or address.

Another exploitable factor regarding prepaid phones is the financial records associated with them. A large number of prepaid phones (Trak Phone, Go Phone, Virgin Mobile) are sold via retailers such as Wal-Mart, Target, and 7-Eleven. Depending on the service provider, the phones that are sold through these retailers must be activated at the point of sale. In other words the phone must be scanned during the checkout in order for it to be activated on the provider network. This serves as a loss prevention feature and keeps thieves from stealing the phones. Think about your own jurisdiction and your own experiences. Have you ever been called to a retailer for a report of a shoplifted prepaid phone? The crooks are aware that they can't activate a boosted prepaid cell phone so they don't bother.

It is possible to get the purchase/activation location information and method of payment information from the provider. But just because it is possible does not mean it is easy. Many investigators will be told by the cellular service providers they are not able to retrieve activation information. They will also say they are unable to provide information about the method of payment information. Method of payment information is especially important because it can reveal additional investigative leads such as where the suspect is paying their bill at, where they are purchasing their rechargeable minutes from, and if they are using a credit or debit card to pay their monthly bill. When law enforcement investigators run into this problem they are being told a partial truth by the subpoena compliance department.

For most providers it is true the subpoena compliance personnel do not have access to the method of payment information but that does not mean the information does not exist. Rather the information is retained by the revenue department of the company and is compartmentalized to prevent unauthorized access to the financial information of the customers. To get activation and method of payment information from a provider, you will usually need to elevate your conversation to a supervisor or department head. On more than one occasion I've been told by the subpoena compliance personnel the information does not exist. My response is something to the effect of, "Then how does your company stay in business? If you do not track billing information your company would not survive. Therefore, the information must exist somewhere in the organization. Who do I need to talk to get it?"

Remember, the search warrants, court order, and subpoenas most law enforcement agencies use contain a demand for method of payment information but it is not routinely provided. Because the vast majority of law enforcement agencies demand, and then do not follow up when the information is not received, the companies have not developed a formal mechanism for the subpoena compliance personnel to fulfill the demand. The more law enforcement agencies demand the complete fulfillment of information contained in the legal process the more likely it is the providers will integrate method of payment information into the subpoena compliance process.

Activation information is especially important in the case of multiple phones. For example, many correctional agencies are struggling with the issue of cell phones inside an institution. Tracking the activation information can provide investigative leads to combat the two most common methods of contraband cell phones inside a correctional institution; staff compromise and bulk phone smuggling.

Staff compromise is self-explanatory. The most common method of bulk phone smuggling is for an accomplice to toss a container or package containing multiple phones and other contraband items over a fence or conceal it in a location where an outside work crew can recover it. Activation information is useful for these investigations because once the location where the phone was activated has been identified; the investigator can go to the loss prevention department of the retailer and recover video surveillance and other records of the transaction.

Also, method of payment information is another method of investigating contraband phones in a correctional environment. Getting the phone into the institution is the only aspect for the contraband smuggler. They must also keep the phones activated by paying the bills. Remember, even though it is a prepaid phone, someone has to pay the bill to keep it going. Tracking this information can lead to the person who procured the phones and coordinated the smuggling into the institution.

The loss prevention departments for many large retailers are an often underutilized source of information for law enforcement. Too many investigators believe the people who work in loss prevention are "wanna-be" cops who couldn't make it in law enforcement. While this is undoubtedly true in some cases, most large retail chains make a great effort to hire former law enforcement or former military personnel to work in loss prevention. Additionally, most large retailers have more sophisticated investigators assigned to roles such as organized retail theft and cargo theft. The investigators who work in these assignments are usually the best in the organization and can be a critical component of tracking down purchase and activation information. Don't overlook them just because they are not "real" cops.

Handling the Records

Now that you have the records in hand and have or haven't paid for them, let's discuss methods for examining them. In this section, you will learn how to handle originals, transfer to electronic files, and finally how to examine the information.

Never Work with the Original Data

Frequently, cellular service providers return the requested files via electronic mail. I know most investigators' first impulse is to immediately open the documents and start looking around. *A word of caution: the files you receive are evidence and need to be treated as such.* Numerous service providers only store their records for three to five years. In fact, one large prepaid provider now only stores call detail records for only six months. It is possible that a complex investigation or lengthy trial process may take longer than that and the original records can be lost forever. If you have not saved an original copy of the records and placed them into evidence, you risk opening the door for defense lawyers to challenge the validity of your evidence. Upon receiving an electronic mail return, I strongly urge you to immediately copy those files to a disk and book the disk into evidence before proceeding with any analysis.

Most court order or search warrant returns will come in an electronic format. The provider may send a compact disk (CD) or digital video disk (DVD). Other providers will send the records to you via email or allow you to download them from a secured website. It is extremely important to preserve the records you receive and treat them like any other form of traditional evidence. If you do receive the records via email or download, you should save the record by transferring it to a disk and booking the disk into evidence. Do this before you open the files and start browsing the returns. Never, ever work with the original records.

Most service providers retain their records for a minimum of five years, but there is no standard for how long they must retain them. Given the glacial pace some courts move at, your investigation and prosecution of the case may take several years. It is possible the original records will be deleted or destroyed during that time, so records you received from your court order or search warrant may be the only ones that remain.

Hash Function/Verification of Files

A cryptographic hash function is an algorithm, which when applied to a block of data such as a file, will return a fixed-length string consisting of numbers or letters. This fixed length string of numbers and/or letters will reflect a change in the underlying data if any is made to the file or the information within it.

Popular forensic tools and applications incorporate various hash algorithms. For example, Cellebrite's Universal Forensic Extraction Device (UFED) generates two hash algorithms for data extracted from using the device.

Examination Report | Page 29 of 33

13	File Source: Phone File Size: 104251 Bytes File Date/Time: 08/25/11 11:40:12 MD5: 477242DFA0E34D35B492 2C7B01249828 SHA256: D652C4B6 7D92ABB A046F8D 15CABB6 1BC7F95 D3C9A3E 0C232F9 E98530E 2BE2B45	Pixel Resolution: 573x768	
14	File Name: Photo0243.jpg File Path: /Photos File Source: Phone File Size: 445670 Bytes File Date/Time: 08/25/11 11:35:54 MD5: EA47979535845ACFA5DB 6E504053ACDB SHA256: C52E2482 2288137 F890C4F 0FBBC68 16C0CE2 680EC6F FFF3271 E30E901 7170AB5	Resolution: 72x72 (unit: inch) Pixel Resolution: 1280x960 Camera Make: SAMSUNG Camera Model: SGH-T259 Date/Time: 2011:08:25 11:35:53	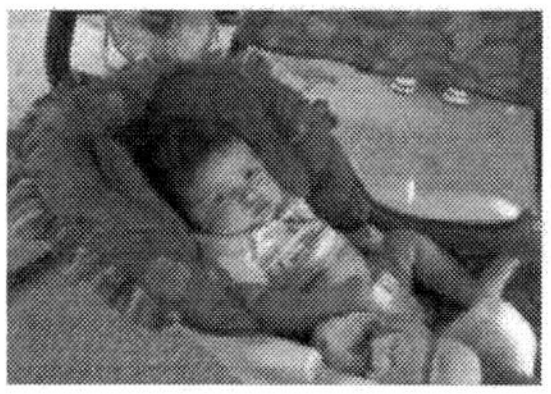

MD5 and SHA256 are commonly used cryptographic hash functions to insure evidence integrity. There are multiple online and downloadable programs to verify the hash of evidence obtained using forensic tools as well as data received from electronic means such as court order/search warrant responses.

Securing Search Warrant/Court Order Returns Sent Via Email

Upon receipt of a court order or search warrant return from a cell service provider an investigator should immediately take steps to preserve the evidence's integrity. The following example is shown using Microsoft Outlook 2007.

Before opening the email message or any of the attachments, highlight the email message and click the **File** located in the upper left corner of the screen. Select **Save As** and choose the option **Outlook Message Format**.

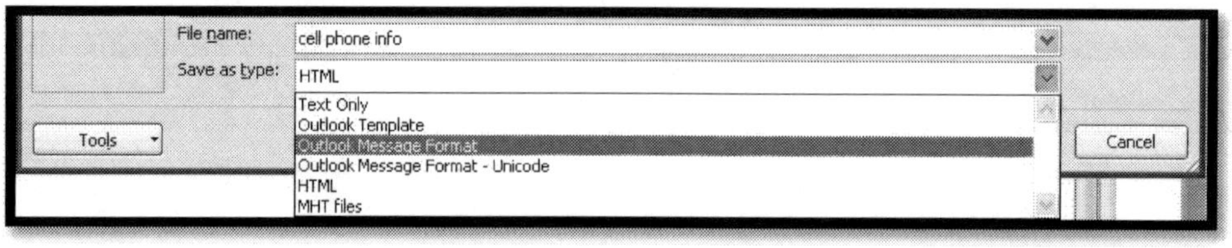

Selecting this option saves the message, the files, and the header information that documents the Internet protocol address(es) of the sender as well as the route the message took to get to your department's server. The IP information can be found by selecting the small drop down arrow in the **Options** box.

From there, select **Message Options** and you will be able to view the IP and routing information which can help verify the integrity of the message.

Choose a location to save the file to, preferably in a folder with the case number or other identifiers.

The next step is to verify the hash of the email message. There are a number of free online services that will calculate the hash value of documents, but law enforcement officers and their agencies are leery of letting an anonymous third party web application access their data. You may choose to use one of these services or to download and install an application to calculate the hash value. *[NOTE: Some of these programs are designed to run from the command line prompt. If you are not familiar with command line, make sure the program you select has a graphic user interface (GUI) and is designed to run on your department's operating systems (Windows).]*

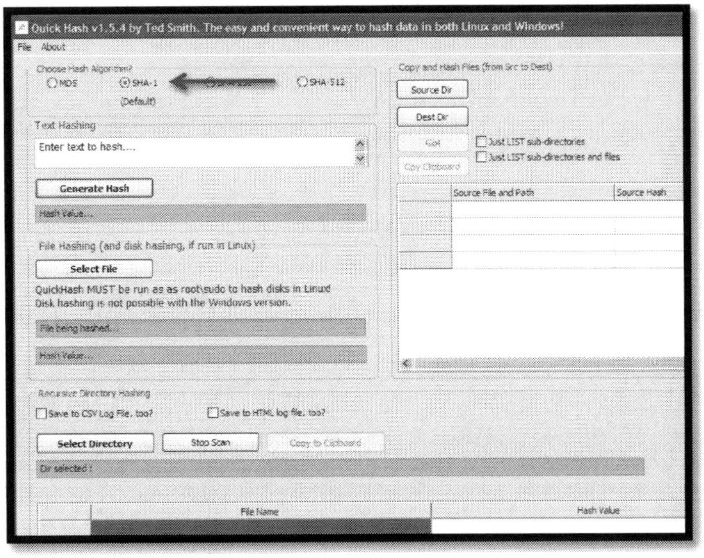

This example illustrates the use of a program called Quick Hash available for free from Source Forge. Open the program and note the default hash algorithm is set at SHA-1 (Secure Hash Algorithm-1). You can choose from several different options in this program, but for illustration I selected a more comprehensive algorithm, **SHA-256.**

Navigate to your save Outlook file using the **Select File** function. The software will automatically calculate the hash value using the algorithm selected. From here, you can **Copy** and **Paste** the hash value into a report or other document for preservation.

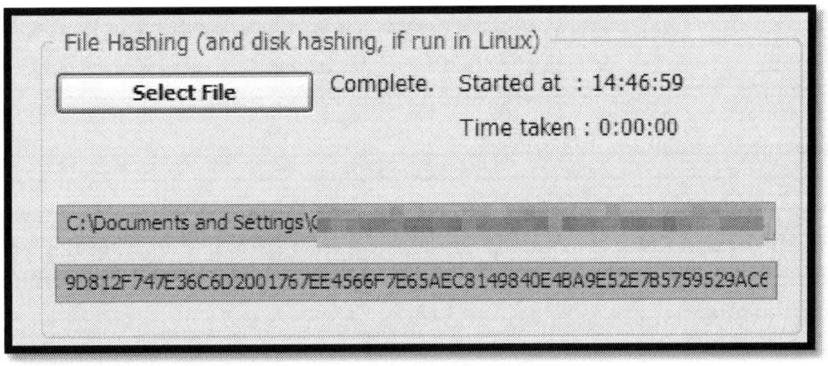

Quick Hash also has a function that allows you to calculate the hash value of files whenever you move them or transfer them to another storage medium. While doing so the program automatically calculates the hash value of all of the files contained in the folder and then creates a .csv (Comma Separated Value) file, which is readable using ExcelThen you can **Copy** and **Paste** the hash value into a report or other document for preservation.

If you already opened the email, it's not too late. Just don't open the files. Within Outlook press the Windows button in the upper left corner of the message box. Select **Save As** and save the records to your computer. Then follow the steps outlined above to calculate the hash value of the files. It would be preferable to hash the files before saving them to your computer, but I haven't found a hash calculator yet that will work inside Microsoft Outlook.

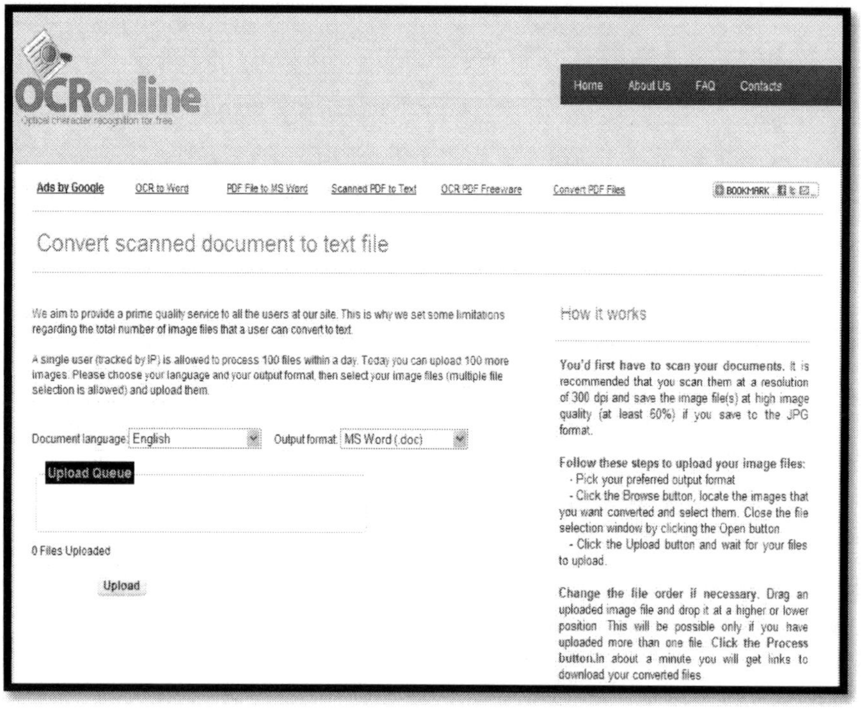

Sometimes a cell service provider will return the records in paper. I always recommend requesting the records in electronic format, but sometimes it's just not possible. If you do receive a large amount of paper returns, call the subpoena compliance department back and see if they can provide them to you in electronic format. If you are unable to get the records from the provider in an electronic format, you are will have to convert them into an electronic format in order to use them for data analysis. One alternative to re-typing all of the data into a spreadsheet is using optical character recognition (OCR) software.

OCR software takes a scanned image of a document and translates it into an electronic format of your choice (word processing, spreadsheet, etc.) OCR software can cost anywhere from $80 to $500. The software's success sometimes depends on the quality of the scanner. When using a scanner, set the dots per inch (dpi) to its highest setting. This may slow the process down when scanning a large amount of documents, but it does assist the software with recognizing the images. Also, keep in mind that every item on the page will get scanned into the document. Logos, hand written notes, etc. will all get scanned into the electronic document and will need to be manually cleaned up.

I have used Omnipage by Nuance Software for several years. While the program is not foolproof, it does significantly reduce the amount of time to translate paper phone records into an electronic format. Most law enforcement agencies that utilize Microsoft Office have a software program included called OneNote. This software has an optical character recognition element that will extract text from image files such a .pdf and .jpeg. OneNote can be used as an improvised OCR program to convert scanned files into useable text.

There are also free online services that will allow you to upload your documents, perform the OCR process, and return the files to you electronically in a format you specify. I don't really care to use them since they require me to upload my file over the Internet and I lose control over where it may go. Also, many of these free services have limits as to the size of the file you can upload.

Reading a Call Detail Record

The response from the cellular service provider can sometimes be confusing or intimidating. While most companies now have instructions for reading the call detail records, the records do not always make sense. If you have any questions or are unsure as to what exactly something means in the response from the cell service provider, make sure to call and ask them directly.

For your reference, I've included a table that outlines and defines some of the typical terminology used in call detail records. Please refer to this table when reading the following fictional example in the next section.

Common Detail Record Terminology

Billed MDN	Stands for Mobile Directory Number and is the industry term for the 10-digit number phone number we use when dialing calls.
Date	Date the call was initiated. (In the next example, the records were requested and returned in Pacific Standard Time.)
Time	Time the call was initiated.
Duration	Duration of calls in minutes and seconds.
Dialed Number	Actual number dialed by the phone's user. The dialed number may not always reflect the true number called. For example, the caller may not enter the area code if he is calling from the same area in which case the investigator must assume the number dialed is in the same area code as the target number. Also, if the caller uses a special feature such as *67, these digits may be reflected in this column.
Destination Number	Actual number called.
Direction	Our Urban PCS example uses codes to abbreviate the direction of the call
ML	Outbound call from Urban PCS mobile.
LM	Inbound Call to Urban PCS mobile.
MSE	Mobile Security Entry Record (This is an internal event for Urban PCS and not a user generated activity.)
CPD	Cellular Packet Data Originated Record – Data event and not a voice record.
Special Features	Unanswered incoming calls will be forwarded to voicemail. This will reflect as an outgoing phone call from the target number. This is because the call is forwarded from the target number to the voicemail system. This can result in duplicate and confusing data when analyzing call detail records and calling patterns.

Using Excel and Add-On Tools

In this section, we will explore how to use Microsoft's Excel to format data as well as some downloadable add-ons that will make life easier for you. As we move through these products we'll use a fictional cell service provider called "Urban PCS" to examine a call detail record file.

Sometimes, Urban PCS provides their records in a confusing format. The column headers in the following example list the data and codes. The information to read these CDRs is usually included with the returns but not always.

	BILLED MDN	DATE	TIME	DURATION	DIALED NUMBER	ATION NUMBER	DIR	STATUS	SF	CALLER ID	CELL
2	4155711687	04/21/2010	00:02:56	00000:02	4155711687	4155711687	CPD	0	0		G2023
3	4155711687	04/21/2010	08:27:00	00000:02	*777		CPD	0	0		G2026
4	4155711687	04/21/2010	08:27:05	00000:02	4155711687	4155711687	CPD	1	0		G2026
5	5107766577	04/21/2010	09:57:46	00000:48	4155711687	4155711687	ML	0	0		G6030
6	4155711687	04/21/2010	09:57:49	00000:46	4155711687	4155711687	LM	0	0	5107766577	G2026
7	4155711687	04/21/2010	11:08:34	00009:27	4155711687	4155711687	LM	0	0	4158676584	G2032
8	4155711687	04/21/2010	11:29:51	00001:25	4155711687	4155711687	LM	0	0	9253357169	G2032
9	4155711687	04/21/2010	11:35:51	00001:10	14153689734	4153689734	ML	0	0		G2026
10	4153689734	04/21/2010	11:35:57	00001:04	4153689734	4153689734	LM	0	0	4155711687	G2338
11	4153689734	04/21/2010	11:38:31	00001:33	4155711687	4155711687	ML	0	0		G2338
12	4155711687	04/21/2010	11:38:36	00001:30	4155711687	4155711687	LM	0	0	4153689734	G2026
13	5107763545	04/21/2010	11:40:42	00003:18	14155711687	4155711687	ML	0	0		G4044
14	4155711687	04/21/2010	11:40:46	00003:12	4155711687	4155711687	LM	0	0	5107763545	G2026
15	4155711687	04/21/2010	11:59:49	00001:23	5107763545	5107763545	ML	0	0		G2026
16	5107763545	04/21/2010	11:59:55	00001:18	5107763545	5107763545	LM	0	0	4155711687	G4044
17	4155711687	04/21/2010	12:03:47	00001:06	4155711687	4155711687	LM	0	0	4159480975	G2032
18	4155711687	04/21/2010	12:05:07	00000:57	15107766577	5107766577	ML	0	0		G2032
19	5107766577	04/21/2010	12:05:20	00000:43	5107766577	5107766577	LM	0	0	4155711687	G3026
20	4155711687	04/21/2010	12:06:15	00000:37	4154105865	4154105865	ML	0	0		G2026
21	4154105865	04/21/2010	12:06:22	00000:25	4154105865	4154105865	LM	1	0	4155711687	G4221
22	4154105865	04/21/2010	12:06:42	00000:04	3114154105865	4154105865	ML	0	1	4155711687	F0000
23	4155711687	04/21/2010	12:07:30	00000:33	14155715629	4155715629	ML	0	0		G2026
24	4155715629	04/21/2010	12:07:34	00000:25	4155715629	4155715629	LM	1	0	4155711687	G5030
25	4155715629	04/21/2010	12:07:54	00000:03	3114155715629	4155715629	ML	0	1	4155711687	F0000
26	4155715629	04/21/2010	12:12:14	00000:56	4155711687	4155711687	ML	0	0		G4088
27	4155711687	04/21/2010	12:12:19	00000:52	4155711687	4155711687	LM	0	0	4155715629	G2026
28	4155711687	04/21/2010	12:13:39	00001:40	13233631929	3233631929	ML	0	0		G2026
29	4155711687	04/21/2010	12:16:47	00000:34	4154105865	4154105865	ML	0	0		G2032
30	4154105865	04/21/2010	12:16:51	00000:25	4154105865	4154105865	LM	1	0	4155711687	G4221
31	4154105865	04/21/2010	12:17:11	00000:03	3114154105865	4154105865	ML	0	1	4155711687	F0000
32	4154105865	04/21/2010	12:17:45	00000:04	4155711687	4155711687	ML	1	0		G4221
33	4155711687	04/21/2010	12:17:48	00000:02	4155711687	4155711687	LM	1	0	4154105865	G2026
34	4155711687	04/21/2010	12:18:07	00000:20	4154105865	4154105865	ML	1	0		G2026
35	4154105865	04/21/2010	12:18:13	00000:15	4154105865	4154105865	LM	1	0	4155711687	G4221
36	4155711687	04/21/2010	12:22:44	00000:31	14155853944	4155853944	ML	1	0		G4026
37	4155711687	04/21/2010	12:23:53	00000:38	7753428272	7753428272	ML	1	0		G4026

NOTE: The value of 0 represents an answered call. The value of 1 represents an unanswered call.

In this example, you might notice different phone numbers in the "Billed MDN." This particular provider creates a double record when two of its subscribers are in contact. As you can see on lines nine and ten our target number 4155711687 initiated a call to 4153689734 but both call records are listed on the target phone numbers CDR. Urban PCS does this to account for the billing activity between subscribers. As Urban PCS is a particular favorite of criminals due to its anonymity there are several thousand duplicate records in this data set.

The first step is to sort the records by your target number and remove all "Billed MDN" that are not your target number. At the beginning of this data set, the CDR consists of 7,232 records. It may not seem like a good idea to delete several hundred or even thousands of calls from your data set, but by leaving those records in your file, it can seriously alter the results of your analysis. In this case, removing the duplicate records between Urban PCS subscribers reduces the data set to 4,774 records.

Using Microsoft Excel to Substitute Codes

It can be cumbersome to refer back to Urban PCS's guide to reading their records so I will use a feature of Microsoft Excel to substitute common verbiage for their codes.

1. Select the column you wish to change the information in.
2. In the Home screen of Excel, select **Find** and **Select** and then **Replace**.
3. Substitute common language for the codes used by Urban PCS.

In the example below, the coded Direction column was replaced with common language, as were the Status, and Special Features fields. When I am done replacing the codes, I usually sort the data by date and time.

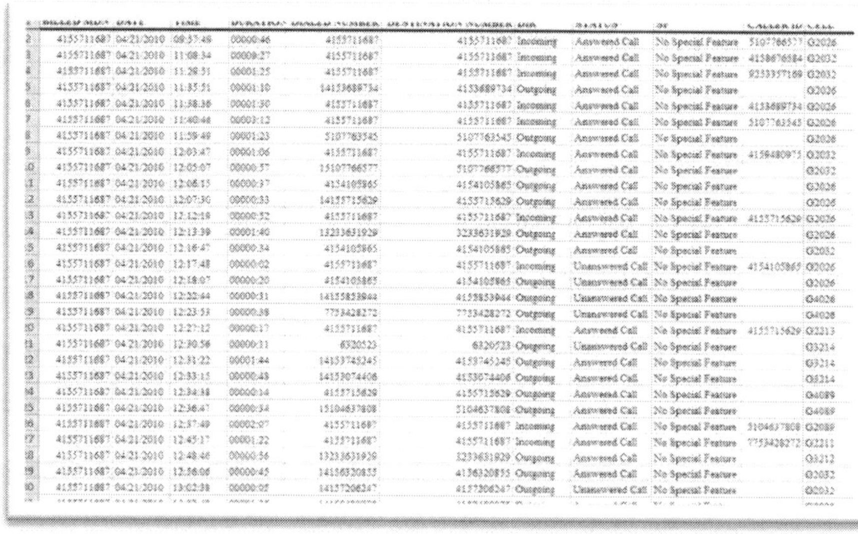

If the timeframe of the crime under investigation is known you can likely stop at this point. There is no need to do further analysis if you believe the suspect associated with this phone number committed the crime between 0900 and 1200 on 4/21/2010. The rest of the records are not likely to be relevant. But what if this is a conspiracy investigation such as narcotics or gangs and the crime is ongoing over an extended period of time? How can you use these records to assist you?

An investigator could manually review the records for interesting calls, phone numbers, or try to identify relevant patterns. For instance, you might note the incoming call on line 27 from a Nevada area code and begin to attempt to identify the subscriber. Each case is unique and the relevance of a phone number may have significance only to a particular investigator or analyst. I recommend a systematic method of evaluating call detail records, particularly in narcotics and gang investigations.

Reading Two-Way Radio Network Call Details on a Record using UFMI

In relation to mobile phone investigations, UFMI stands for Universal Fleet Mobile Identifier. The UFMI can be thought of as an additional phone number used to identify push-to-talk transmissions made on iDEN networks. iDEN, or Integrated Digital Enhanced Network, is a mobile telecommunications technology developed by Motorola and a two-way radio network. Communications made over the iDEN network have a UFMI associated with the mobile handset in addition to a telephone number.

Communications made between iDEN phones will display the UFMI on call detail records in a format similar to 117*68*44556. The first three digits of the number are the Urban Identifier that describes the carrier and the general geographic area the UFMI is assigned to. Call detail records with UFMI do not display a direction such as incoming or outgoing because these are two-way radio communications. You can determine the initial direction by looking at the call detail records of both subscribers. The first communication made by selecting the appropriate contact and depressing the push-to-talk button, will be several seconds before the receiving contact responds. This time difference is seen in the call detail records. Communications made during the same time frame will not be differentiated in the call detail records. So a three-minute back and forth conversation over the iDEN network will not display how long each handset was transmitting. It will only display the total time of the communication.

Mobile phones equipped with iDEN technology are typically identified with a model number that begins with an "I," for example an i465 is a Motorola manufactured iDEN equipped model of phone. iDEN technology is primarily associated with Boost Mobile phones, the pre-paid division of Sprint Nextel Corporation.

Cleaning Calls to Voicemail

No matter what tool you are going to use to analyze your CDR, they all have a universal requirement. Specifically, the data needs to be in a uniform format, or "cleaned." Pen-Link and i2 Analyst Notebook are among the two most common tools for analysis of call detail records. Unfortunately, these software programs are expensive and frequently have a steep learning curve to use them effectively.

With our fictional Urban PCS records, we are going to reduce the amount of data and standardize the remaining information. This will also eliminate the unwieldy amount of calls to voicemail you will likely encounter. In the example below, we are going to completely remove the Dialed Number Column. Remember that column refers to the actual digits entered by the phone's user. The Destination Number is the actual number dialed and normally free of extra digits such as *67 or the number 1 in front of the area code. Next, sort the records so we can review the data.

LLED MIN	DATE	TIME	DURATION	DESTINATION NUMBER	DIR	STATUS	SF	CALLER ID CELL
4155711687	05/05/2010	12:01:36	00000:22	1	Outgoing	Unanswered	No Special Feature	G4258
4155711687	04/30/2010	18:06:59	00000:03	3	Outgoing	Unanswered	No Special Feature	G2026
4155711687	05/17/2010	00:34:50	00000:02	3	Outgoing	Unanswered	No Special Feature	G2026
4155711687	05/17/2010	00:34:53	00000:02	6	Outgoing	Unanswered	No Special Feature	G2026
4155711687	04/28/2010	20:15:12	00000:14	67	Outgoing	Unanswered	No Special Feature	G4202
4155711687	05/11/2010	18:22:46	00000:43	82	Outgoing	Unanswered	No Special Feature	Q2588
4155711687	05/12/2010	13:40:17	00000:24	5556	Outgoing	Unanswered	No Special Feature	G4061
4155711687	05/12/2010	13:41:29	00000:18	5889	Outgoing	Unanswered	No Special Feature	G4061
4155711687	04/25/2010	20:31:12	00000:10	2402029	Outgoing	Answered C	No Special Feature	G3061
4155711687	05/05/2010	15:36:55	00000:32	2402029	Outgoing	Unanswered	No Special Feature	F4039
4155711687	05/08/2010	21:10:21	00001:58	2864390	Outgoing	Answered C	Three Way Call	G2026
4155711687	05/05/2010	15:31:02	00000:09	3746727	Outgoing	Unanswered	No Special Feature	F2039
4155711687	04/23/2010	09:06:29	00000:19	4106589	Outgoing	Answered C	No Special Feature	G2026
4155711687	04/23/2010	11:23:25	00000:19	4106589	Outgoing	Answered C	No Special Feature	G2026
4155711687	04/23/2010	12:55:40	00000:34	4106589	Outgoing	Answered C	No Special Feature	F4039
4155711687	05/04/2010	23:02:17	00004:22	4106589	Outgoing	Answered C	No Special Feature	S3250
4155711687	05/03/2010	09:53:32	00000:27	4106589	Outgoing	Answered C	No Special Feature	G2026
4155711687	05/05/2010	09:58:54	00000:20	4106589	Outgoing	Answered C	No Special Feature	G3026
4155711687	04/24/2010	22:31:49	00001:57	4240311	Outgoing	Answered C	No Special Feature	G3208
4155711687	05/06/2010	16:35:02	00003:10	4240311	Outgoing	Answered C	No Special Feature	G3204
4155711687	05/08/2010	16:42:24	00004:23	4240311	Outgoing	Answered C	No Special Feature	G4061
4155711687	05/07/2010	15:13:32	00002:02	4240311	Outgoing	Answered C	No Special Feature	G2026
4155711687	05/11/2010	10:45:11	00000:14	4240311	Outgoing	Unanswered	No Special Feature	Q4476
4155711687	04/27/2010	15:17:56	00002:27	4249351	Outgoing	Answered C	No Special Feature	G3204
4155711687	05/13/2010	14:30:40	00001:55	4249351	Outgoing	Answered C	No Special Feature	F3039
4155711687	05/13/2010	14:39:46	00000:27	4249351	Outgoing	Answered C	No Special Feature	F4039
4155711687	05/13/2010	14:40:26	00000:48	4249351	Outgoing	Answered C	No Special Feature	F4039
4155711687	05/13/2010	14:41:18	00001:35	4259351	Outgoing	Answered C	No Special Feature	F2039
4155711687	05/03/2010	14:55:10	00000:18	5711687	Outgoing	Answered C	No Special Feature	G2026

Using Add-On Tools with Excel for Cleaning Records

You can see in Figure 9 that there are numbers that do not reflect a phone number in the Destination Number column. These are the numbers a user accidentally enters when keys are pressed inadvertently. They do not indicate an actual phone number dialed. We also run into an issue with phone numbers that are only seven digits. If you are going to do data analysis, your phone numbers need to be standard 10-digit numbers including the area code. Working with a combination of seven and 10-digit phone numbers can alter the results of your analysis. As a reminder, if the phone number is a seven-digit phone number, you will have to assume it is in the same area code as the target phone number unless there is evidence to the contrary.

For basic analysis it does not matter if you want to format the numbers as 4155551212, 415-555-1212, or (415) 555-1212. However, some sophisticated software programs, as well as databases maintained by other law enforcement and intelligence agencies, require the data to be in 10-character format with no spaces, dashes, or parenthesis.

ASAP Utilities

I've attempted to reformat seven-digit phone numbers into ten-digit phone numbers with varying degrees of success. If you anticipate working heavily with phone records, such as in a pen register or wiretap case, I definitely recommend downloading a free utility for use with Excel. The product is called ASAP utilities and it has created an interface for automating the more complicated functions in Excel that would otherwise require writing formulas and macros within the program.

ASAP Utilities has a free version of the download that installs a tab at the top of the toolbar in Excel. In order to convert the seven-digit numbers into 10-digit numbers with the area code, I am going to demonstrate using ASAP Utilities. After installing the program go to Text, then selection one (Insert before current value), which will allow us to insert the three-digit numbers area code in front of the seven-digit number and create a standard formatted 10-digit number telephone number.

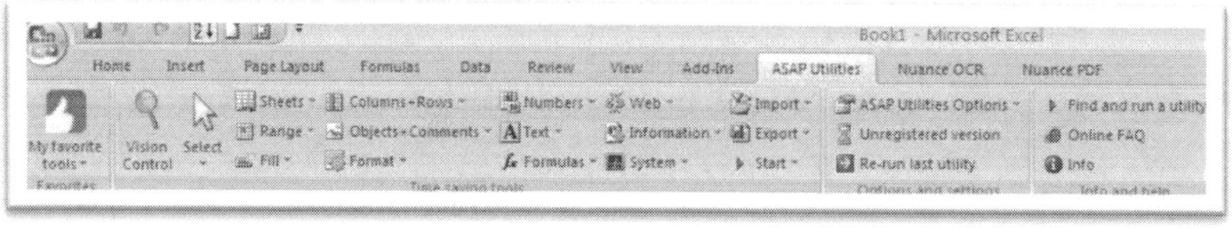

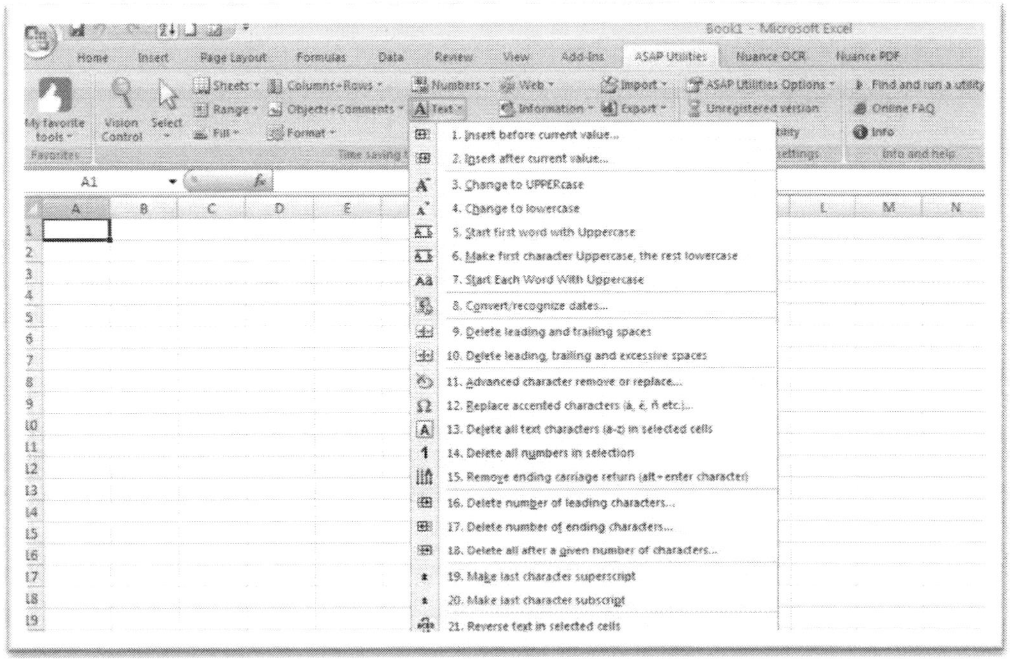

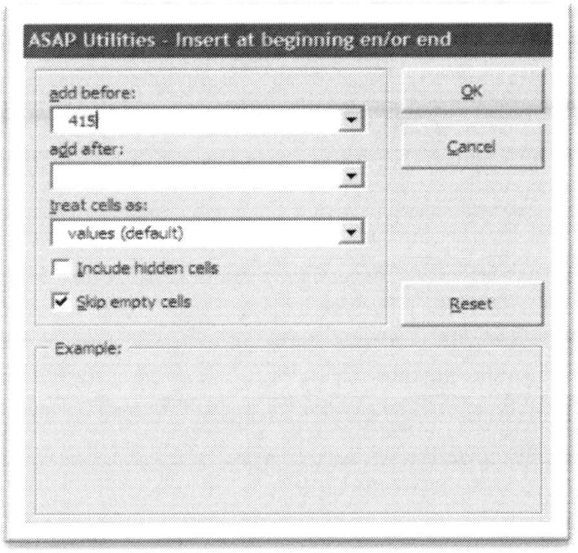

There is a way to create standard formatted 10-digit telephone numbers in Excel without add-ons, but I've had so many challenges trying to do this that I chose to use ASAP. Difficulties were changing the format of the cell, creating custom formats, and writing formulas for the range of cells, which often becomes problematic when you are working with large data sets from different sources. ASAP proved to be the easiest and simplest way, and you can't beat the price. To learn more, about ASAP go to www.asap-utilities.com.

Even a simple analysis of telephone records can get complicated when the call detail records are from more than one provider. They need to be formatted the same, or "cleaned" to effectively analyze these records. This entails creating a standard set of data fields and ensuring the data conforms to the same format and standards. Failing to create standard data fields and uniform formatting of the data itself can adversely alter the finished product.

Proprietary Software with Automated Processes

Software programs such as i2 Analyst Notebook and Pen-Link have automated the process for inputting data. Using their proprietary tools, the software programs can clean the data automatically, which greatly reduces the time an investigator or analyst spends preparing the records into a standard format. However, these software programs can be expensive for some agencies.

Call Detail Record Analysis

The detail records of a suspect's phone holds an abundance of valuable information. We need to apply some analytical tools to this data, specifically frequency analysis, to unpack this data. Additional analytical techniques involve Sequence Analysis, which refers to the identification of repeated calling patterns within sets of call detail records. For example, if a confidential informant places a phone call to a dealer to purchase narcotics and the dealer subsequently makes a phone call to another phone number, you have identified a sequence. If the sequence repeats every time the confidential informant attempts to purchase narcotics, you have an identifiable pattern that could possibly reflect the dealer's source of supply. Sequence Analysis infers certain possibilities but does not necessarily prove something to be true. In this example, the drug dealer could be calling an associate to give him a ride to the meeting location and not necessarily to re-supply his narcotics.

Another interesting aspect of Sequence Analysis relies in the presumption that the higher up a subject is in an organization, the fewer outgoing calls they make to subordinates. To put it in a law enforcement context, picture the chief executive of your organization. In any given day the Chief or Sheriff is likely to receive incoming calls from subordinates reporting on current situations, requesting resources, or seeking permission for certain activities or projects. The subordinates are reporting upwards to the commander of the organization. The Chief or Sheriff is likely to place less outgoing calls to subordinates than the number of incoming calls from the subordinates. Those calls will be to direct the operation and to seek further information regarding a topic.

Most criminal organizations have some sort of structure, formal or informal, which loosely parallels many law enforcement and military organization charts. Sequence analysis can be used to attempt to identify the upper hierarchy of a criminal organization.

Once your records are in a standardized format you can begin your own basic frequency analysis using another tool found in Excel, the Pivot Table Wizard. Pivot tables are a powerful function built into Microsoft Excel which can be used to translate call detail records (CDRs) into something useable. It is a data summarization tool that can automatically sort, count, and total the data stored in one table or spreadsheet and create a second table, called a pivot table, displaying the summarized data.

Most investigators will acknowledge that in their personal lives they only call a very few phone numbers every day or even several times a day. Typically these calls are to significant others or spouses, employers, and close friends. Criminals are no different. They will also be in regular contact with their significant others and close associates and retain those contacts after they change their cell phone numbers.

We are going to use this tool to help us understand 1) who our target contacts the most, and 2) who the target is in contact with the longest. Law enforcement uses this most frequently when they are trying to identify the user of a particular phone number by attempting to identify their close associates.

Investigators also use this when they are trying to locate fugitives by determining whom they called the most prior to fleeing.

The Pivot Table Wizard can be found under the Data tab in Excel. Remember: *never work with your original data!* You should only use a copy of the files and not the originals you received from the cellular service provider.

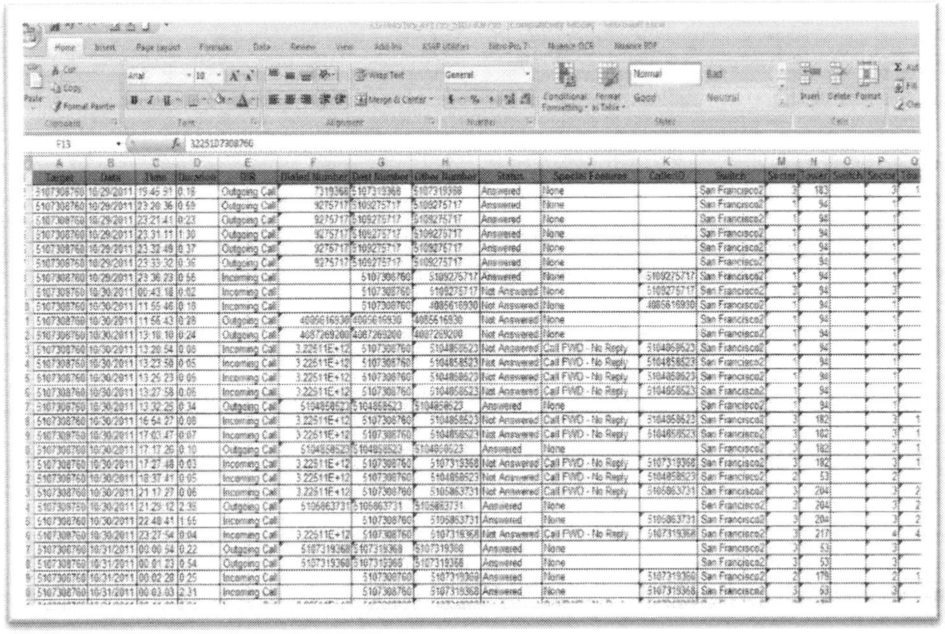

Step 1: Clean Your Data

In order for a pivot table to work effectively the call detail record data must be standardized. Some cell service providers include routing codes, extra digits, and additional information in the same text box as the phone number. These extra digits will need to be cleaned from the data. You may also see additional data in the Dialed Number column; this does not reflect an actual phone call, as is the case when someone enters an incomplete phone number before pressing send. Since those are not actual calls, but records of incomplete calls, I delete them. Depending on the provider you may also see extra routing numbers or extra digits, such as *67 when someone blocks their phone number from appearing on caller ID. There are a number of methods for removing those leading digits including the Find and Select function from the Home tab of Excel. However, after trial and error, I have become a faithful user of a free suite of tools called, "ASAP Utilities." These tools make it very easy to format all of your data quickly.

Formatting Data into a Standard Formats

As you do more work with Excel, you will likely run into the same problems time and again. I've prepared the short list in the following section and provided you with a few workarounds.

Converting Call Durations

A problem encountered when trying to format the data into a standard scheme involves converting the durations of calls. Most investigators and analysts prefer to view call duration in a format that is easily understood. For example, a call duration of 3,427 seconds is obviously a lengthy time, but it makes the data more intuitive and presentable to refer to it as 57:07.

A simple formula for converting call durations that are measured in total seconds is:

=A1/86400

This presumes in this case that the duration is listed in column A. You will need to change the reference to A1 to reflect the column the data is in. The end result will still not look like it is usable, but right click on the cell or column, select **Format Cells**, select **Custom**, and finally select **h:mm:ss**. This tells Excel to display the data in Hour:Minute:Second format.

Text to Columns

Another issue can be records returned with a number of leading zeroes in front of the call duration such as those found in the returns from our fictional Urban PCS cell phone company (back a bit in Figure 5.5). As with everything else, there are many ways to remove the leading zeroes. You can use ASAP utilities to remove a predetermined number of extra characters from in front of your duration. I like to use a feature built into Excel called "Text to Columns." One of my favorite analysts has a saying regarding this feature: "Text to columns can cure almost anything in Excel!" Simply put, this feature can rapidly clean up a bulk of different formatting issues, especially extraneous digits.

Insert a blank column next to your call duration column to use this feature. If you don't, the data in the adjacent column will get written over when the process is completed. Highlight the **Call Duration** column. Next, click the **Data** tab at the top of the toolbar and select **Text to Columns**.

Next, select the **Fixed** width option in your Wizard Box and click **Next**. This will bring up a **Data Preview** window. From within the Data Preview window you can select where you would like to tell Excel to split the data. Insert the line break next to the numbers you want to separate. In this case the line is placed between the last zero in front of our call duration and the first number of the call duration.

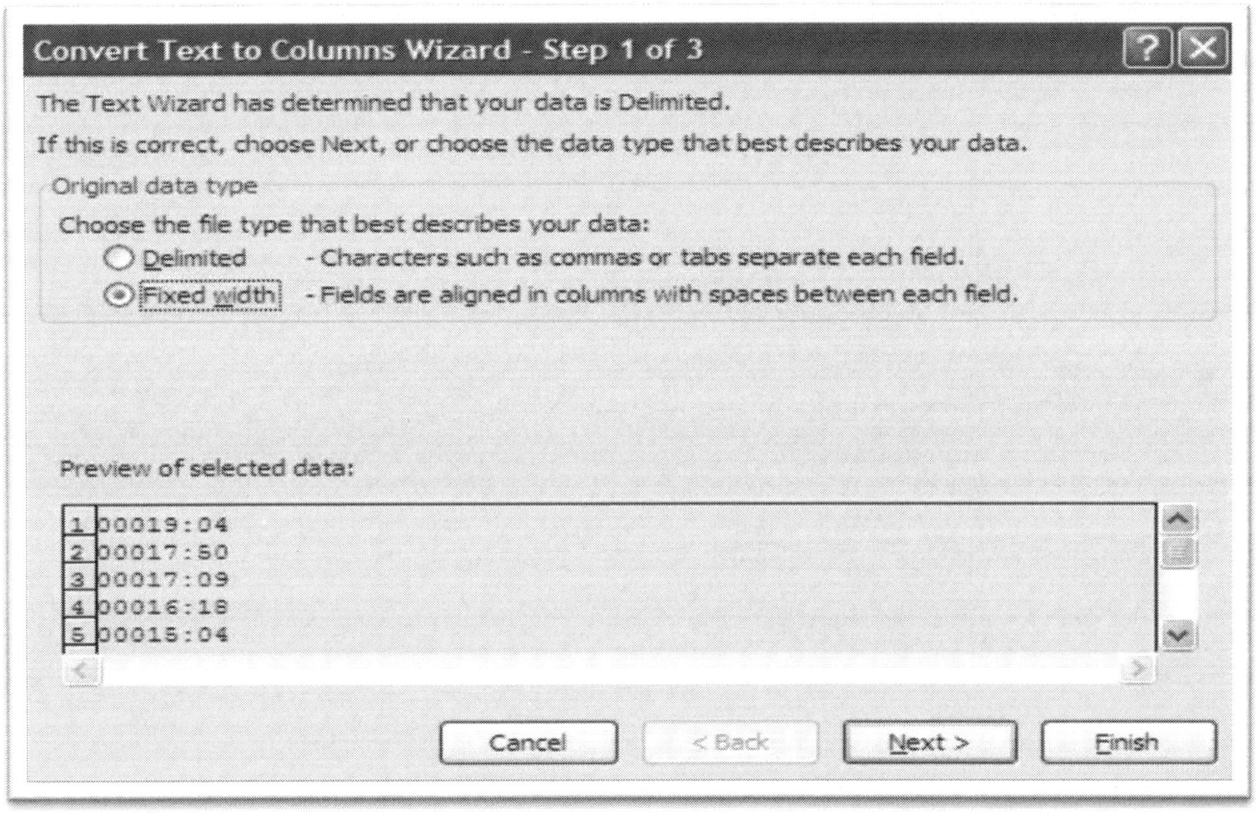
Convert Text to Columns Wizard - Step 1 of 3
The Text Wizard has determined that your data is Delimited.
If this is correct, choose Next, or choose the data type that best describes your data.
Original data type
Choose the file type that best describes your data:
Delimited - Characters such as commas or tabs separate each field.
Fixed width - Fields are aligned in columns with spaces between each field.
Preview of selected data:
1 00019:04
2 00017:50
3 00017:09
4 00016:18
5 00015:04
Cancel
< Back
Next >
Finish

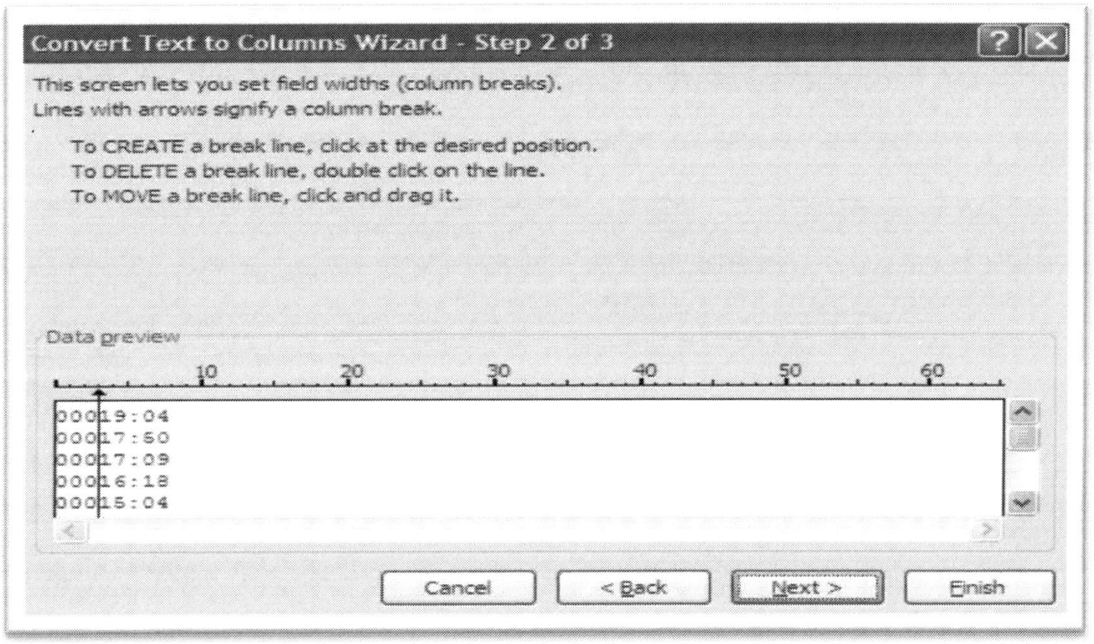
Convert Text to Columns Wizard - Step 2 of 3
This screen lets you set field widths (column breaks).
Lines with arrows signify a column break.
To CREATE a break line, click at the desired position.
To DELETE a break line, double click on the line.
To MOVE a break line, click and drag it.
Data preview
10
20
30
40
50
60
00019:04
00017:50
00017:09
00016:18
00015:04
Cancel
< Back
Next >
Finish

After you have set your line break, select **Finish** and your data should look something similar to the table below.

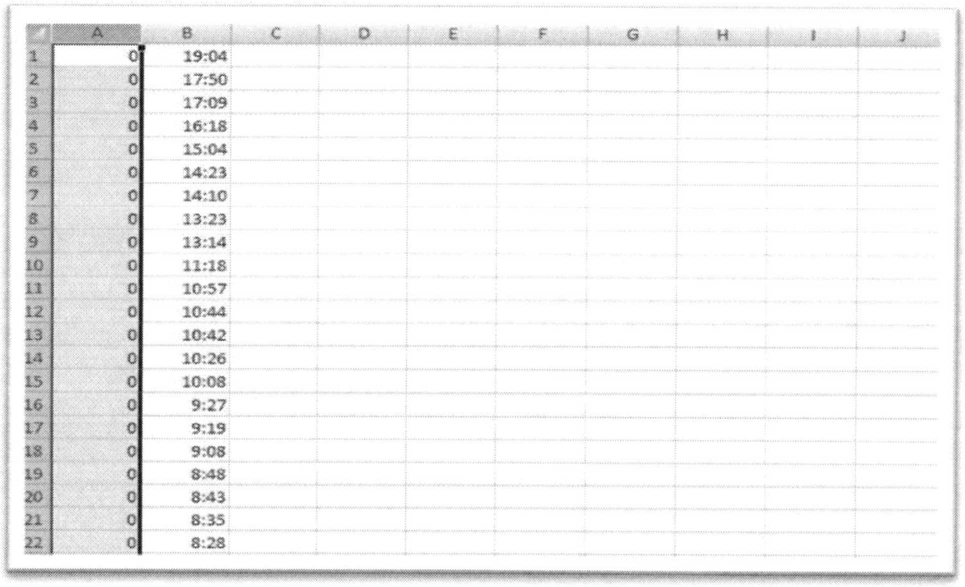

You'll note that all of the leading zeroes have been removed and you are left with the data in the M (Minute): S (Seconds) format.

You may also note that Excel interprets your data in call detail records as text and not numbers. When Excel detects numbers stored as text, it places a small green triangle in the upper left of the cell.

You could click on the green triangle and manually change each cell of text into a number, but that would take forever.

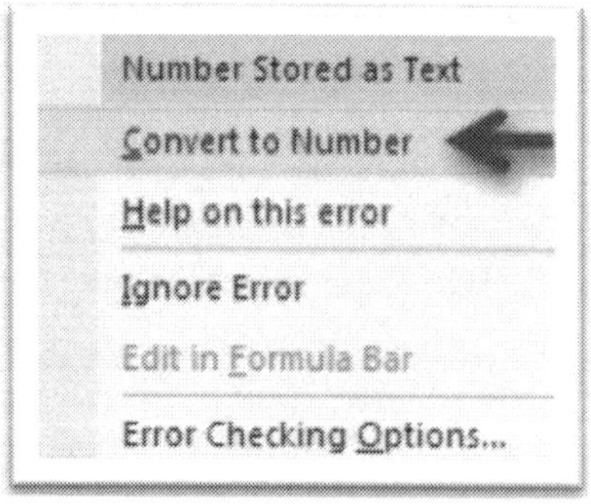

To quickly convert all text to numbers highlight the entire column by clicking the cell with a letter in it at the top of the column. Right click and select **Format Cells**.

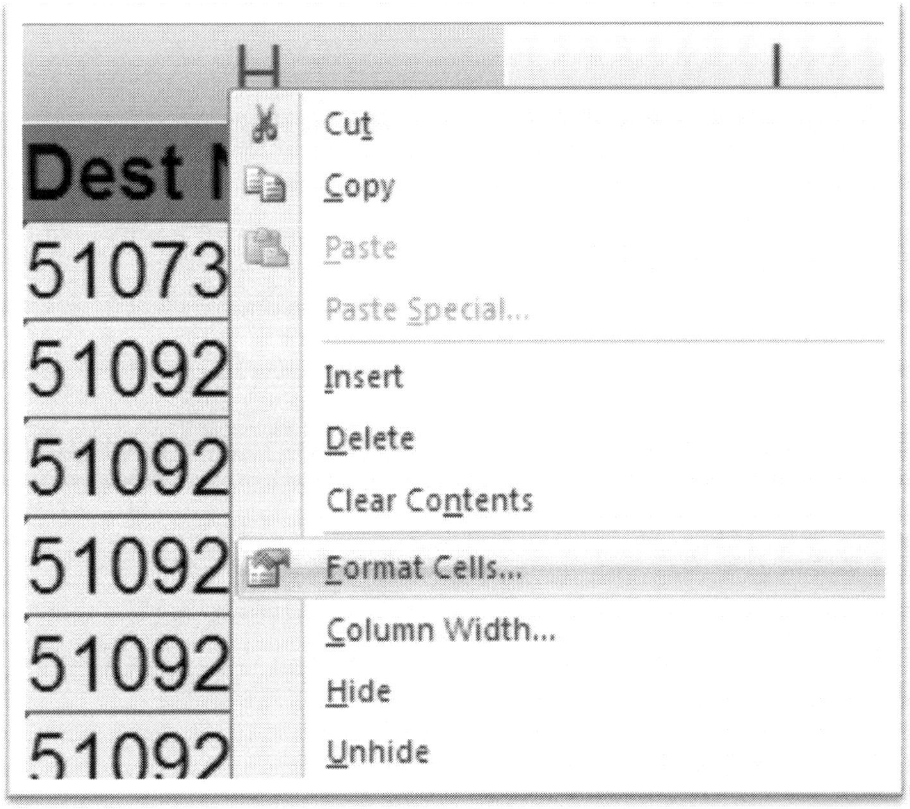

Select **Number**. By default you will see the number **2** in the box marked **Decimal Places**. Change that to **0**.

Step 2: Insert New Columns

If your call detail records do not have a separate column for the target phone number, you will need to create one. The easiest way is to right click your mouse button on the column header **A** and select **Insert**. This will automatically insert a column and shift the data over.

The A1 cell type is something to help you remember what this data is, such as "Target Number." This will be important later when there will be multiple columns to choose from and you want to make sure you select the proper one.

Insert the target number and then fill the entire column with your target telephone number. Copy the phone number three times into the first three cells. Then highlight all three cells by left clicking the mouse button and dragging it downward until all three cell are highlighted. If you hover your mouse cursor over the lower right corner of the last cell what looks like a small plus sign + will appear. Left click the plus sign + and Excel will automatically populate all of the cells below it, with the target number.

You will also need to insert another column for your "Other Numbers." This data already exists in the call detail records, but some providers display this information in multiple columns. Right click your mouse button, select **Insert** again to place a new blank column in your data, and label it with something you will remember such as "Other Number."

The other phone numbers need to be consolidated in one column. Depending on the provider, you may have to sort your data by the "Direction" of the call or the "Caller ID" column to obtain this information. Copy the data from those rows into your new column so that all of the other phone numbers are in one place.

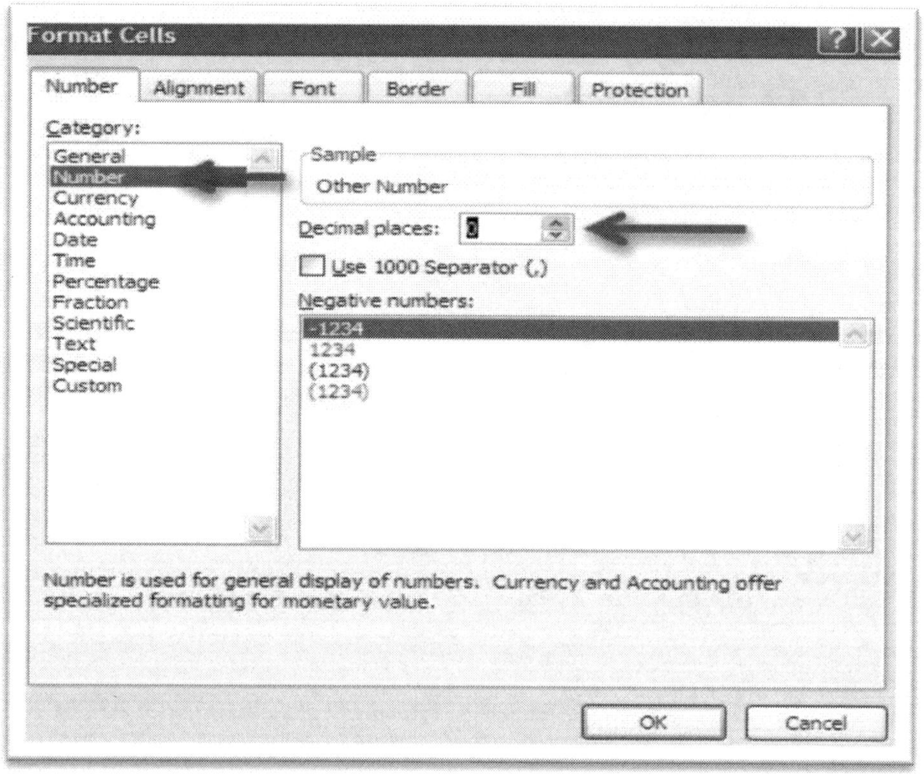

Step 3: Finding the Pivot Table

The **Pivot Table** function is located in two places in Excel. The first is in the **Insert** tab located at the top of the Excel workbook.

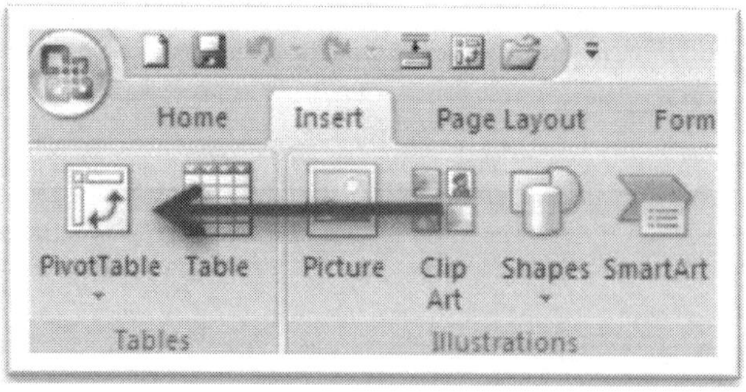

You may also wish to add the Pivot Table to the **Quick Access Toolbar** located at the upper left corner of the Excel workbook. In order to add the Pivot Table to the toolbar, select the drop down error to the right of the toolbar. You will need to select **More Commands** to locate the Pivot Table function.

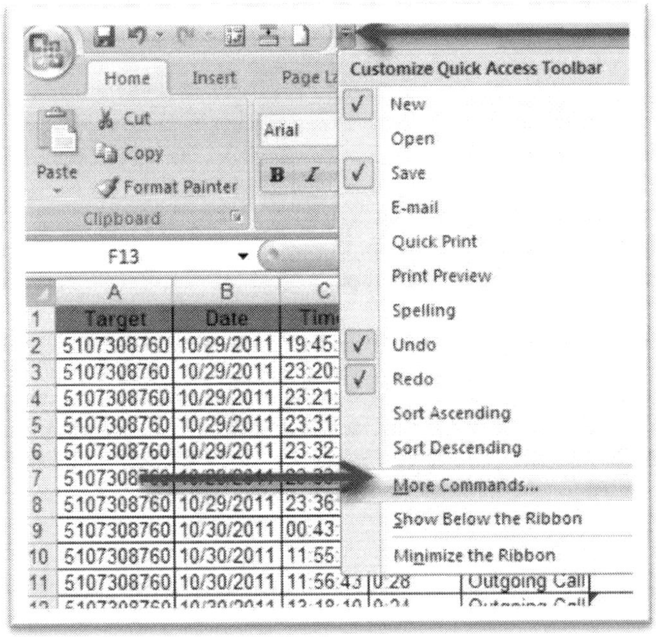

Next select **All Commands** and scroll down until you see the **Pivot Table Wizard**. Select the **Add** button to include the Pivot Table Wizard in the toolbar.

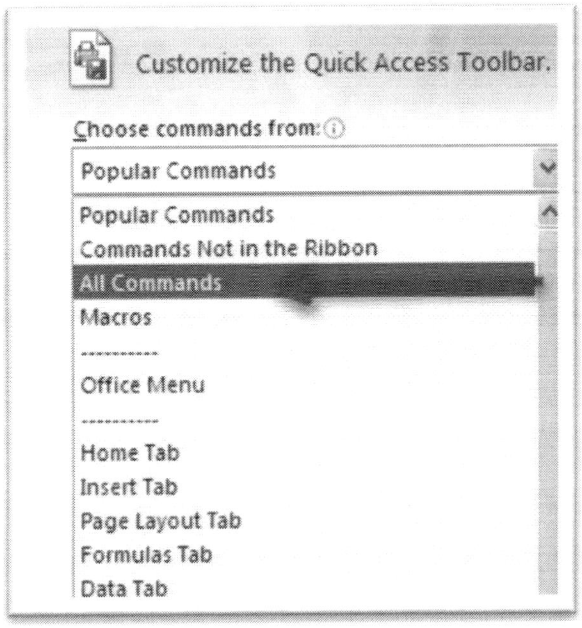

When you are done you should see this icon in your Quick Launch Toolbar.

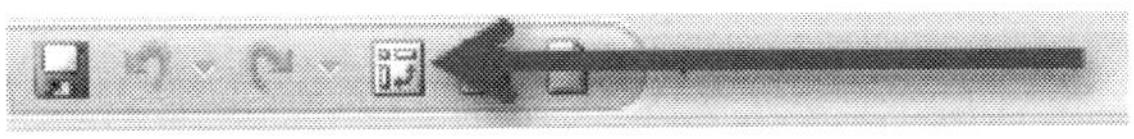

Step 4: Using the Pivot Table Wizard

Select **Pivot Table** from either the **Insert Tab** or the newly created icon in the Quick Launch Toolbar. This will start the program and automatically bring up this screen:

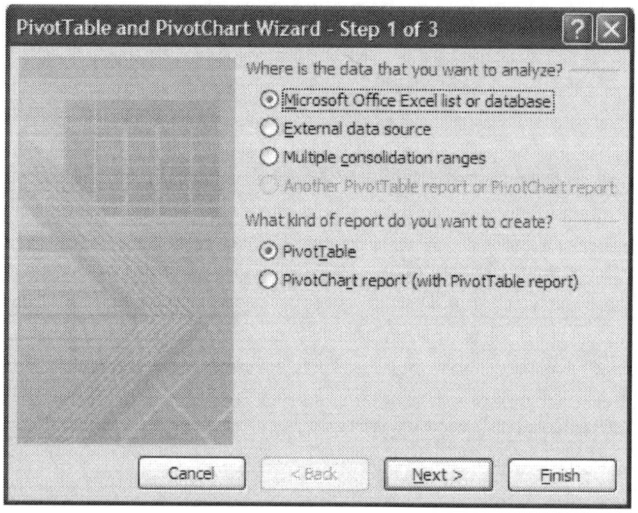

We are not going to do anything fancy from this screen so select **Next**. You could select **Finish** from this screen, but there are some different options in the following steps.

The Step 2 screen tells the Pivot Table Wizard where the data is going to come from.

Before selecting **Next**, make sure all of the data in your sheet is selected. The data in the sheet should be surrounded by a moving row of dashes sometimes referred to as "marching ants."

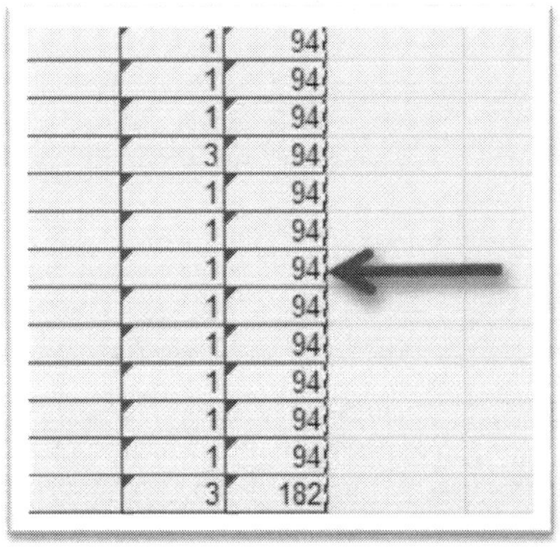

All of the data inside the highlighted area and surrounded by the marching ants is going to be imported into the Pivot Table. If for some reason all of the data is not selected, or if the entire work sheet is highlighted including columns and rows with no data in it, select **Cancel.** Click on any cell containing data on the sheet and re-run the Pivot Table Wizard.

To complete the setup of the Pivot Table select the **Finish** button from the Step 3 screen. By default this will create the Pivot Table in a new worksheet that will be added in front of your existing worksheets.

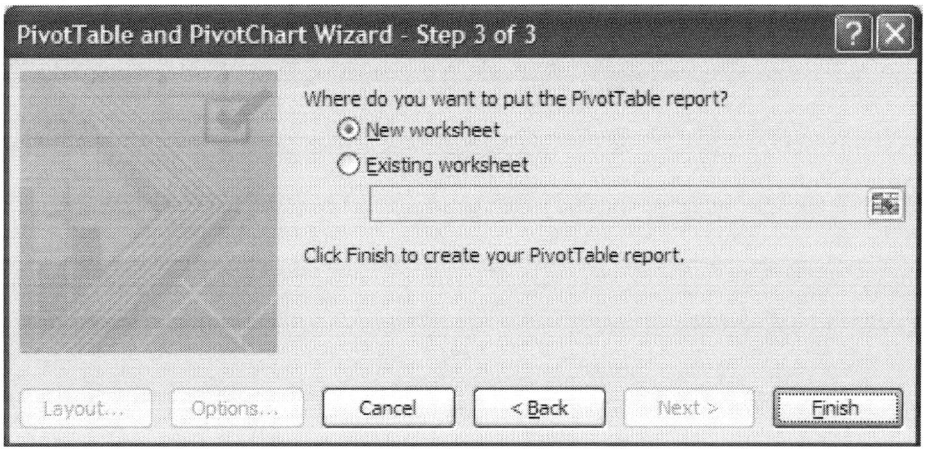

Step 5: Using the Pivot Table

Once the Pivot Table process has been completed you should be presented with the following in a new sheet added to your Excel workbook:

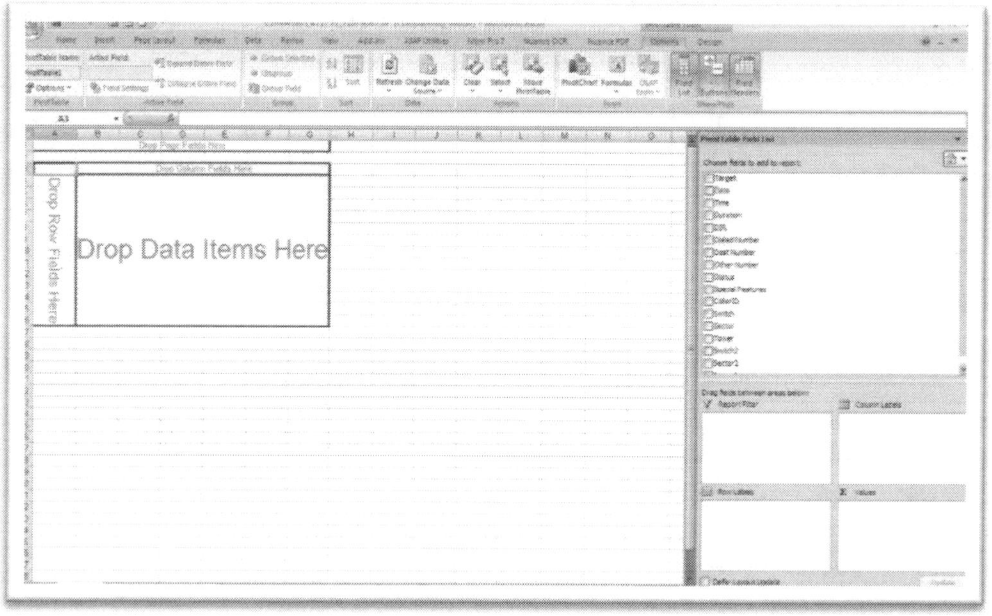

The entire column headers, including the **Target Number** and **Other Number** should appear in the **Pivot Table Field List** located in the upper right of the screen. From this screen you can perform basic call frequency analysis using the data from the call detail records.

Select **Target Number** and **Other Number** by placing a check mark in the boxes. Your data may appear in the boxes labeled **Row Labels** and **Values**.

If the data is not automatically in those boxes, click and drag the data so it appears as below:

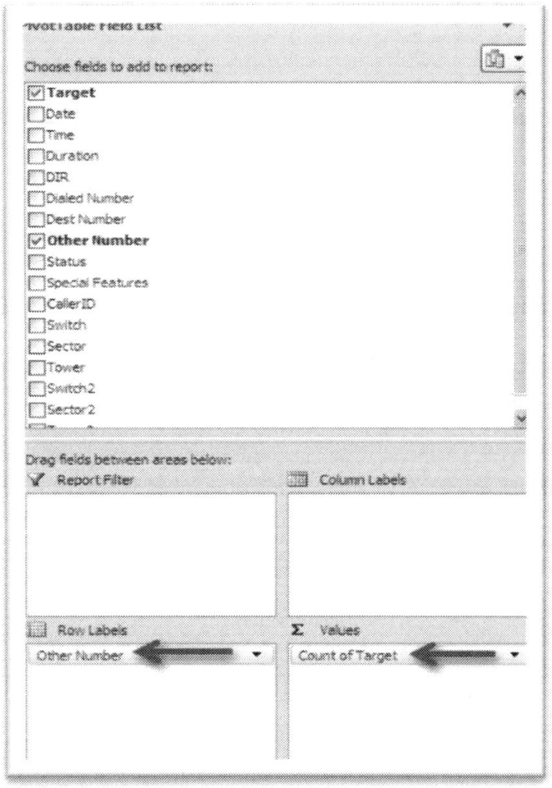

Sometimes the data in the **Values** box will be different and will say **Sum of Target** and not **Count of Target**. If this happens you will see strange numbers in the **Pivot Table** because the software has decided to add up the total of all of the phone numbers instead of count how often they occur. If that happens select the drop down menu in **Sum of Target**.

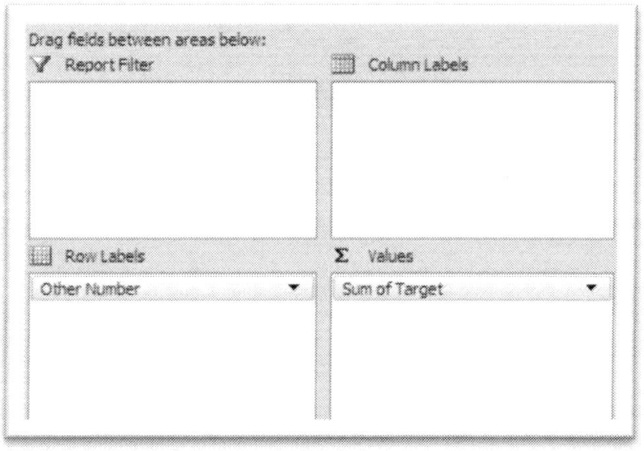

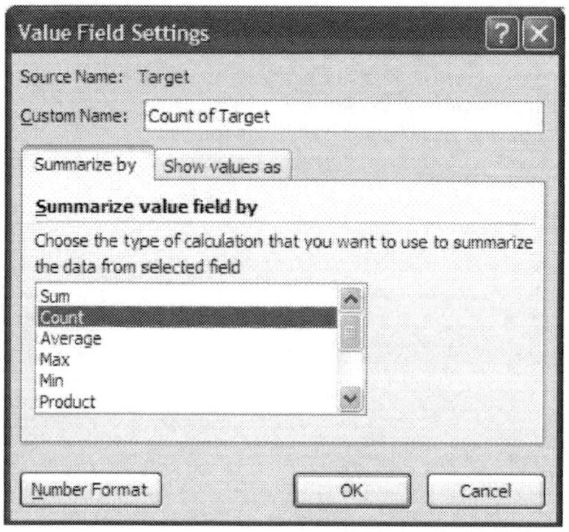

Once the **Value Field Setting** has been properly configured you should see each unique phone number listed with a total representing the number of times the target phone number was in contact with the other phone number.

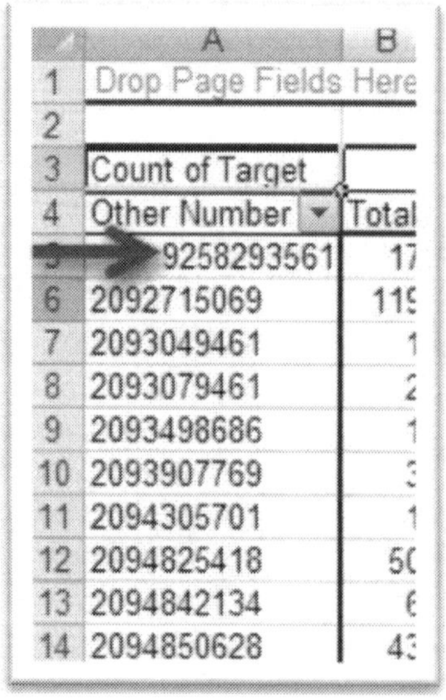

However, in the example listed above you can see that the first phone number is not listed sequentially. In this case, the number was not properly formatted and is still viewed by Excel as text and not a number. If this happens to you go back and reformat the other numbers so they are all represented as a number.

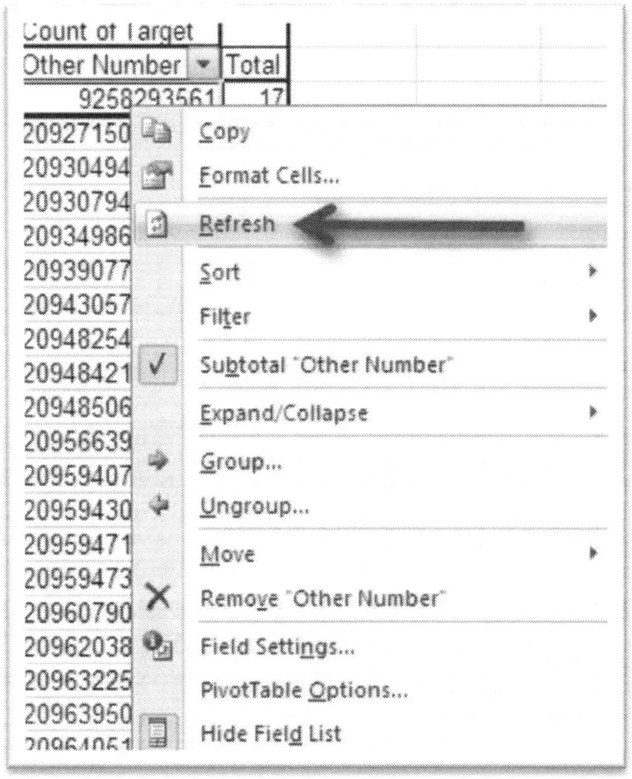

It is not necessary to re-run the Pivot Table. Select the sheet containing the data and make the necessary corrections. Then select the drop down box located above the phone numbers and select **Refresh**.

If everything looks good, move on to the next step.

Step 6: Sorting the Numbers

The Pivot Table has given us some good information. You should have a list of unique phone numbers and a sum of the occurrences the target contacted. However, by default, the Pivot Table is sorted by the phone number and not the total number of contacts. In order to find the phone number the target was in contact with most often, we must change the **Sort** order of the Pivot Table. Click the drop down box located at the top of the column and select **More Sort Options**. In order to see the target number's most frequent contacts in order, select the **Descending** option from the **Sort** menu and then select **Count of Target**.

This will re-sort the data by the highest number of contacts to the lowest and allow you to visualize the most frequent contacts. Note that the target number will usually be the largest number of contacts. Depending on the service provider, these calls are actually incoming calls that were routed to voicemail or they were the user, checking his voicemail messages.

In most analysis, the top 10 to 20 phone numbers will reflect a statistical majority of the phone numbers your target is in contact with. In this example, the top 20 phone numbers reflect approximately 25 percent of the total numbers in contact with the target. Within this top 20, will be the close inner circle of your target. However, you may make a huge error by only focusing on the high frequency or high duration calls. Think about a mid to upper level narcotics investigation – how often do you believe your target will call the head of the organization? These calls could be very infrequent and basic call detail analysis may miss them.

The next steps are basic police work. Identify the subscribers and users of the high frequency contacts.

Other Uses for Pivot Tables

Another use for the Pivot Table is basic cell tower analysis. A Pivot Table can quickly add up the most frequent cell towers that were in communication with the phone. This can provide clues to a suspect's location or other areas he was known to frequent. Visualizing cell tower data in a Pivot Table will require a key from the cellular service provider or from the FBI's CALEA (Communications Assistance to Law Enforcement Act) Unit. However, the basic Pivot Table functions are identical to those used in determining high frequency contacts.

Follow the steps listed above until the Pivot Table is generated. However, instead of listing the **Target Number** and the **Other Number** you will highlight the **Target Number** and the **Tower**. Note that the provider in this example provides both beginning and ending cell tower data, so you may need to run the Pivot Table on both sets of data to get an accurate idea of the handset's location.

By changing the **Sort** order in a similar fashion to the one illustrated above, you can effortlessly sort the cell towers by their frequency of use.

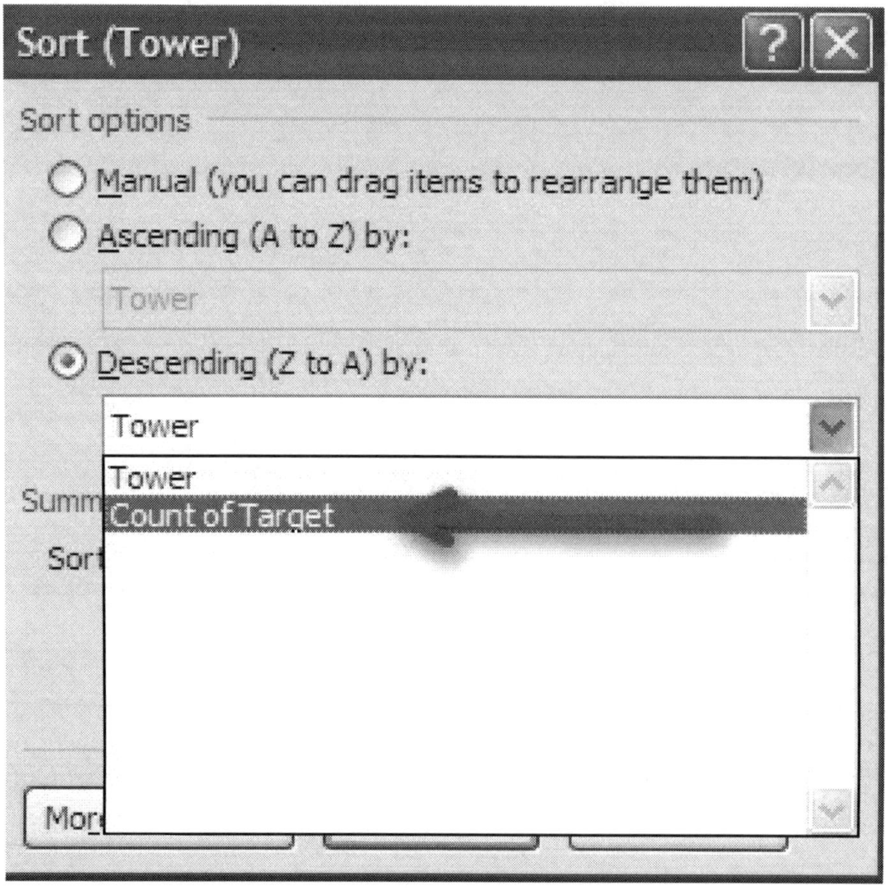

In this example you can see there are only a few towers with a significant number of hits. You might also note the second most common data point is listed as **Blank**. This is because the provider in this example does not record cell tower information when the call is forwarded to voicemail. Most of the major cellular service providers do not do this nor do they give cell tower data for text message or data events. However, AT&T is an exception to that rule as they capture cell tower data during phone calls, text messages, and data events such as Internet browsing.

	A	B
1	Drop Page Fields Here	
2		
3	Count of Target	
4	Tower	Total
5	88	1900
6	(blank)	1216
7	587	790
8	182	726
9	8	172
10	452	153
11	165	141
12	226	96
13	257	90
14	60	80
15	18	80
16	179	76
17	183	75
18	117	55
19	53	54

AT&T Records

For being such a large, profitable company you would think AT&T would be able to help out the law enforcement community by sending their search warrant, subpoena, and court order responses back in accessible format. But for some reason, AT&T does not (or will not.) Unfortunately, AT&T sends their responses back as Adobe Portable Document Format (.pdf) files. This requires us to go through a little bit of work in order to make them usable.

09/27/2011
SCAMP

MOBILITY USAGE
(with cell location)

Run Date: 09/27/2011
Run Time: 20:07:04
SMS Usage For: [illegible]
Account Number: [illegible]

tem	Conn. Date	Conn. Time	Originating Number	Terminating Number	IMEI	IMSI	Description	Cell Location
32	08/01/11	01:04A	[illegible]	[illegible]	[illegible]	[illegible]	IN	[56897/28572:-121.25523:38.1216:320]
33	08/01/11	01:07A	[illegible]	[illegible]	[illegible]	[illegible]	IN	[56897/28572:-121.25523:38.1216:320]
34	08/01/11	01:12A	[illegible]	[illegible]	[illegible]	[illegible]	OUT	[56897/28572:-121.25523:38.1216:320]
35	08/01/11	01:13A	[illegible]	[illegible]	[illegible]	[illegible]	IN	[56897/28572:-121.25523:38.1216:320]
36	08/01/11	01:14A	[illegible]	[illegible]	[illegible]	[illegible]	OUT	[56897/28572:-121.25523:38.1216:320]
37	08/01/11	01:14A	[illegible]	[illegible]	[illegible]	[illegible]	IN	[56897/28572:-121.25523:38.1216:320]
38	08/01/11	01:28A	[illegible]	[illegible]	[illegible]	[illegible]	IN	[56897/28201:-121.25181:38.14161:80]
39	08/01/11	01:30A	[illegible]	[illegible]	[illegible]	[illegible]	IN	[56897/28201:-121.25181:38.14161:80]
40	08/01/11	01:33A	[illegible]	[illegible]	[illegible]	[illegible]	IN	[56897/28201:-121.25181:38.14161:80]
41	08/01/11	01:33A	[illegible]	[illegible]	[illegible]	[illegible]	OUT	[56897/28201:-121.25181:38.14161:80]
42	08/01/11	01:34A	[illegible]	[illegible]	[illegible]	[illegible]	IN	[56897/28203:-121.25181:38.14161:200

There are two primary methods to convert AT&T call detail records into something usable: free and paid. The free method is to open a copy of the file (because we are never working with the original evidence files) using a pdf reader. Most law enforcement agencies have access to pdf readers installed on their departmental computers. If you don't have one, you will need to get one and install it. After opening the document go into the **File** menu usually located in the upper left of the screen. Select **Save As** and save the document as a .txt (Text) file. This copies the data and saves it in a format readable by text editors such as WinPad and Notepad. Open the resulting file.

997474
09/27/2011
SCAMP

MOBILITY USAGE
(with cell location)

at&t

Run Date: 09/27/2011
Run Time: 20:06:18
Voice Usage For: [illegible]
Account Number: [illegible]

Item	Conn. Date	Conn. Time	Seizure Time	Originating Number	Terminating Number	Elapsed Time	Number Dialed	IMEI	IMSI	Description	Cell Location
1	07/27/11	12:32A	0:01	[illegible]	[illegible]	0:24	[illegible]	[illegible]	[illegible]	M2m_VMC	[56897/16407:-121.25181:38.14161:80]
2	07/27/11	12:40A	0:07	[illegible]	[illegible]	0:25	[illegible]	[illegible]	[illegible]	m2M_DIR	[56897/16407:-121.25181:38.14161:80]
3	07/27/11	01:00A	0:02	[illegible]	[illegible]	0:03	[illegible]	[illegible]	[illegible]	M2O_DIR	[56897/16351:-121.28967:38.11753:80]
4	07/27/11	01:02A	0:05	[illegible]	[illegible]	1:28	[illegible]	[illegible]	[illegible]	m2M_DIR	[56897/28293:-121.28967:38.11753:200]
5	07/27/11	01:04A	0:28	[illegible]	[illegible]	0:05	[illegible]	[illegible]	[illegible]	M2M_DIR	[56897/28293:-121.28967:38.11753:200]
6	07/27/11	01:05A	0:28	[illegible]	[illegible]	0:04	[illegible]	[illegible]	[illegible]	M2M_DIR	[56897/28293:-121.28967:38.11753:200]
7	07/27/11	01:07A	0:26	[illegible]	[illegible]	0:16	[illegible]	[illegible]	[illegible]	M2M_DIR	[56897/16507:-121.34919:38.10106:80]

From inside the txt file select all of the data using **Control** (CTRL) and the letter **A** button, and copy the data using Control (CTRL) and the letter **C** button. Next, open a new workbook in Microsoft Excel. Paste the copied data into cell A1 using the Control (CTRL) and the letter **V** button.

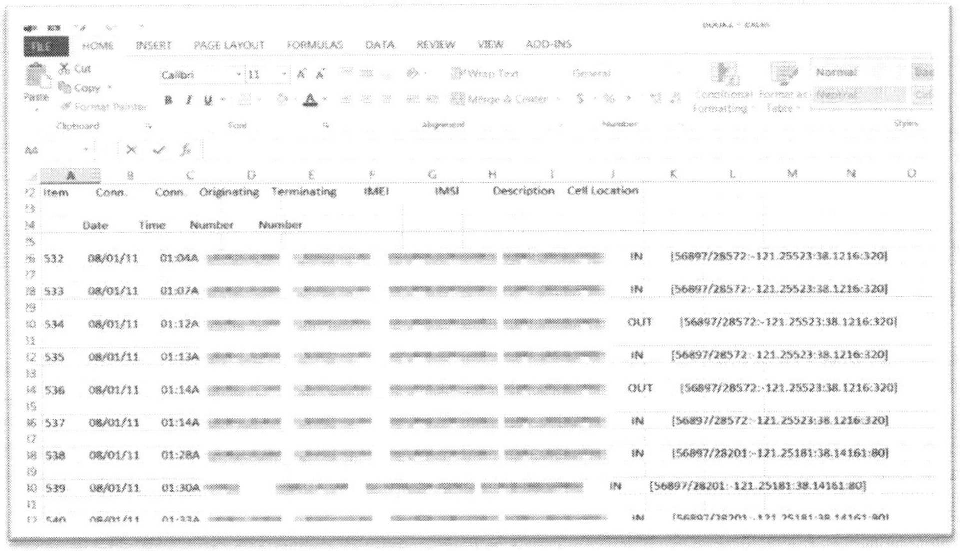

There is still a bit of work to do after you have pasted the records into Excel. When the data was pasted into the cells it brought over everything from the text files-page numbers, header information, and basically any text that was on the sheet. We need to get rid of the extra data. Also, even though the data looks like it is in separate columns, it's not. All of the data is actually in column A and to make use of it you will need to separate each piece of data into its own column.

One final note: AT&T puts all call detail records, text message records, and data records in one pdf file. If you try the following steps on all of the consolidated files, you will create more work for yourself down the road. Take the time to copy the data from each section into its own spreadsheet – e.g. call detail gets a spreadsheet, text messages get a spreadsheet, and data usage gets a spreadsheet. To add additional sheets to the workbook, go the bottom of the page and click on the plus sign (+) next to the active work sheet. You can rename each worksheet by right clicking on the worksheet tab and typing in the names to match the data.

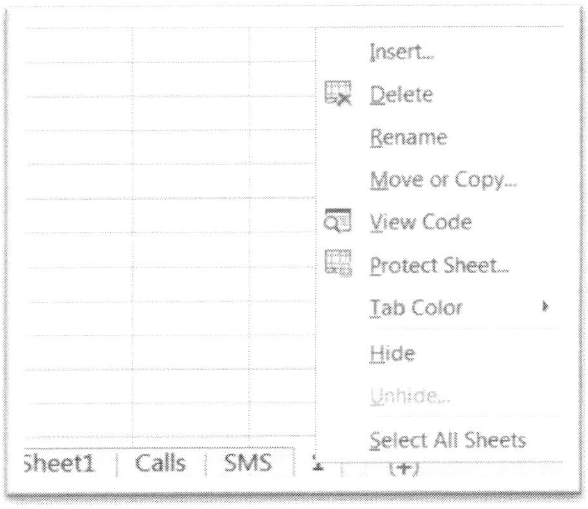

The next step is to get rid of all of the extra junk. Use the **Sort** function found in the **Data** ribbon on the upper part of the screen. Then highlight the data you do not need and delete it. You should be left with only the actual transactional records and all of the extraneous information should be gone.

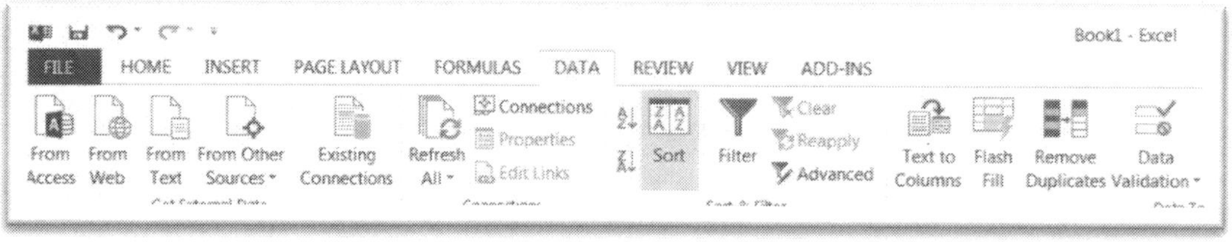

The final step is to get each item of data into its own column. To do this, highlight column A by clicking the **A** at the top of the data.

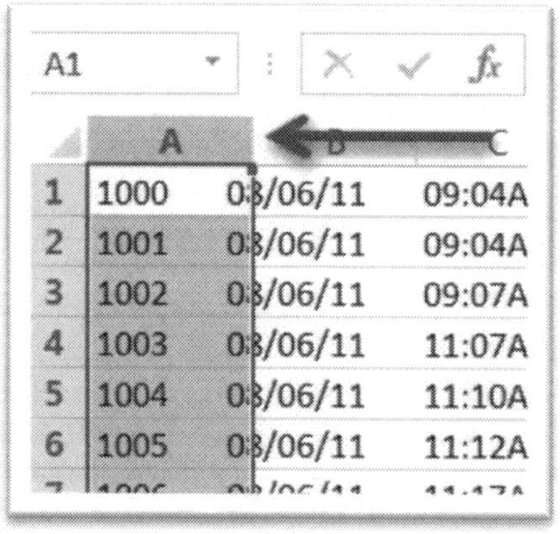

Next select go to the **Data** tab in the ribbon at the top of the screen and select **Text to Columns**. You should see a dialogue box similar to the following graphic. It is important to select **Delimited** from the two options before pressing **Next**.

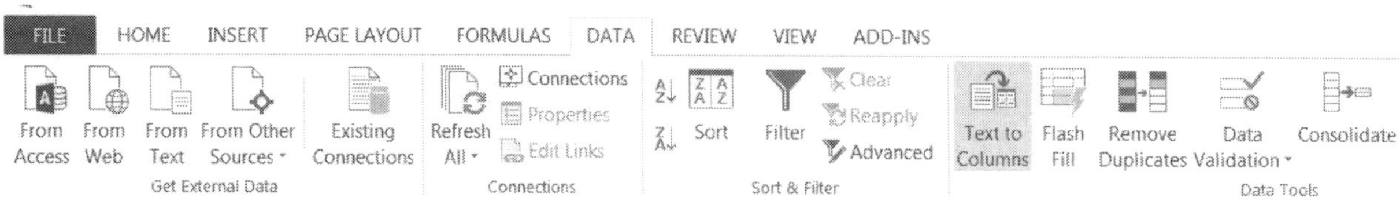

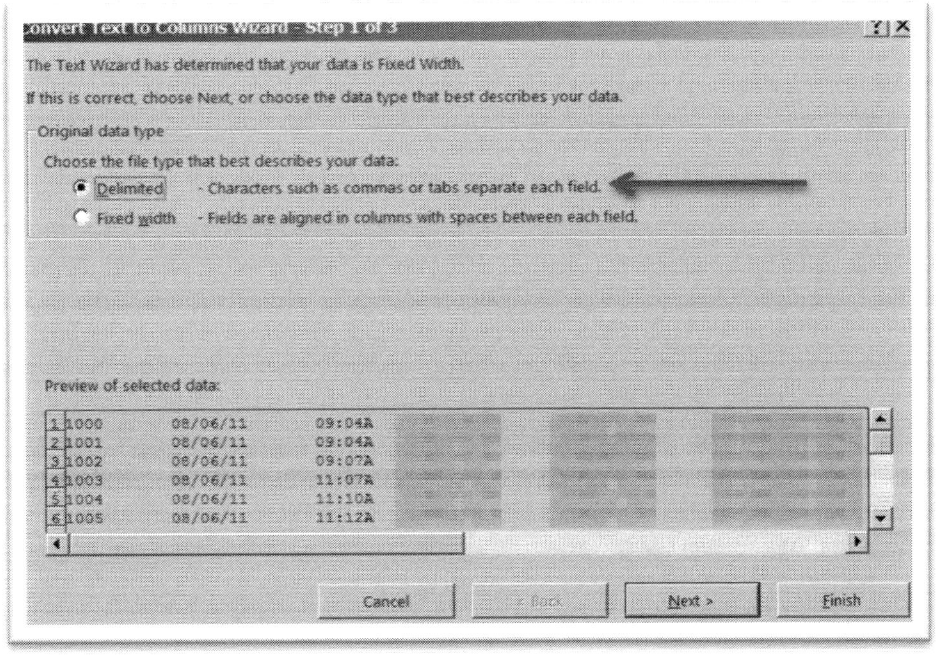

This dialogue box gives you options for how it will separate the data. If you have followed all of the steps outlined above you will need to put a check mark in the box next to **Space** and a check mark in the box that reads **Treat consecutive delimiters as one**. You know you are on the right track if you see the black lines separating the data fields in the Data Preview window as shown below.

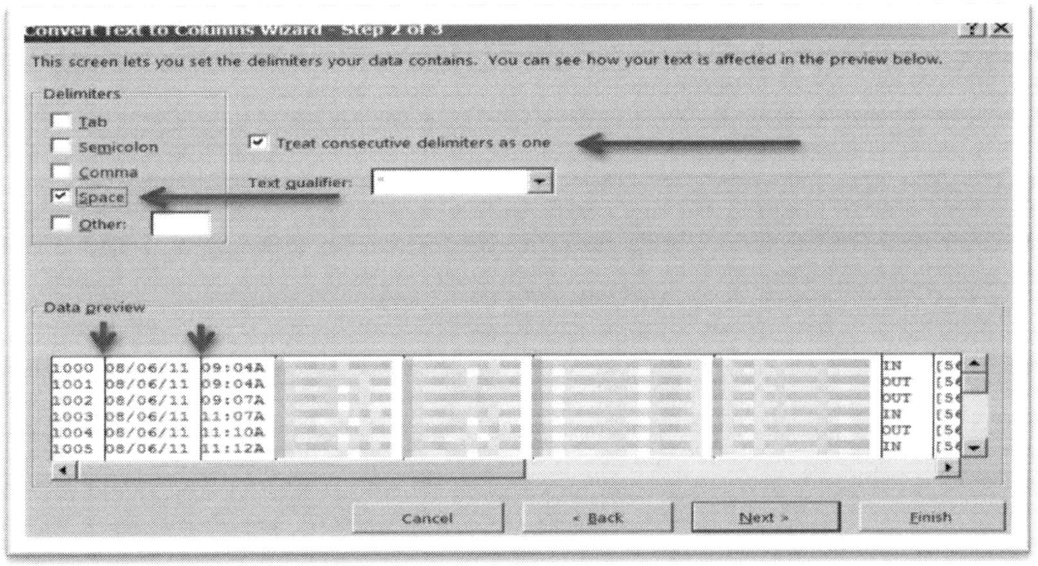

From here, there is no need to go to Step 3 of the process. Press **Finish** to complete the process. This process had separated all of the data which had been in column one so that each element is in its own column and row. This will be important if you want to use any of the data-analysis steps to be discussed later in this chapter or if you want to do any basic sort functions for preliminary analysis.

Presenting Your Data Effectively with RFFlow

Law enforcement officers or analysts often have to brief their co-workers, superiors, or other agencies with their investigative findings. A frequency analysis done in Excel may lack the visual impact the presenter desires. As stated, some of the existing software programs are too expensive for many agencies.

A basic alternative can be found in a software program called "RFFlow." The software program is designed for charting functions and has a template for telephone analysis. This template can produce a graphic chart illustrating call frequency that helps an investigator focus on the target's close associates.

The software can be found at www.rff.com and offers a 30-day fully functional download for evaluation purposes and is economically priced at $49.00 for a single license. Once the program has been installed it is simple to import your data for analysis. Prior to using the software, your data must be formatted so the program can understand it.

Make a copy of your formatted call detail records. Delete the non-essential columns of data so the only remaining fields are your target number, the other phone number, and the direction of the call (e.g. incoming versus outgoing). Any other data fields can skew the results. The file must be saved in a format RFFlow recognizes. These file formats are comma separated value (.csv) or text (.txt). Select the **Save As** function from the main menu to convert your file into the proper format.

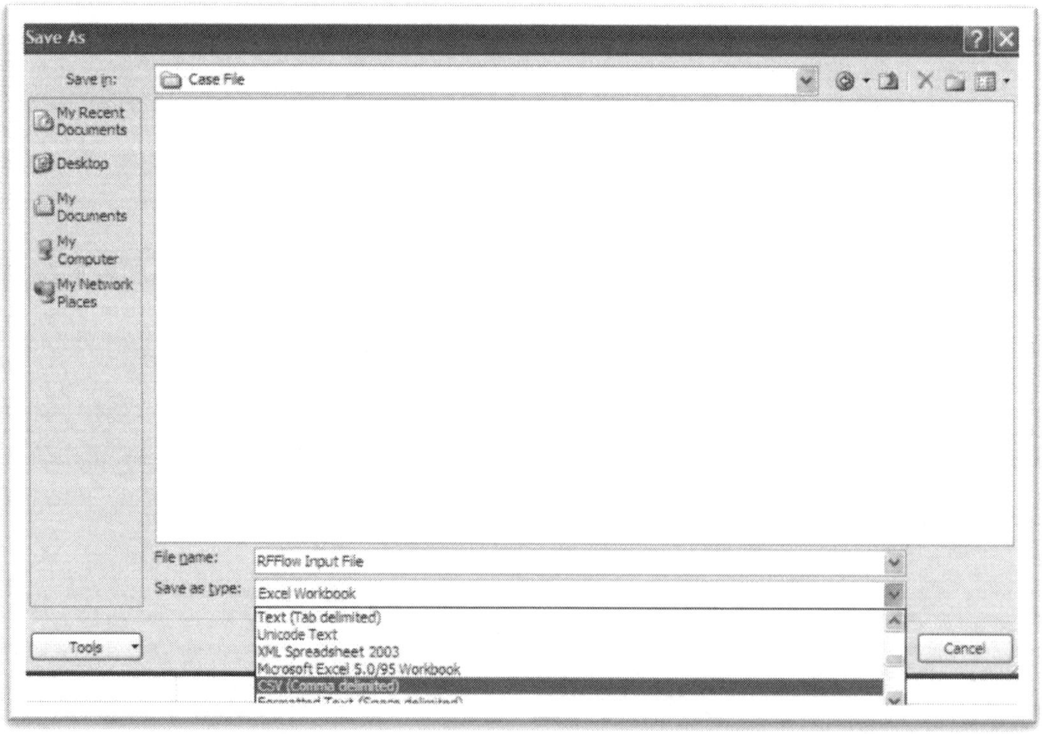

Open RFFlow and select **Tools**, **Auto Charts**, and then **Telephone Analysis**. Navigate to your data file and select **Open**.

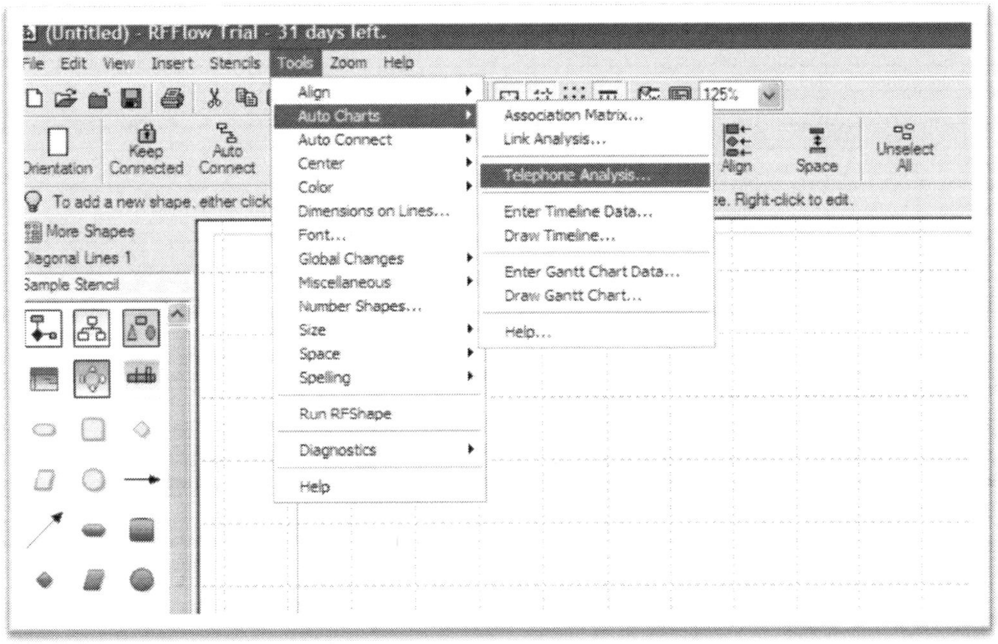

The next screen will give you an option to select the **minimum number of records** to illustrate. If your intention is to identify and illustrate the high frequency phone numbers in contact with your target, you should use this function to filter some of the infrequent contacts out of the data set. In this case, I selected a minimum of 10 records.

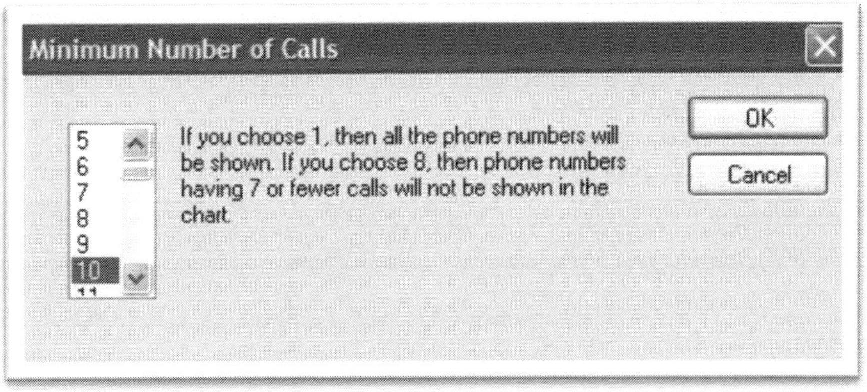

RF Flow will automatically generate a spoke diagram illustrating all phone numbers that are in contact with the target number more than 10 times.

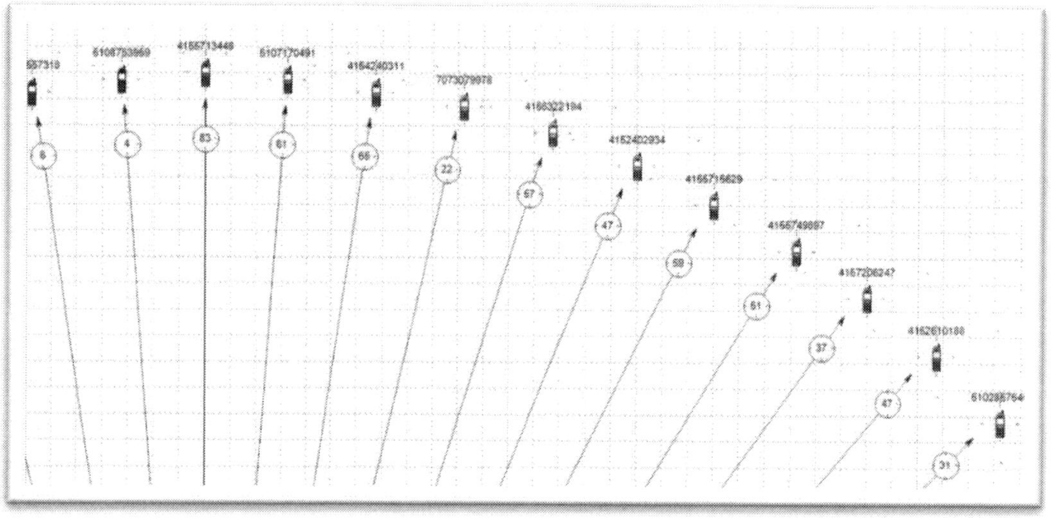

It is easy to zoom in, examine, and identify the high frequency numbers in this format.

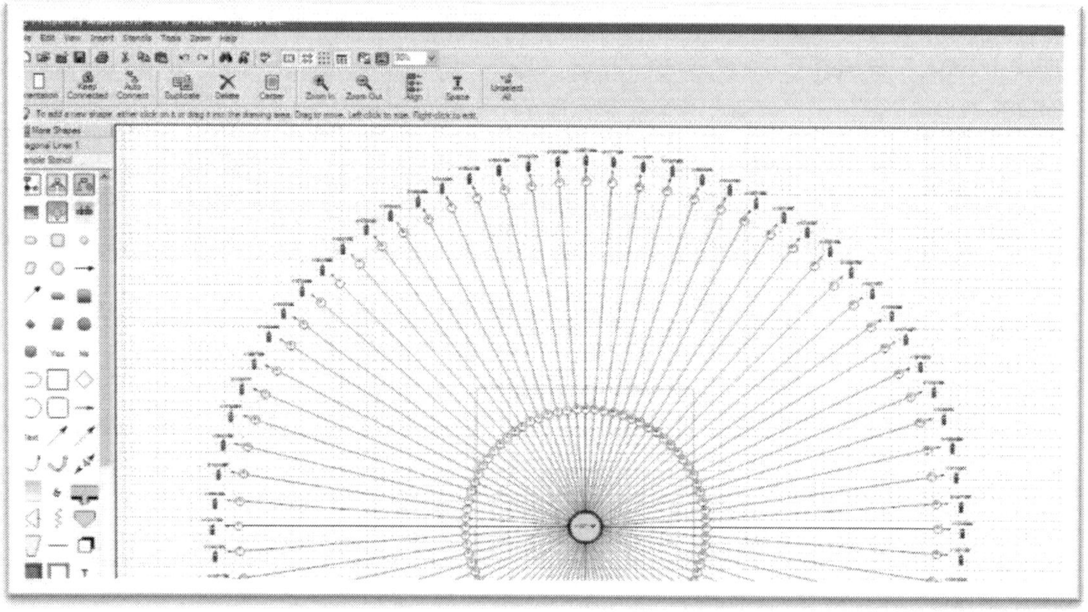

Using Cell Phone Data to Reveal Patterns

When debriefing an informant, expect him to have little – if any – identifying information on his dealer. Because what an informant considers as identifying information is not usually the same as what a law enforcement officer does. Our Ozzie the Ounce Level Dealer is an example of an informant whose identifying information doesn't deserve a search warrant. On the upside, most informants can quickly pull out their cell phone and give you the phone number of their dealer.

If you have an informant who wants to work off his case, I highly recommend downloading the contents of his phone. To demonstrate his willingness to cooperate, ask him for consent to search the phone. By examining the call logs you can validate how often he is in contact with his supplier that may establish or demolish his credibility. You can also discover connections that the informant may not have been willing to give you at the beginning of your "relationship."

Now you can take it one step further. In examining the larger set of information contained in your suspect's cell phone records, you can establish patterns of behavior. When those patterns are combined with your surveillance techniques, you gain a better understanding of your suspect's actions. Now you are able to create a detailed diagram of the structure of a criminal organization. This diagramming process is time consuming, but there are software tools available to speed up the process.

Analysis Software Tools

Analysts and investigators use software tools to effectively study call detail records for patterns and sequences. These tools feature visualization software, such as Microsoft PowerPoint that illustrate the findings of the investigation, sophisticated analytical tools such as i2 Analyst Notebook, or Pen-Link to automate the analytical process. These programs also link charts from other programs, which is effective for revealing patterns when delivering your presentations. Linking charts is the process of fusing together data from different programs. Then different formats and sources can be merged into a common model (i.e. taking information from Excel and including it in a PowerPoint presentation but being able to update Excel data from both places).

I worked with a variety of software programs to analyze and illustrate the results of my investigations counting those I've mentioned. Each software program has its pros and cons, such as:

Microsoft Excel and PowerPoint

- Pros: Almost universally available, simple to use, performs basic analysis and illustration
- Cons: Limited ability to detect patterns and sequences, difficulty handling large data sets, requires data to be in a uniform format

i2 Analyst Notebook

- Pros: Capable of handling large data sets, able to import data with minimal or no "cleaning," automatically generates easy to understand link charts
- Cons: Cost per license can be prohibitive (particularly for small to medium-sized agencies), requires training to be used effectively

Pen-Link

- Pros: Designed to work with the common wiretap and pen register equipment, capable of handling large data sets, able to import data with minimal or no "cleaning," specifically designed to support telephonic investigations
- Cons: Cost per license also can be excessive (particularly for small to medium-sized agencies), user-interface can be confusing, requires training to be utilized effectively

Threads

While most analysts and investigators are familiar with i2 and Pen-Link, they might not be aware of a new product on the market. I used this product extensively and, in my opinion, it is one of the best on the market today.

- Pros: Sophisticated and automated reporting features, can double as a case management/intelligence database, specifically designed for telephonic investigations, short learning curve to use, less expensive than i2 and Pen-Link
- Cons: Newer company so there are still some program bugs, graphical link charts are somewhat primitive, less expensive than the other programs, but cost can still be prohibitive for smaller law enforcement agencies

A Free Alternative

A free alternative to some costly commercial programs is available from a company called yWorks. Its product, yEd Graph Editor, is a charting program that makes a variety of charts and diagrams. The software allows you to manually create link charts and also can automatically import data from Excel. It is intuitive and very easy to use when manually creating charts. In fact, I find it easier to manually create a chart in this program than in other programs, which costs thousands of dollars and requires weeklong training to use them. Two of the features I especially like are the ability to import custom graphics for use as icons and a hyper-linking function that lets you link your icon to a specific folder or file for later reference.

yEd does not have a dedicated telephone template, so working with a large data set will likely require a lot of manual editing, but for basic link charting and case organization yWorks is hard to beat.

Text Message Retention

Contrary to what privacy advocates and defense attorneys might believe, most of the major cell service providers do not store text message content once the message has been delivered. This may not be the case for national level intelligence agencies, but it is for state and local law enforcement officers. However, there are a couple of important distinctions contained in the opening sentence. For example, text message content can be recovered from the provider if the message was never delivered to the handset. As long as the phone remains off, the message remains in the delivery server waiting for communication with the handset. This is helpful during missing persons' investigations where the phone is turned off and investigators are interested in the most recent text message sent to the target phone number.

Another important distinction is the difference between short message system (SMS) text messages and multimedia message system (MMS). At least one of the major providers retains MMS content consisting of the media file and accompanying text content. Additionally, larger cell service providers have an online backup capability from which text message may also be recovered with the proper legal demand. Law enforcement investigators miss this data in their legal demands because they have been using the same search warrant, court order, or subpoena template for the last 30 years and have not updated it to reflect services the provider offers. If MMS and/or data stored in online backup services are not specified in the legal demand, the companies will not provide data. In most cases they also will not tell you the subscriber has access to those services in the first place.

The only providers who routinely store text message content once it has been delivered are Verizon and MetroPCS. Verizon stores three to seven days of text message content based on their server storage capacity. If the data is not preserved using 18 USC 2703(f) then they are not recoverable if the legal demand is received after the message has been deleted from the server. MetroPCS routinely stores text message content for substantially longer – between 30 and 60 days. As with all good things this comes with a cost. The company demands payment of $50.00 for stored text message content. Sprint Nextel is one of the major providers who stores MMS content but not SMS content. Its retention period appears based on server storage capacity and if the data is not preserved it is not recoverable.

Specialized Records Searches

Criminals to change their phones and/or phone numbers frequently. Sometimes this is done because of real or suspected law enforcement surveillance, law enforcement seizure of the device, non-payment of their bills, or the fact that criminals lose their phones (sometimes to other criminals). However, criminals don't actually abandon the device or throw it away. This is especially true with high-end smart phones that may contain a large amount of content the criminal wants to keep such as pictures, videos, music, games, or other applications.

The most popular of "dropping" a phone is to change the number. Remember a suspect, who has an expensive smartphone loaded with apps his personal media files, does not want to have to repurchase paid programs or transfer and reload his media. Additionally, a high-end phone can be a status symbol for the criminal.

A criminal may call his provider and change the number, usually after telling the provider he is receiving harassing phone calls. If a suspect discards his device or loses it through other means, he will stay with the same service provider and report the handset as lost or stolen. He will then re-establish service using the same subscriber information.

In situations where you have a "cold" or older phone number of a suspect, call the cell service provider to see if the suspect changed his number. Some providers require a search warrant or court order to do so. If that is the case, include language in the search warrant or court order directing the provider to search by the prior account holder's name and request any other phone numbers on the same account. In conspiracy investigations, one person could be responsible for multiple phone numbers on the same account. This is especially true in human trafficking and prostitution investigations, where a suspect maintains high control over his subordinates.

Some criminals believe that by changing the Subscriber Identity Module (SIM) card on the phone they have defeated law enforcement's ability to track them. Switching the SIM card associates the phone number from the device it came from with the new phone. What many people do not know is that when a SIM card enabled phone makes or receives a phone call it broadcasts several identifying numbers to the provider, which are subsequently recorded. The information transmitted includes the serial number of the SIM card, known as the International Mobile Subscriber Identity (IMSI) as well as the serial number of the phone itself, or the International Mobile Equipment Identifier (IMEI). The IMEI is roughly analogous to the vehicle identification number of a car and the IMSI can be thought of as the license plate number. While it is relatively easy to switch the license plates on a car, it takes much more skill and sophistication to change the VIN number. With the service of a court order or search warrant, it is possible to check with the provider to see if the IMEI is still active with a different SIM card.

The license plate and VIN analogy brings up another important point. It is possible to change the IMEI of a phone using specialized equipment. This is usually done to conceal a stolen phone and reintegrate it into the marketplace. Investigating stolen phones and fencing operations is covered in more detail in a subsequent chapter.

Identifying a Suspect's New Phone Number Using Call Detail Records

A method to determine a suspect's new phone number is to examine the suspect's call detail records using the Excel pivot table method outlined earlier in this chapter. Using the suspect's most recent call detail records, examine the pivot table for records of high frequency and duration calls to identify the target's friends and family. Common sense tells us that a suspect who places hundreds of calls a month to a friend, spouse, or significant other will continue to do so with his new phone number. Generate a court order or search warrant to the providers of those high frequency numbers and identify the new number from them.

Another relatively easy way to identify a suspect's new phone is to examine his old phone records for calls to voicemail. When a suspect is trying to evade law enforcement the first step is to obtain a new phone and turn off his old phone. This complicates the law enforcement investigation, but it also grants an opportunity to determine his new number. In most cases the suspect is not going to actually toss his old phone out the window or into a river. He still needs it. The old phone contains his contacts and other data he wishes to keep like image and video files, games, and applications. In addition to containing the information he needs, the old phone serves as the last point of contact for the suspect's friends and acquaintances.

The suspect needs to retain his old phone, or at least keep the service active for a short while, to receive voicemail messages. An investigator can leverage this fact by checking the suspect's most recent call detail records for calls to voicemail with a longer than average duration. Compare the suspect's likely actions by examining the average person's behavior as it relates to voicemail messages. Most incoming calls to voicemail where the phone is powered on but no message is left will likely be 30 to 35 seconds in duration. This accounts for four incoming rings before the call is forwarded to voicemail and the caller hangs up before leaving a message. Incoming calls to voicemail where the phone is powered on and a message is left will likely be between 35 to 55 seconds in duration. So any incoming call to voicemail where the duration is over 60 seconds will probably indicate one of two things: a) the incoming caller left a lengthy message (which would probably warrant additional follow-up, especially in the case of fugitive investigations); or b) the suspect was calling his old phone number to check the voicemail messages and listened to one or more of them.

See how many prior phone calls the long duration voicemail caller had prior to dropping his old phone. By examining the pivot table data generated from the call detail records, you can see if the incoming caller had any prior calls to the target number. If there are no prior calls to the target phone number, you might want investigate that number further. The first piece of information you should seek from the provider of the incoming voicemail caller is the activation date. If the activation date is at or near the same time as when the suspect dropped his other phone/number, you probably have his new phone number – all because he had to check his voicemail messages.

There is another method for quickly obtaining a subject's new phone number, which is provided by a law enforcement entity. However, this method is too sensitive for publication as I have no control over where this book ends up. If you are in need of this service contact me via email at aedens@policetechnical.com from a government address. List your agency information so I can verify your employment before sending it to you.

Calls -to -Destination Searches

Cell service providers can also do a "calls-to-destination search." This is a specialized computer search of their records for everyone else who has called a target phone number during a given time frame. It's helpful if your crook has blocked his caller ID or called a number where the caller ID is not displayed or captured.

Let's assume your agency receives a threatening phone call on the main business line. The caller specifically threatens to assassinate a particular officer in a specialized unit. Criminals know that calls to 911 or the non-emergency line are automatically displayed, even if they have caller ID blocking on their number. In this case, the suspect presumably knew this and called the business phone number for the department in order to make his threats. The caller ID of the suspect was not displayed or captured during the call, even when he was transferred to the dispatch center. Based on the specificity of the threat, the department believes a credible threat exists against the officer. Traditional investigative methods would likely involve researching the officer's previous arrests and significant incidents in an attempt to identify likely suspects. However, in this instance the officer is assigned to a specialized enforcement unit targeting street gang members so the suspect list is long and distinguished.

It seems impossible to attempt to identify the phone number of the suspect, but this is not the case. The first step is identifying the specific time frame when the call was received. Even a small law enforcement agency receives a large number of phone calls on its main business line, so narrowing down the time frame is essential. However, that can be a double edged sword. Too narrow and you may miss the incoming call. Too broad and you may comprise multiple results that will all need to be researched. In this instance the approximate time the call was received is identified and an additional minute is added in the legal process to be on the safe side.

The next step is to identify *all* of the communications service providers, both cellular and landline that cover your geographic area. In a vast metropolitan region, this can be intimidating with a massive amount of companies. But effort is worth it in this case, especially if you are the officer who received the threats. Once you have identified all of the companies, you will send them all a court order directing them to perform the calls-to-destination search. In essence, the court order directs them to search their entire call record database for any of their subscribers who called the particular phone number (in this case the main business number for the police department) during the specified time frame. For example:

Conduct a calls-to-destination search for the period 1/1/2013 1400 PST to 1/1/2013 1405 PST, of Urban PSC call detail records (CDRs) (date, time, duration, originating number, destination number) to identify any and all Urban PCS mobile telephone numbers used to place calls to the following numbers: 510-555-7272.

Be sure to include additional language directing the provider to list the subscriber information and call detail records for any user they identify as making a call to the target number. This language is part of the standard court order and search warrant language found in the Appendix of this book.

Calls-to-destination searches can be done anytime – the date, time, and receiving phone number can be identified even if the calling phone number is not known. Other uses for calls-to-destination searches can include:

- A large scale drug dealer will not give an informant his phone number and only makes outgoing calls to the informant from a caller ID blocked phone number
- Fugitive investigations in which a family member or friend acknowledges receiving a phone call from the suspect but is unable or unwilling to provide the suspect's phone number
- Calls to homicide or narcotics tip lines wherein the caller's information is so specific he must have firsthand knowledge of the crime
- Harassing phone calls made to a phone number from caller ID blocked numbers
- Bomb threats to schools involving a caller ID blocked number

Calling Cards

There is no two ways about it – if your suspect is using calling cards, then you have significant hurdles you will have to overcome to advance your investigation. As you probably know, a prepaid calling card is a small paper or plastic card, sized and shaped like a credit card and used to pay for telephone services. A purchaser may select from thousands of different denominations and brands. Ethnic communities will have calling cards advertising to their particular region of the world. The card itself is purchased and activated at the point of sale. The user's phone calls are deducted from the prepaid amount depending on the length of the call and the destination country. To place a call, the user must first dial an access number, typically a toll free line. He then enters the personal identification number, found on the back of the card, and the phone number he would like to dial. Having the physical card itself is unnecessary because knowing the access telephone number and the PIN is sufficient. Standard cards, which can be purchased and used without any sort of account, give a fixed amount of credit and are discarded when used up; rechargeable cards can be "refilled."

Herein lies the problem from an investigative standpoint. When examining your suspect's call detail records, you will only see the local access or toll free number on the record and not the number dialed. Once the call is completed you will be unable to see any of the digits entered. The digits entered once a phone call is completed, whether it's to a local access number for a calling card or the toll free number to a financial institution, are commonly referred to as "post cut-through dialed digits" or PCTDD. The legal issues surrounding the collection and admissibility of PCTDD are discussed in the chapter regarding dialed number recorders/trap trace devices also known as, "pen registers." Suffice it to say there is a significant degree of legal dispute regarding PCTDD, but there are few other methods for obtaining calling card information without using it.

You must obtain the card that has been used to make the call, attempt to identify the calling card company, and use a calls-to-destination search for the target phone number or install a pen register on the suspect's phone lines.

Acquiring the actual card used to make the call requires old-school investigative methods such as informants and digging through the suspect's garbage to retrieve the card. Both of these methods have an uncertain degree of success and rely heavily on luck. However, these methods can result in getting the name of the calling card company and the personal identification number (PIN) assigned to the card. If you are able to determine the PIN you can submit a Court Order or Search Warrant to the provider using the PIN number in place of a phone number – a pen register.

A calling card string (pictured above) obtained from a pen register will appear pretty obvious in comparison to other calls. An example of a PCTDD string might appear similar to:

18002339008011551163467721123311120409011

It is relatively easy to break apart the string and decipher the call:

Access number International number	PIN number
18002339008 0115511634677	2112331120409011

Deciphering the access number is the first step. It is the first 10 or 11 digits of the string and is usually in the North American Plan (NAP) format ### (area code)-###-prefix. No NAP number can begin with a one. This designates a long distance or toll free call.

The international number is also easily identifiable as it is prefaced by 011. This alerts the system to prepare for an international call and remaining digits are the actual phone number.

Calling Card Stings

Some law enforcement agencies have had success introducing known calling cards with pre-recorded PIN numbers to suspects using informants. You can set your department up in the prepaid calling card business, as well. There are re-sellers who will supply you blocks of PIN numbers or the complete calling card at a discount. Once the PIN numbers are known and recorded, an undercover officer or informant sells or gives the card to the suspect for his use. The drawback to this method is, most calling cards have a maximum denomination of $20.00, which does not last a long time. You might only be able to capture a few phone calls before the prepaid minutes expire. The suspect is then forced to look for another card.

There is another variation of this method using prepaid calling cards or prepaid phones. For technical assistance contact Orion Systems regarding their ULETS system and other investigative assistance. They can be reached at www.ulets.com or 925-556-5600.

International Phone Records

If your suspect is using calling cards, there is a high probability he is using those cards to make international phone calls. Many calling cards are directed toward immigrants and advertise a particular country or region such as Mexico, the Philippines, and the Middle East. In addition to the problems associated with investigating calling cards, comes the difficulty of deciphering international phone numbers.

One of the primary challenges in deciphering international phone numbers is determining the format for the particular country or region. The common North American Plan format for phone numbers of 10 digits is far from universal. Countries may have eight- to 12-digit phone numbers while some have only six. The key to determining the appropriate format is to focus on the first digits following the international access number 011. International country codes are between one and three digits and it can be extremely difficult to correctly identify which part of the number dialed is the country code. For example, international calls to North American Plan numbers are preceded by the number one. Similarly international calls to the nations of the former Soviet Union are preceded by the number seven. From there it can be a little more confusing since country codes for other nations are either two or three digits, making it hard to determine which one to use. Calls to Europe can use a two-digit county code if the call is destined for England (44) or a three-digit country code if the call is destined for Ireland (353).

To correctly decipher the international number, you can manually parse the telephone number and separate the country code from the phone number. An easier way is to use one of the Internet tools available such as numberingplans.com or fonefinder.net. These services are free and will reveal not only the country dialed but also whether the phone number is a landline number or cellular as well as the geographic region the phone number is associated with.

When examining call detail records where the phone number was directly dialed without a calling card, the international number called will be designated with a + symbol. The 011 digits dialed will not appear on the call detail records.

Obtaining International Call Detail Records

There is no easy way to get international call detail records or subscriber information. Various federal law enforcement and intelligence agencies have some capacity in this area, but it is rare that they share this information with state and local law enforcement officers.

The Federal Bureau of Investigations maintains Legal Attaches, commonly referred to as "legats," at the embassies of most countries. These Legal Attaches are tasked with establishing and maintaining liaison with the main law enforcement and security services in foreign countries. The relationship is designed to be a two-way street. Federal law enforcement agencies can make requests for information to these foreign law enforcement and intelligence agencies and those same agencies can request cooperation from the FBI and other U.S. law enforcement agencies for assistance. This cooperation is dependent on the both countries being party to a Mutual Legal Assistance Treaty or MLAT. A country being a signatory to a MLAT is great, but not all countries have signed the agreement with the United States. While the absence of a MLAT is not a show stopper, it means there is no formal mechanism in place for law enforcement cooperation. Also keep in mind an MLAT does not necessarily guarantee cooperation. For example, the MLAT signatories include countries problematic for law enforcement like Nigeria, the Cayman Islands, and Venezuela.

It is possible to obtain international call detail records and subscriber information through the Legal Attaches program. However, foreign governments are under no obligation to provide that information.

Also, it may be necessary to disclose the details of your investigation to the FBI and the foreign government to get those records. Additionally, this process is not likely to be quick. A request must be made through the local FBI who routes it to its headquarters. The request is then routed to the appropriate Legal Attaché, who, in turn, submits the request to the foreign law enforcement agency. The foreign law enforcement agency then has to undertake whatever procedural and legal review it is mandated to follow. The results are then sent back through the reverse process. If you think you need international assistance, contact your FBI Field Office or Resident Agency and ask for help.

Again, there is another method for obtaining international call detail records, but I could not include it in this book. Please contact me via a government email account for further assistance.

Mutual Legal Assistant Treaty (MLAT) Signatories

- Antigua and Barbuda
- Argentina
- Australia
- Austria
- Bahamas
- Barbados
- Belgium
- Belize
- Brazil
- Bulgaria
- Canada
- Cayman Islands
- Cyprus
- Czech Republic
- Dominica
- Egypt
- Estonia
- Finland
- France
- Germany
- Greece
- Grenada
- Hong Kong
- Hungary
- India
- Ireland
- Israel
- Italy
- Jamaica
- Japan
- Latvia
- Liechtenstein
- Lithuania
- Luxembourg
- Malta
- Mexico
- Morocco
- Netherlands
- Nigeria
- Panama
- Philippines
- Poland
- Portugal
- Romania
- Russia
- Slovakia
- Slovenia
- South Africa
- South Korea
- Spain
- St. Christopher and Nevis
- St. Lucia
- St. Vincent—The Grenadines
- Sweden
- Switzerland
- Thailand
- Trinidad—Tobago
- Turkey
- Ukraine
- United Kingdom
- Uruguay
- Venezuela

Caller ID Spoofing

Caller ID spoofing is using technology to cause a telephone network to display a number on a receiving phone's caller identification (caller ID) system that is not the true originating phone number.

The advent of voice over Internet protocol (VOIP) systems have made spoofing easy and inexpensive. A popular method for spoofing caller ID involves a subscription-based service or a web-based interface. Anyone can purchase a number of prepaid spoof minutes, which are similar to those on a calling card. The purchaser is assigned a PIN number and an access number. The user contacts the access number from any landline or mobile phone, enters his PIN number and the phone number to be called, followed by the phone number they want displayed. Some services offer additional features such as voice alteration and recordings of the calls emailed to the user at the conclusion of the call.

Caller ID spoofing has been used by criminals in harassing and stalking type crimes. Several high profile instances involved reporting false crimes to 911 to elicit a response from law enforcement. The action has been coined "SWATing." According to published reports, some of these instances have resulted in tactical units responding to reports of violent crimes only to find unsuspecting subjects who have been victimized by this "prank."[8]

How it works

You must understand what is transmitted when you make an outgoing phone call to know how caller ID spoofing works. When you make an outgoing phone call a packet of data is transmitted that contains the following data fields:

- **Automatic Number Identification** (ANI): The phone number you called from
- **Automatic Number Identifier II** (ANI II): A two-digit code containing information about the type of phone number you called from, such as landline, coin operated payphone, correctional institution, and mobile phone
- **Calling Party Number** (CPN), also referred to as CID-Caller ID, CLID-Calling Line Identification, and/or CNID-Calling Number ID: This is also the phone number you are calling from. However, the data in this field can be manipulated unlike the ANI. CPN data is designed to be manipulated for a number of legitimate reasons. For example, CPN data contains a marker that designates if the calling number is private or can be displayed. If the number is flagged as private, the system

[8] http://www.911dispatch.com/911/swating_pranks.html

displays "PRIVATE" in lieu of the number. CPN data is also changed on outgoing phone numbers originating from a private branch exchange (PBX) a private office or organization uses. Law enforcement organizations may use a PBX, which displays a general outgoing phone number and not the desk number of the originating officer or investigator. Caller ID spoofing takes advantage of the ability of CPN information to be manipulated. Using VOIP technology, caller ID spoofing providers can change the CPN data to display the data entered by their customer.

Costs for these services vary depending on the total number of minutes, also called credits, purchased. One popular Internet vendor offers plans consisting of anywhere from 25 credits for $4.95 to 2,500 credits for $299.95. A credit is equal to one minute of talk time and five credits for a spoofed text message. Payment is accepted in the form of credit card number or by using PayPal.

The Truth in Caller ID Act of 2009

In December 2010 the President signed the Truth in Caller ID Act. The Act makes it a federal crime "...to cause any caller identification service to knowingly transmit misleading or inaccurate caller identification information with the intent to defraud, cause harm, or wrongfully obtain anything of value..."

This law is a watered down version of other legislation which would have essentially stopped caller ID spoofing. Violators of this law face the possibility of fines for every instance they misuse caller ID. There is an exception in the law for law enforcement officers conducting "any authorized activity."

Investigating Caller ID Spoofing Cases

To investigate a caller ID spoofing case you will need to use calls-to-destination search for all of the major providers. An Internet search can make it appear there are several thousands of these providers, which would normally make this type of search impractical. However, many do not know if the majority of the providers found during a search are actually affiliates of several large providers such as SpoofTel, SpoofCard, and Telespoof. Start with them as they are likely to populate the first several pages of an Internet search.

Spoofing service providers maintain information about their customers including the numbers they call, the Internet protocol address used to subscribe to their service, method of payment information, and the number they called from. All of this data gives law enforcement investigators avenues to investigate with further legal process.

Working the caller ID spoofing service providers will likely capture the unsophisticated teenager making a bomb threat to his school or the harassing stalker. More sophisticated users could utilize other caller ID spoofing technologies such as Skype and cloud based PBX services. Skype users have the ability to change the displayed number on caller ID systems. Along with other reasons, since Skype is based in Luxemburg, cases involving the company were notoriously difficult to investigate. Skype did respond to law enforcement requests but since its acquisition by Microsoft, it has become much easier to deal with. Skype generates user logs on the host computer that can show calling activity using the service. If a suspect in a caller ID spoofing case is known or believed to be using Skype, a forensic examination of his computer may reveal evidence contained in the activity log.

Cloud based private branch exchange, or PBX services, are another method used to spoof caller ID. PBX services are designed to serve as a central telecommunications hub for a business and used to require the installation of physical equipment. PBX has transformed to a cloud based service and this method is mostly used by hackers. For example, one posting placed on a hacker forum in mid-2013 discussed using the cloud based PBX service Vitelity to spoof caller ID. The service providers will change depending on the vulnerabilities discovered in the hacking communities, but using legitimate companies to facilitate this activity may leave traces behind.

Some law enforcement agencies have explored using caller ID spoofing in their investigations with mixed results. There are a number of drawbacks for using caller ID spoofing during law enforcement investigations. Some agencies have reported running out of minutes or credits in the middle of a call, causing some obvious difficulties. The other challenge involves returned calls. If the spoofed call is unanswered, attempts to return the call to the spoofed number may reveal that the phone call did not actually originate from that number.

Caller ID Spoofing Weaknesses

Caller ID spoofing suffers from a couple of weaknesses that can be exploited during law enforcement investigations. Since commercial caller ID spoofing services first started, the number of companies have dwindled and consolidated into a few large providers. These companies faced the possibility of prohibitive regulation in the Truth in Caller ID Act of 2009 and seek to avoid any further legal restrictions that may force them out of business.

This might have caused some of the larger companies to cooperate with law enforcement legal process requests. In fact, in some of the "SWATing" stories discussed earlier, it appears those companies did cooperate with law enforcement investigations. However, somewhat complicating the issue is that some of these companies are based in Canada. This would likely involve seeking assistance from the FBI Legal Attache for assistance. Check the caller ID spoofing providers' websites to determine their willingness to cooperate.

You can get a general feel for their corporate attitude by looking at their privacy policies and legal disclaimers. For example, SpoofTel, the self-proclaimed "World's Leader in Caller ID Spoofing," has a security department listed on their contact page (see below) and states in their privacy policy they will keep customer information private unless given legal process from a law enforcement agency.

One of the largest providers of caller ID spoofing is Tel Tech Systems, which operates under the names PhoneGangster, SpoofCard, Telespoof, Trap Call, and Stealthcard. Their subpoena compliance contact information is:

Tel-Tech Systems, Inc.
Legal compliance:
Tel-Tech Systems, Inc.
1433 Hooper Avenue Ste #129
Toms River, New Jersey 08753
Send service: Fax or Hard Copy
Phone: 732-276-1300

Fax: 888-689-2671
E-mail: legal@spoofcard.com

Legal contact: Jessica Thorn
Phone: 732.276.1300 X305
Fax: 877-533-4717
E-mail: jthorn@teltechcorp.com
Emergency Contact: legal@spoofcard.com (after business hours)
Website: www.teltechcorp.com

For other providers you may want to try the consolidated subpoena compliance contact list at http://www.search.org/programs/hightech/isp/.

Caller ID spoofing using a commercial service requires the suspect to access the provider from either his phone or computer in order to use the service. As these are paid services the suspect must use either a credit card or PayPal. All of this activity leaves traces and the caller ID spoofing company can provide you with the information with proper legal process. Undoubtedly, this will require additional court orders, search warrants, or subpoenas to search the suspect's phone number, Internet protocol (IP) address, and method of payment information, but these leads are more likely to resolve to an actual cell phone subscriber, the physical location where the computer was located, and his financial information.

Trap Call

The ability of toll free numbers to reveal the ANI is the basis for the service known as Trap Call. Trap Call is a service one of the major caller ID spoofing companies offer. Trap Call deals users a subscription service that allows them to reject unknown or caller ID blocked incoming calls. Those rejected calls are then routed to Trap Call's service, which is an enhanced toll free number. As the subscriber of a toll free number is paying the bills he is entitled to see the incoming numbers and caller ID blocked is stripped off these calls. The call is then routed back to the subscriber's phone and can be answered or sent to voicemail. Trap Call then sends the subscriber a text message containing the ANI of the incoming call. Trap Call also offers a premium service that lets a subscriber run the incoming phone number through public records databases for subscriber information, digitally recording and transcribing incoming conversations, and black lists of prohibited incoming callers.

Trap Call does not work with all cell service providers. According to its website, Trap Call only works with AT&T, T-Mobile, Verizon, and Sprint. Prepaid providers such as Metro PCS and Boost are not supported. Trap Call also has applications that can be installed on phones using Blackberry, Apple, and Android operating systems.

I learned about Trap Call first hand during a wiretap investigation. The case agent in the investigation had the main suspect under surveillance. In order to confirm the suspect was still using the phone number which was the target of the pending wiretap, the case agent made a phone call to the suspect using his department issued phone. The suspect was observed answering the phone and the surveillance was terminated. Several minutes later the case agent received an incoming call from the suspect asking why he had called him. The case agent was able to convince the suspect he had dialed the phone number in error but was unable to figure out how the suspect had obtained his caller ID blocked law enforcement mobile phone number. It was not until the wiretap was operational that we were able to see the incoming text messages from Trap Call regarding the unblocked incoming calls.

Most government agencies obtain their cell phone contracts under a broad umbrella contract by city, county, or federal entity. Fortunately the suspect in the above case did not have the public records check feature which would have likely indicated the subscriber of the phone was the State of California. Even if the records check did not reveal the name of the investigative agency, it would have seriously hampered the investigation if the suspect became concerned and dropped his phone.

Trap Call has some obvious advantages for law enforcement investigators. However, I see no reason why we should patronize a company that also offers caller ID blocking services and thrives by allowing criminals to mask their identities. The Law Enforcement Telecommunications System from Orion Systems offers the same services. They are also heavily involved with local, state, and federal law enforcement agencies in a variety of investigations. Consider Orion Systems before shelling out personal or department money to a caller ID spoofing company.

The ability to see the underlying phone records of incoming calls to a toll free number is the reason I recommend tip lines be 800 numbers. If someone calls a tip line and references information that only the suspect could have known, his phone number can be quickly identified by reviewing the incoming calls record.

Chapter 4: Cell Towers and Cell Sites

Before we examine the intricacies of cell sites, we should address the proper vocabulary. The reader should be aware that the proper term for a cell site or cell tower is "Base Transceiver Station" (BTS), as it is used by the telecommunications industry. However, in the law enforcement community, we commonly refer to the BTS as a "cell tower" or "cell site," but for simplicity I am going to refer to them henceforth as "cell sites." But why "cell sites" and not "cell towers"?

While all cell towers are cell sites, not all cell sites are towers. Many believe the actual structure is the cell site, but since cell site antennas can be mounted on countless structures and not just at the tops of towers – particularly in urban locations – we should aptly refer to all of them here as "cell sites." Although it is normal to see 100-foot tall towers dotting the landscape in semi-rural and rural areas, this is not true in every location. These metal monstrosities are an eyesore and an affront to anyone who has to live near them, and, as a result, cell service providers go to great lengths to integrate cell sites with existing infrastructure by mounting antennas on the sides of tall buildings or on church steeples. Providers also try to camouflage cell sites by adding plastic branches and leaves to hide the outline of the tower and make it appear like a tree.

Cell Site Infrastructure

While the antenna mast – or tower – is what most law enforcement officers think of when discussing cell sites, it consist of several components. In addition to the BTS, which facilitates communication between the tower and the provider's network, cell sites also have an emergency power supply such as a backup generator. The BTS and power supply are usually located in proximity to the cell tower or mast in an enclosed structure.

One may find multiple antenna arrays on a single mast. In most cases, this is to accommodate a cell service provider's need to increase its coverage in a certain geographic region. However, a provider may lease space on a mast to other companies. This is particularly true in areas with restrictive zoning regulations and communities that do not want cell masts dotting the horizon.

Operational View of the Cell Site

There is no hard and fast rule regarding a cell site's coverage area. Some of those in the law enforcement community circulate the concept that an "average" cell site range is two miles. Two miles is a "safe" measurement for use in mapping cell site activations of a suspect's phone. Unfortunately, "safe" is not always correct, and misinformation is damaging during investigations and testimony.

The effective range of a cell site is not a fixed figure and is impacted by factors including:

- The frequency used
- The transmitting cell phone's power and size
- The amount of transmissions sent or received at any given time.
- Geography, such as adjacent buildings and hills
- Weather conditions
- Routine or emergency maintenance may shut down a cell site
- Large crowd events or unusual occurrences may overwhelm a particular site.

In most instances, a cell site's range will be set to overlap with another site so there is enough coverage for a "handoff" between the two. This allows a mobile phone call to be maintained when the handset is moving, such as in a car or train.

Service providers place cell sites in areas where they can provide the most effective coverage to the subscribers. In high-density urban areas, where buildings and other structures can adversely affect the cell signal, sites can be as close as several hundred meters. In low-density suburban areas, sites are placed 1 to 5 miles apart. Cell sites can be placed 10 to 15 miles apart in rural areas without obstructions. The maximum range of a cell site antenna or mast can be limited by the underlying technology. For example, Global System for Mobile Communications (GSM) systems have a fixed maximum range of 40 km, or 23 miles, based on technical limitations of the equipment. While there is no designated maximum range for Code Division Multiple Access (CDMA) and iDen (push-to-talk) systems, the effectiveness is limited by the low-power transmission of the mobile phone.

Depending on the operating frequency of the site and its height, the effective range of a site can be anywhere from 1.5 to 5 miles. Law enforcement agencies use an average cell site radius of 1.5 to 2 miles to demonstrate the possible area where a mobile device was located. This estimated radius is helpful for general estimation and illustration purposes only but should not be used as evidence. Consider the information regarding a cell site not as final proof but as a pointer system, which gives you some basic information on how to proceed.

In addition to conditions that impact which cell site a mobile device communicates with, an investigator must understand that a mobile phone may connect to anywhere between two to six cell sites simultaneously and must select the best signal for communication. The best and strongest signal, however, is not always from the closest site.

A detailed examination of the operating characteristics of the site is needed to form an educated opinion on the location of a mobile device within a cell site's range

To address this, some law enforcement agencies are using outside contractors to evaluate a crime scene's cell site coverage. Other law enforcement agencies are training their personnel to use cell site surveying equipment and techniques. This evolving field takes into account factors that may impact which cell sites were in communication with a suspect's mobile device. Based on their evaluation, they are able to make an educated opinion.

An investigator should understand the basic principles behind the entire cellular communications network – not just the cell site locations. This can help in weighing cell site evidence in the investigation and prosecution.

The central computer that controls phone connections, also known as the Base Station Controller, and the hardware and software of the mobile phone, assists in a device's switch from one cell site to another during conversations. As a mobile phone moves, the controller and the device try to select the strongest signal and free up the weaker signal, which becomes available to another user. Remember that the mobile phone will attempt to connect to the cell site with the strongest signal. Again, this is not always the closest cell site, since the closest cell site may actually have a weaker signal due to geographic or environmental factors, or heavy use.

In general, each site has three sets of directional antennas aimed in separate directions and receiving/transmitting into three different cells and frequencies. The coverage areas of these directional zones are commonly referred to as "sectors." A mast may have less than three directional antennas/sectors if there is no need to provide coverage to a certain area. For example, a mast located at the base of a hill, or covering an unpopulated area, may only provide coverage to the freeway that passes it. Consequently, that mast may have only one or two sectors. There are also masts that feature omni-directional antenna and do not provide sector information. Additionally, there are masts that feature as many as six directional antennas/sectors.

Each standard mast provides 360 degrees of coverage. The three antennas each provide 120 degrees of coverage (See image below). It is assumed, unless noted otherwise, the site is oriented north, or zero degrees, and the first sector, sometimes referred to as "alpha," covers 60 degrees on either side of zero; 300 to 60 degrees.

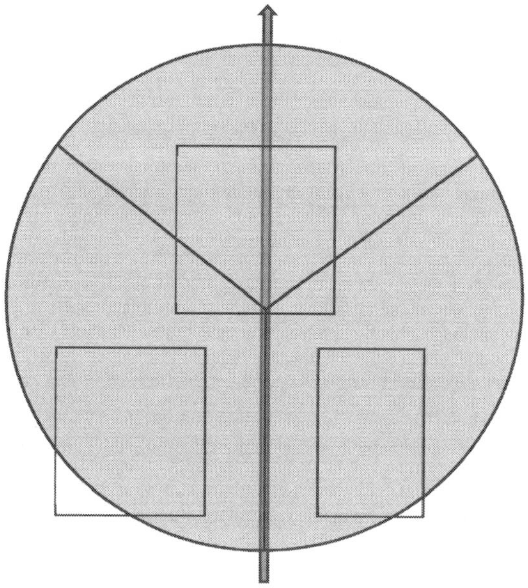

Pitfalls to Avoid

There are five errors law enforcement officers make when using cell site information in their investigations.

1. Some fail to remember that the cellular handset will *not* necessarily communicate with the closest site but with the cell site from which it receives the strongest signal. Be prepared to address which factors may have impacted a particular cell site – or group of cell sites – such as rush-hour traffic, large public gatherings, and sporting events.
2. Investigators will commonly refer to the cell phone and the target of the investigation interchangeably. I *strongly* recommend you avoid this dangerous habit. Your credibility will be destroyed during testimony if you refer to the records of cell site activations as movements of a suspect and not the movements of the phone – unless you can definitively state, through direct observation or admission, that the target was using the device.
3. Without sophisticated survey equipment, it is not possible to state that a cell phone was at a precise location. Rather, cell site information tends to show the general geographic location of the phone at the time a call was made or received as well as the sequence of the cell site usage. While it is possible to proactively determine in real time the location of a cellular device using GPS or cellular geolocation, it is not easily feasible to do so historically. There are specialists who, with the assistance of cellular engineering maps and signal measurement equipment, are able to calculate the probability of the subject being within the coverage area of a particular cell mast and sector during a particular time. Despite claims from the American Civil Liberties Union (ACLU) and the Electronic Frontier Foundation (EFF) and media perceptions, the technology does not

exist to precisely determine the historical location of a cellular device based on cell site and sector information.

4. An important – and overlooked – aspect of investigations of cell sites relates to the azimuth of the site. As noted above, unless specified otherwise, it is presumed the alpha sector is oriented toward 0 degrees. However, the azimuth may be oriented away from true north (See Figure 16), which will change the orientation of the sectors. As noted by Paul Bedell in *Wireless Crash Course*, "The grid angle/azimuth is the orientation in degrees of the first face of all sectorized cells from true north." In English, this means the site may be orientated away from true north. If the azimuth of the site is 90 degrees instead of 0 degrees, you can see the changes in the orientation of the sectors (See the image below). Information on the site's azimuth is contained in the cell site key, which often accompanies the call detail records provided by a cell service provider in response to a court order or search warrant. If they are not included with the returns, contact the provider and make sure it sends you the key
5. The fifth error occurs when analyzing and illustrating cell site coverage areas of the sectors. Remember, cell sites and mobile phone handsets are essentially parts of a sophisticated radio system. The transmissions are radio waves and are not confined to the circles or lines often used to illustrate them on a map. There can be significant overlap between the signals from one sector to another. This overlap can be as much as 20 to 30 percent on either side of the sector. Failing to factor this overlap can adversely impact the investigation and testimony.

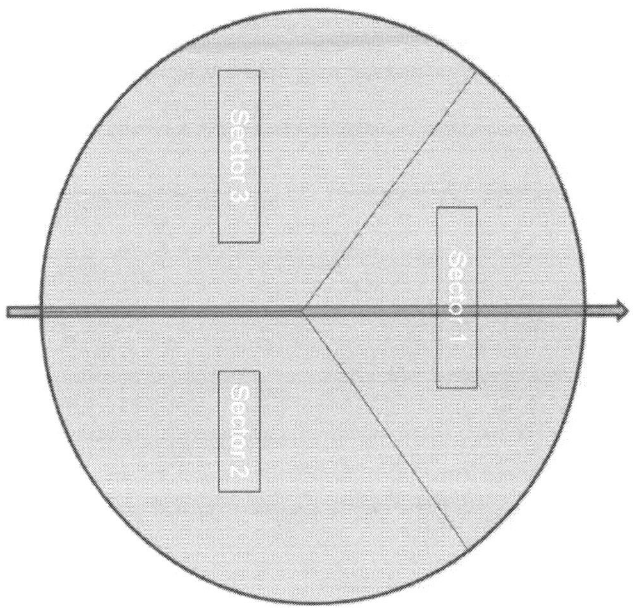

Site Identification

Cell service provider can identify their sites in different formats. Some providers use their own site number format, including their internal numerical designation for the site and an indication of which sector was activated for the particular transmission. Other cell service providers format their site identifiers and sectors using Location Area Code (LAC) and Cell Site IC (CID). Most providers will include a key to decipher the location, sectors, and azimuth of their sites.

Femto cells

A femto cell is a small, low-power cellular base station, usually designed for a residence or small business and intended to provide increased coverage for the area. Similar to Wi-Fi, the femto cell connects to the cell service provider using a broadband connection such as DSL or cable. A femto cell is capable of supporting up to four active phones in a residential setting, and up to 16 in a small business.

Pico cells

A pico cell is similar to a femto cell but is designed to cover a larger area such as an office building or a shopping center. As with a femto cell, the pico cell is designed to extend cell coverage to an area where signals from the network do not reach well or in areas of high concentration of phone users such as a public transit station.

COWs

COWs – or cells of wheels – are trailer deployed cell sites used to augment existing service during highly attended sporting or concert events and provide coverage during disaster recovery operation.

CALEA's Database of Cell Sites by Provider

In October 1994, Congress took action to protect public safety and ensure national security by enacting the Communications Assistance for Law Enforcement Act of 1994 (CALEA). The law clarifies and expands the existing statutory obligation of telecommunications carriers to assist law enforcement in executing electronic surveillance pursuant to court order or other lawful authorization. CALEA seeks to guarantee that after law enforcement obtains the appropriate legal authority, telecommunications carriers will have the necessary capability, and sufficient capacity, to assist them regardless of a carrier's specific systems or services.

The Federal Bureau of Investigation established the CALEA Implementation Team to preserve law enforcement's ability to conduct lawfully authorized electronic surveillance (wiretaps). The CALEA team maintains a website with restricted law enforcement access. Any law enforcement officer is able to gain access to the site after application and verification of his credentials. The website has several extremely useful features, including an archived database of cell site locations and information by provider. The database is searchable through a variety of methods, including by city name and zip code and/or latitude and longitude. This can assist law enforcement officers in determining which cell service providers are active in their market area. The CALEA website also has a historical archive of cell sites, which is helpful if you are examining an older case and need to see sites in existence at the time. The length of how far back the archived sites goes depends on the provider, but, in most cases, it is at least several years' worth of data.

Accessing the CALEA Database

The CALEA database was once easy to find. It also used to contain a lot more information, such as meeting notes and presentations from cell service providers and law enforcement guides for the individual companies. To get to the CALEA site, you need to enter this address in the navigation bar of your favorite Internet browser: **https://208.250.68.210**. New users will need to click on the **Request Access** link at the bottom.

The CALEA group takes a bit of time to verify a prospective user's law enforcement affiliation. If you anticipate using this database, I recommend registering before you would actually need it because it would take days to check your background. You don't want to try and register at 0300 on a Sunday and expect access to it. Fill out the appropriate information in the boxes and wait for the email granting you access.

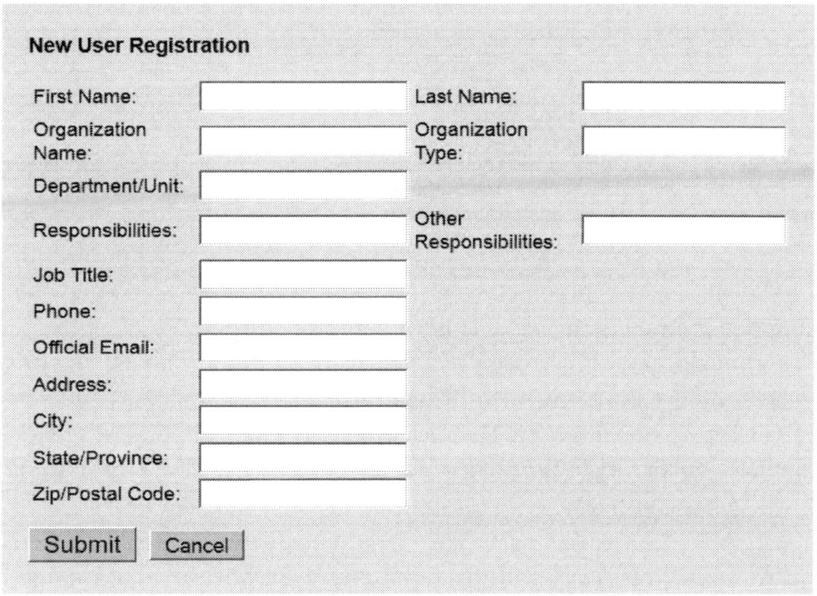

New User Registration

First Name: Last Name:
Organization Name: Organization Type:
Department/Unit:
Responsibilities: Other Responsibilities:
Job Title:
Phone:
Official Email:
Address:
City:
State/Province:
Zip/Postal Code:

Submit Cancel

Mapping Cell Site Information

Cell site location and sector information is easily visualized on a map and can aid investigators and prosecutors. There are mapping software tools and programs available for little or no cost, such as Google Earth, batchgeocode.com, and Microsoft Streets and Trips. Each service or program comes with its own pros and cons. For example, entering multiple cell sites in Microsoft's Streets and Trips is easy but not so fun in Google Earth. Batchgeocode.com is an online service and some law enforcement officers have concerns about putting their information out on the Internet. None of these tools have features for feasibly illustrating azimuth or coverage areas.

The cell site keys obtained from the provider or from www.askcalea.net will provide the location information in two formats: street address of the site and the latitude and longitude. I prefer to use the latitude and longitude of the site, as the street address may not reflect the site's precise location. Also, it is easier to have a consistent standard for data importation using latitude and longitude as opposed to trying to format multiple different street addresses. Another reason for using a cell site's latitude and longitude of a cell site is there are instances where the data was incorrectly entered by the provider. In the example listed below, the provider lists the street address of the cell site as "19020 North Indian Avenue in Riverside." However, the latitude and longitude place the cell site correctly at 19020 North Indian Canyon Road in Desert Hot Springs CA – approximately 50 miles away from the listed street address.

LA-ONTARIO 2	Whitewat	19020 N. Indian Ave.	Riverside	Riverside	CA	92240	33.906111	-116.541389

To map all of these locations at one time, I suggest placing the relevant information into an electronic spreadsheet such as Excel. Once in an electronic format, it is possible to upload the files using a data importer feature found in mapping programs. I also recommend displaying all of the sites in the area and not just the ones the phone accessed. This displays the relativity of a particular site or series of sites.

Additionally, you should map relevant locations from the investigation, such as the known crime scene and suspect's residence or work place. Stay away from speculating about "facts" that are not established. For example, by using the times the cell sites below were activated, it might be reasonable to assume the suspect's phone was traveling southbound on Highway 101). It is a mistake to testify to this as a "fact." It is reasonable to use this as an investigative tool to assist law enforcement officers in identifying *potential* routes of travel or locations, but it falls well short of factual evidence.

Proactive Mapping

With a little bit of effort, it is possible to obtain the locations of every provider's cell sites in a particular geographic region. This information can be acquired from the CALEA Unit's cell site database or cell service providers. By proactively mapping the cell sites in a jurisdiction, an investigator can immediately determine the appropriate cell service provider and the cell sites that cover a geographic area. This can save valuable time in the event of a serious felony criminal investigation. If the methods for creating a comprehensive map are outside of your skill set, look within your city or county government and locate the Geographical Information Services (GIS) specialist or technician. Countless small and medium sized county and local governments have one or more GIS specialists. These specialists have the equipment and capabilities to create comprehensive maps depicting whatever information you provide them. Most GIS specialists are comfortable working with the latitude and longitude of the particular site. Make sure the providers and the site identifying information are included on the map or on an adjoining key for easy reference.

Mapping Tools

Mapping cell sites can be time consuming and challenging, especially when there are multiple sites involved and/or there are a number of azimuth changes. Most of the mapping services let a user represent a cell site or other structure with an icon or user generated designs. These work well with a standard cell site orientation, but what if you are trying to illustrate multiple sites with varying orientations? Other than by estimation it is difficult to accurately represent a section which covers 150 degrees.

One commonly used tool to overcome this challenge is an onscreen protractor. I know most people haven't used a protractor since elementary school, but this program puts a tool onscreen and lets you accurately represent cell site sectors based on their azimuth. The program costs less than $30.00 and has a free trial version. It can be found at www.iconico.com. But if it's free, it's for me. You can also find an onscreen protractor in the suite of tools from picpick.org. The protractor is merely one of a suite of tools that include screen capture and basic image editing programs and is available for non-commercial use for free.

A free tool to annotate and draw on Google Maps is available at www.scribble.com. The website is helpful for manually illustrating a single cell site at a time. There is no batch upload feature, but it does offer tools that are useful for illustrating maps including a full featured set of icons.

Historical Mapping

Sophisticated equipment exists for the precise mapping of the coverage and signal strength of a certain cell site and sector. Technicians can use this equipment to geographically "map" the strength of the site and sector as well as to determine the likelihood a device was in a particular area.

Most cell service providers have already done this and maintain engineering survey maps, also called "propagation maps," of their sites and respective sectors. The cell service providers do not like to divulge this information, even with legal process, as they consider it to be proprietary. However, with persistence, you can request and obtain these maps. Note that the survey map was accurate, as the time the survey was completed and compiled should be on the map. Recent construction and other factors may affect the accuracy of the survey map especially if they were completed after the survey was and a subsequent survey was not done afterward. An investigator seeking to introduce survey maps into evidence should prepare to address this issue with defense counsel under cross examination.

Cell Site Dumps

It is possible to access the information stored by a cellular service provider's system particular to a device's registration with a specific cellular site. You have to go back to the basics of the underlying cell site technology for this statement to make sense. Imagine you have just landed at an airport away from your home. You turn the phone on to call your loved ones and let them know you've landed safely. After what seems like a long wait during for your phone to find a network, you finally get a signal and start making calls. Here is what happened behind the scenes: When you turned the phone on, it started listening for a System Identification Code (SID), which is transmitted from the command channel of a nearby cell site. The site, or Base Transceiver Station (BTS), controls its own functions and antennas. The BTS then communicates with a Base Station Controller (BSC) which, in turn, communicates with a Mobile Switching Center (MSC). The MSC is responsible for routing calls to your mobile device. When a cell phone is turned on, it scans for the cell service provider's System Identification Code (SID). Once the phone detects the SID from the site, it transmits a registration request to the network. As your phone has not previously, or recently, been used in the geographic area covered by the MSC, the MSC queries the Home Location Registry (HLR) of your service. As your phone is new to the area, it makes an entry in the Visitor Location Registry (VLR) and the HLR, so the system knows where to route your calls.

What is a Home Location Register (HLR)?

An HLR is the main database of subscriber information for a cellular service provider's network. The database consists of customers' profiles or account information, account statuses, user preferences, special features subscribed to by the user, and the mobile phone handset's current location.

What is a Visiting Location Register (VLR)?

A VLR is a database similar to a HLR, which the cell service provider utilizes to record the profiles of users who are outside of their home area (roaming). The VLR is based on the information retrieved from the HLR.

What is a Mobile Switching Center (MSC)?

The MSC is a switch used for call processing by routing voice calls and text message to the subscriber. The MSC also handles the hand-off from one cell site to another, records billings information, and interfaces mobile phone calls with landlines.

It is possible to use the VLR and HLR information to determine which cell phones were in a particular area at the time. There are two types of checks that can be made for a specific cellular site. The first is to request the identification of all phones that made or received a call or text message within a specific cell site/sector. This type of search is especially helpful when a) you know or believe the suspect used his cellular device and b) you have a reasonable idea when this communication occurred (e.g. surveillance video during a bank robbery shows the suspect receiving a cell phone call during the commission of a crime.) If these two conditions are met, you must submit a court order or search warrant to all of the providers and the sites that cover the area of the crime (See Proactive Mapping). The court order or search warrant directs the provider to examine its records of calls and data events of a particular cell site, or multiple ones, during a set time frame and give a list of all active calls to you. This will result in a record of all calls made or received on the site in question.

Keep in mind the amount of data you obtain is likely going to be substantial if the cell site is located in a high density area. If the suspect made or received a call during the time frame you requested, it will be reflected in the records. However, there will be additional records that are likely irrelevant to your investigation. Information overload can be a consequence of specifying a wide time frame, say an hour, in the court order or search warrant.

Conversely, it is also important to avoid too small a time frame, such as one minute, because it can be difficult to correlate the times maintained by the cell service provider with the times established by other evidence, such as eyewitnesses or the time stamp on surveillance footage. Most providers charge from $100 to $250 per site to perform this service, so your department can expect to pay in the range of $1,000 to $2,500 depending on how many providers you serve with the order. This search is not considered "records kept during the normal course of business" and is not exempt from the provision outlined in 18 USC 2706 regarding payment. When you do retrieve the results, they can appear overwhelming. The success or failure of this technique is largely dependent on the underlying investigation.

Here is an example of using multiple cell site dumps to attempt to identify a suspect. Let's presume some facts in this scenario. A commercial burglary occurred during which time a security guard was killed. The investigation revealed the crime likely occurred between 0300 and 0500. An investigator retrieved the cell site records for the service providers who covered the area of the crime, which resulted in a record that looks like this:

TCSeizure Dt	Call Type	Calling Number	Called Number	Call Duration	Answer Dt	Disconnect Dt	FirstCell ID	LastCell ID	First LAC	Last LAC
2010-06-30 03:01:22	MOC	5553420681	5557317737	38	2010-06-30 03:01:57	2010-06-30 03:02:00	24952	24952	275	275
2010-06-30 03:01:49	MTC	5558491524	5553420227	104	2010-06-30 03:01:58	2010-06-30 03:03:33	24953	24953	275	275
2010-06-30 03:01:57	MOC	5557710370	5552910811	90	2010-06-30 03:02:33	2010-06-30 03:03:27	24952	40111	275	275
2010-06-30 03:02:14	MOC	5553420681	5557317737	38	2010-06-30 03:02:48	2010-06-30 03:02:52	24952	24952	275	275
2010-06-30 03:03:01	MOC	5553420681	5557317737	36	2010-06-30 03:03:34	2010-06-30 03:03:37	24952	24952	275	275
2010-06-30 03:03:45	MOC	5557460513	5552384285	29	2010-06-30 03:04:08	2010-06-30 03:04:14	24953	24953	275	275
2010-06-30 03:04:20	MOC	5557460513	5552384285	264	2010-06-30 03:04:39	2010-06-30 03:08:44	24953	24953	275	275
2010-06-30 03:05:31	MOC	5553420681	5557317737	41	2010-06-30 03:06:08	2010-06-30 03:06:12	24952	24952	275	275
2010-06-30 03:07:53	MOC	5555340843	5556377244	40	2010-06-30 03:07:56	2010-06-30 03:08:33	24951	24951	40494	40494
2010-06-30 03:08:28	MOC	5553420874	5553420966	225	2010-06-30 03:08:40	2010-06-30 03:12:13	24953	24953	275	275
2010-06-30 03:08:46	MOC	5553420681	5557317737	40	2010-06-30 03:09:22	2010-06-30 03:09:26	24952	24952	275	275
2010-06-30 03:09:37	MOC	5553420681	5557317737	41	2010-06-30 03:10:13	2010-06-30 03:10:18	24952	11951	275	275
2010-06-30 03:11:00	MOC	5553429813	5553429309	41	2010-06-30 03:11:37	2010-06-30 03:11:41	24952	24952	40494	40494
2010-06-30 03:11:51	MOC	5553429813	5553429309	43	2010-06-30 03:12:29	2010-06-30 03:12:34	24952	24952	40494	40494
2010-06-30 03:12:41	MOC	5553429813	5553429309	80	2010-06-30 03:13:19	2010-06-30 03:14:01	24952	24952	40494	40494
2010-06-30 03:12:48	MOC	5555159608	5556558471	65	2010-06-30 03:13:01	2010-06-30 03:13:53	24952	24952	275	275
2010-06-30 03:13:35	MOC	5553420681	5557317737	39	2010-06-30 03:14:09	2010-06-30 03:14:14	24952	24952	275	275
2010-06-30 03:14:07	MOC	5553429813	5553429309	42	2010-06-30 03:14:44	2010-06-30 03:14:49	24952	24952	40494	40494
2010-06-30 03:14:30	MOC	5553420681	5557317737	39	2010-06-30 03:15:05	2010-06-30 03:15:09	24952	24952	275	275
2010-06-30 03:14:56	MOC	5553429813	5553429309	58	2010-06-30 03:15:33	2010-06-30 03:15:54	24952	24952	40494	40494
2010-06-30 03:18:27	MOC	5553420681	5557317737	38	2010-06-30 03:19:01	2010-06-30 03:19:05	24952	24952	275	275
2010-06-30 03:19:24	MOC	5553423697	5556798212	329	2010-06-30 03:19:34	2010-06-30 03:24:53	24953	24953	275	275
2010-06-30 03:19:35	MOC	5553420681	5557317737	40	2010-06-30 03:20:13	2010-06-30 03:20:15	24952	23911	275	275
2010-06-30 03:19:57	MOC	5553429813	5553429309	44	2010-06-30 03:20:35	2010-06-30 03:20:41	24952	24952	40494	40494
2010-06-30 03:20:49	MOC	5553429813	5553429309	43	2010-06-30 03:21:27	2010-06-30 03:21:32	24952	24952	40494	40494
2010-06-30 03:21:39	MOC	5553429813	5553429309	47	2010-06-30 03:22:16	2010-06-30 03:22:26	24952	24952	40494	40494
2010-06-30 03:21:42	MTC	5554004978	5553338075	365	2010-06-30 03:21:51	2010-06-30 03:27:47	24952	24952	275	275
2010-06-30 03:22:33	MOC	5553429813	5553429309	55	2010-06-30 03:23:11	2010-06-30 03:23:28	24952	24952	40494	40494
2010-06-30 03:22:50	MOC	5553420681	5557317737	41	2010-06-30 03:23:27	2010-06-30 03:23:31	24952	24952	275	275
2010-06-30 03:25:37	MTC	5555523846	5553424737	12942	2010-06-30 03:25:45	2010-06-30 07:01:19	24953	24954	275	275
2010-06-30 03:25:50	MTC	5556798212	5553423697	6199	2010-06-30 03:26:00	2010-06-30 05:09:09	24953	24953	275	275

At first glance, the investigator is presented with what appears to be an almost overwhelming amount of data to work with. To simplify the data and make sense of it, the investigator can utilize a tool mentioned earlier in this book: a Pivot Table Wizard found in Excel. Instead of manually sorting the fields, the Pivot Table Wizard can automatically sort the numbers to create a list of unique numbers. Follow the default screens for creating the table by clicking **Next** and then **Finish**. This will insert a new sheet in the Excel workbook, which will look like this:

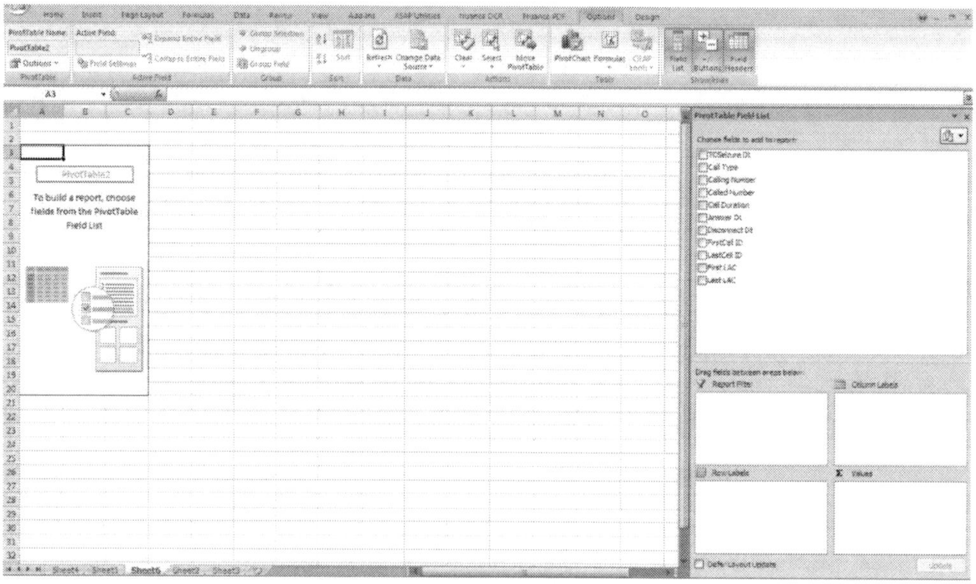

By selecting the **Calling Number** box and **Row Labels**, the table generates a list of phone numbers that can be used for running record checks or bulk uploading into commercial and intelligence databases.

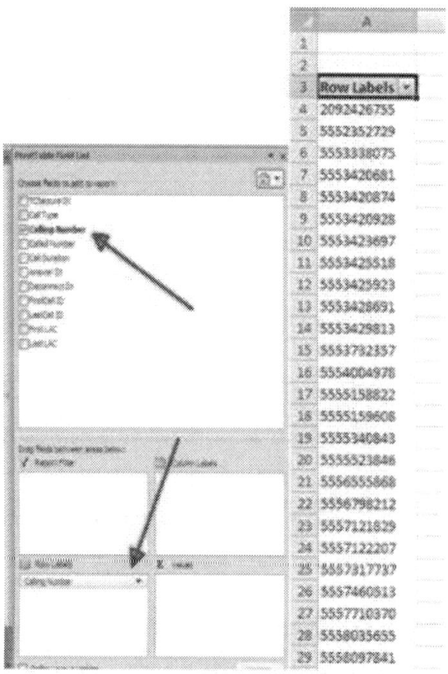

This technique accomplished two things – neither are going to immediately crack the case, but they may provide leads or evidence down the road.

The first benefit of this technique is that the cell site's information is now preserved. As discussed earlier, the amount of data a cell site can retain and its storage duration depends on the storage capacity of the equipment and the volume of data it routinely collects. As newer data comes in, it overwrites the older data, deleting it forever. Criminal investigations can take months, or even years, to solve. By the time a phone number is relevant, the underlying cell site records may not exist.

The second benefit of this technique is that now that the information is preserved, it can be compared to known suspects and used to develop suspect leads.

More than One Cell Site

The real benefit of cell site dumps comes when it is confirmed – or assumed – that the suspect used more than one cell site, which occurs when he moves. The more activity or the farther he moves, the better. Let's take a look at another example:

A bank robbery occurred that matches the modus operandi of several other holdups. In each of the robberies, video surveillance shows the suspect sending and receiving text messages, presumably to a lookout or getaway driver, before displaying a weapon. As the suspect's cell service provider is not known, the investigators request cell site dumps for the six major providers that cover their jurisdiction. The investigators request cell site dumps from each of the cell service providers for the sites that appear most likely to cover the bank. They repeat this for several of the other robbery locations using the same process.

The process reveals two phone numbers communicating with each other in the area of each robbery. The information provided from using this technique is not going to convict anyone by itself. Further investigative work is needed. However, in the absence of other evidence, cell site dumps can give investigations leads to follow up on. I know this technique has been successfully used for numerous types of investigations, including homicides, robberies, drive-by shootings, arson investigations, kidnappings, and home invasions.

Cell Site Information is Perishable

Consider the amount of data a cell site base station must maintain. In addition to maintaining records of every phone call, text message, and multimedia message made and received from a particular antenna, the base station also does this for about every handset in its service area. Remember that the cell service provider needs to maintain general information about a handset's location for timely routing of incoming phone calls. This is accomplished using the Visitor Location Register (VLR). The VLR maintains information about every handset that it communicates with, whether or not the device makes or receives a call or text message. Due to the volume of information the base station stores, it is imperative you preserve that information if it may be relevant to an investigation. Most base stations have a 45- to 90-day storage capacity. If you believe a cell site's information may be important submit a preservation letter to the provider as soon as possible.

Use of Cell Site Information during Interviews and Interrogations

Cell site information can be used for statement corroboration. With it, an investigator is able to prove or disprove alibis regarding suspects' locations before, during, or after the commission of a crime. It is critical during interviews or interrogations to establish dominion and control over the cell device during the timeframe of the crime. Ask the suspect if he has a cell phone, how long he's owned it, and if he allows anyone else to use the phone for extended periods of time. Investigators hesitate to bring up this line of questioning for fear that the suspect will become aware of their interest in the device and destroy or conceal it. Remember, the suspect cannot alter the underlying records the cell service provider maintains. Concealing or destroying the device could result in losing valuable information. However, establishing exclusive or shared control of a device can offer a stronger case for the investigating agency and the prosecutor's office.

Outside of a suspect's statements regarding his ownership of the phone, the other methods available are: possession of the device; physical evidence retained on the phone, such as fingerprints or DNA; direct observation of a suspect using a device; and audio identification through electronic surveillance or monitored phone calls.

Placing the Phone in the Crook's Hand

An essential technique used to establish probable cause for a wiretap involves visual observation of a suspect using a device tied to a certain phone number. An investigator, in conjunction with a surveillance team, will place a pre-text phone call to the suspect's number and observe as he answers it. This technique should not be limited to wiretap investigations and can be used any time an investigator is attempting to determine who is using a particular phone or phone number.

Other Uses for Cell Site Location Information

Cell site investigations are frequently the purview of investigators who are focusing on a single crime with a known time frame or a series of crimes. Many narcotics and fugitive investigators do not use cell site information, which is unfortunate because this data can give an investigator valuable information into the areas where a suspect frequents. In the case of narcotics investigators, instead of spending days to weeks following a suspect, or risk discovery during the installation of a GPS tracking device, they should consider historical cell site location information for leads to physical locations used by the trafficker, including stash houses and grow sites. The steps are relatively easy and involve the use of the same Pivot Table function in Excel as discussed in the Call Detail Records Chapter. In this case open a copy of the Excel file containing the suspect's call detail records.

Make sure a cell containing data is selected. From the **Insert** tab located at the top, select the **Pivot Table** button. Accept the default information on the screen and press **OK**.

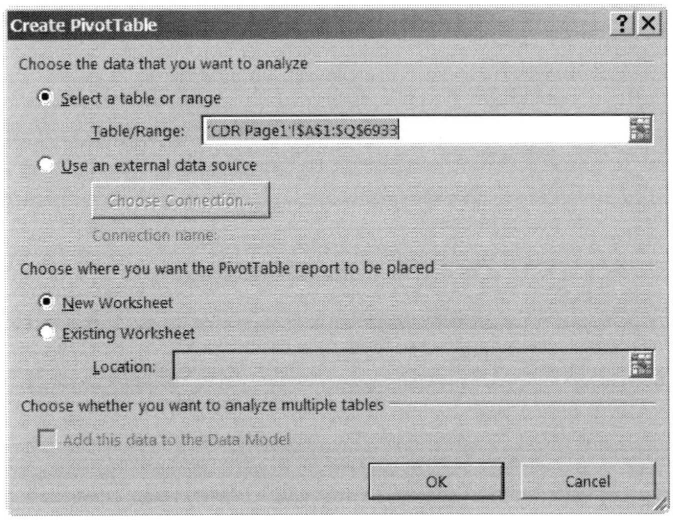

This is the main Pivot Table interface screen. The column headers from the call detail records now populate the fields on the right side of the screen. In this case, selecting **Tower** will create a list of all of the unique cell sites the target's phone was in communication with.

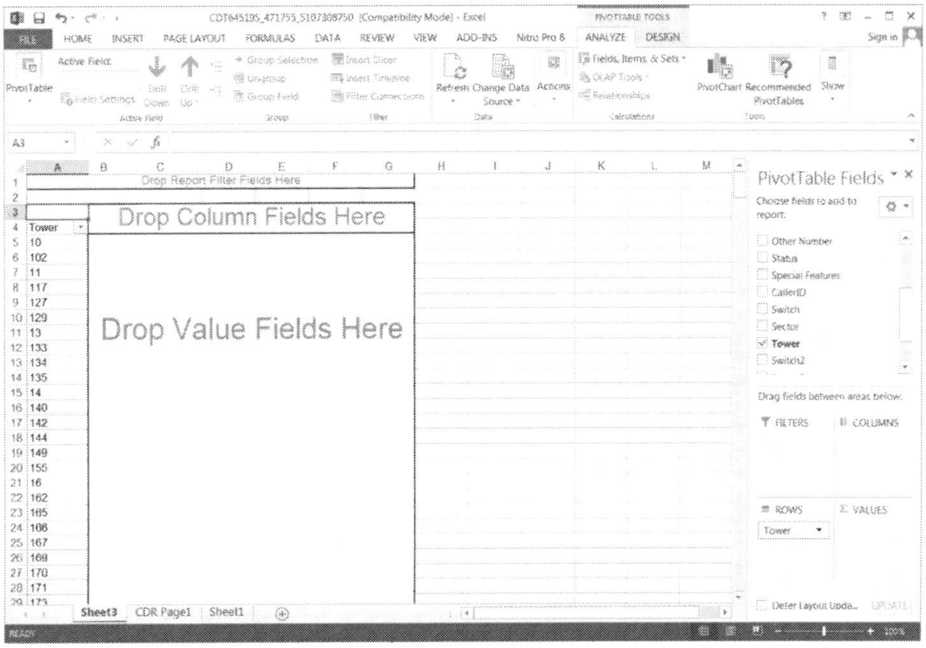

Remember, the appearance of the data in the Pivot Table fields may change depending on how the provider labeled the call detail records. In this example, the first serving cell site is listed as **Tower**. Select the box next to the appropriate label. Next, left-click and drag the tower box, or whatever the provider listed the cell site information as, into the field in the lower right corner of the sheet labeled **Values** and drop it there. Occasionally, Excel will default to the **Sum** function, which adds up all of the numbers in the column. If this happens, select the drop-down box next to the word **Sum** and select **Count**.

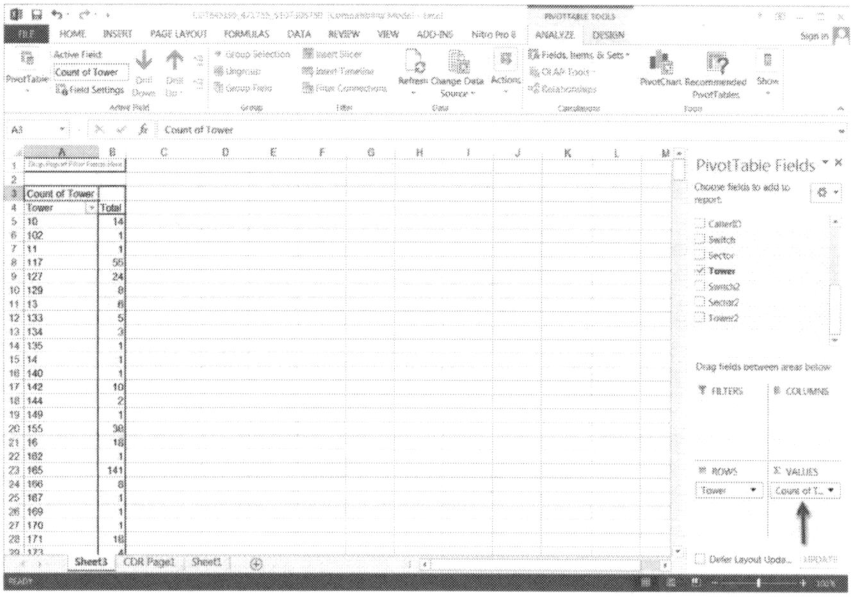

You should immediately see the Pivot Table has added the number of times the call detail records indicated the suspect's phone was in communication with a cell site. By default, the Pivot Table will sort the data in the first column alphabetically or, in the case of numerical data, from smallest to largest. The **Sort** function that you may be used to in Excel does not work the same way in a Pivot Table.

To sort the cell sites in descending order, select the drop-down box with the down-arrow just below the data field. In the example below it is located in cell **A4**.

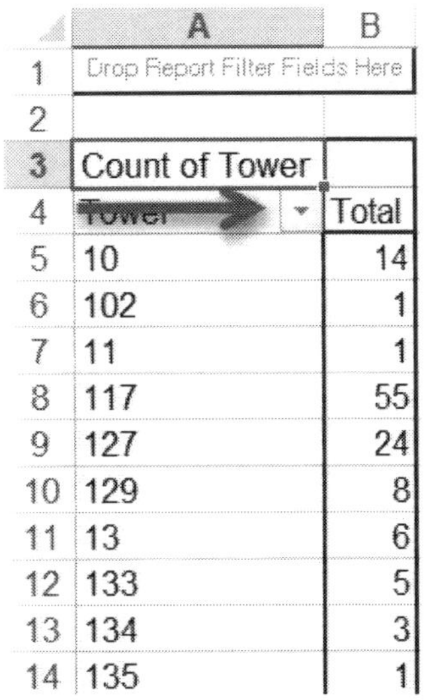

Pressing the button will bring up sort options for the data. Select **More Sort Options**.

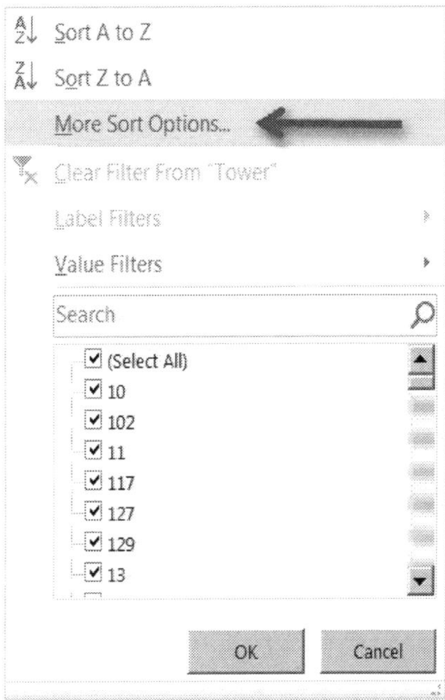

The next dialogue box asks which field of data you want to sort by. Make sure the radio button next to **Descending (Z to A)** is selected and then choose the **Count of Tower**.

The results should appear similar to the example below. Using the key the cell service provider has given, it is now possible to determine the general geographic area where the suspect's phone is. You can see the first areas to start checking are usually the top 10 or so. This technique will not solve all of your problems. You still need to do some basic police work to determine what the relevance of those locations is, such as home and work location. But in the case of narcotics investigations, this can provide clues to the area of stash houses and marijuana grows. Fugitive investigators should combine this data with information developed during the investigation, such as the home addresses of parents and significant others.

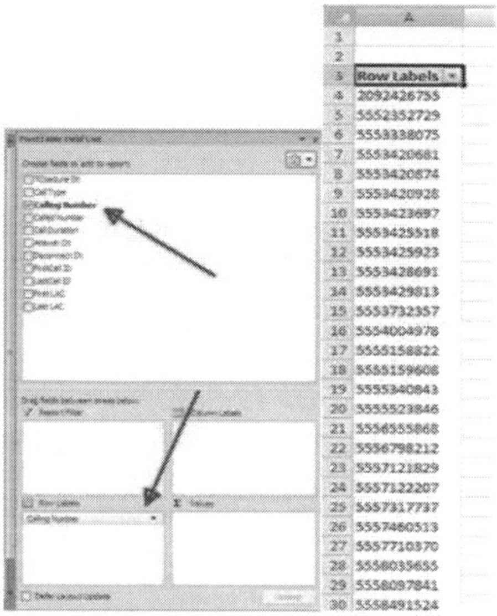

	A	B
1	Drop Report Filter Fields Here	
2		
3	Count of Tower	
4	Tower	Total
5	88	1900
6	587	790
7	182	726
8	8	172
9	452	153
10	165	141
11	226	96
12	257	90
13	18	80
14	60	80
15	179	76
16	183	75
17	117	55
18	53	54
19	219	52
20	224	52
21	453	48
22	89	42
23	221	41
24	94	38
25	155	38
26	2	37
27	25	35

Mobile Phone Tracking

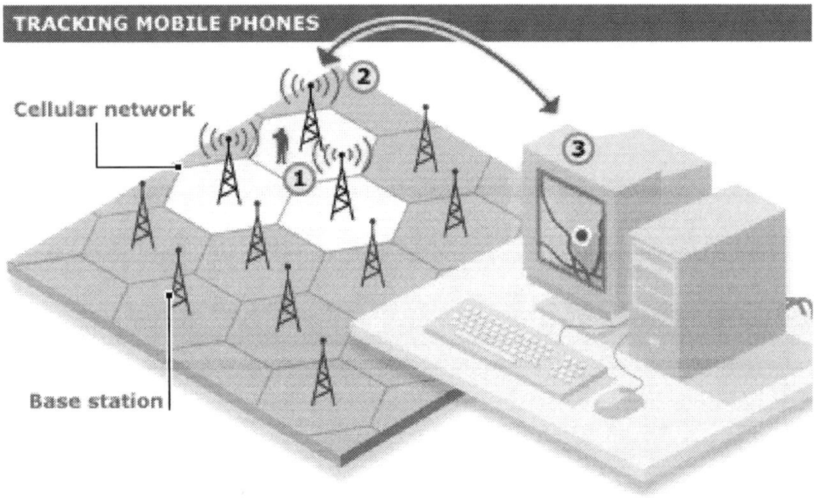

The Federal Communications Commission (FCC) established a requirement for communications providers, known as "Enhanced 911" or E911, to facilitate emergency response to 911 calls. The requirement was intended to ensure cell calls were routed to the appropriate 911 call center and assist rescuers in attempting to locate victims who were not able to determine their location. The FCC Phase II location accuracy, whose deadline to comply was extended to 9/11/2011, is 50 meters for 67 percent of the calls and 150 meters for 95 percent of the calls.

Law enforcement officers can leverage the underlying technology used to facilitate E911 compliance to track a suspect's cell phone in real time/near real time. The two primary methods are using the global positioning system (GPS) embedded in the device and cell geolocation/triangulation.

Global Positioning System (GPS)

Most of today's cell phones come equipped with a global positioning system (GPS) as a standard feature. The equipment can exist in the handset and be used by the provider even if a subscriber has not enabled the service. The GPS receiver communicates with a series of satellites in geosynchronous (stationary) orbit around the world. A device requires unobstructed communication with at least four satellites to be effective. Satellites broadcast signals, which the receiver uses to calculate its location based on latitude, longitude, and altitude. Using exigent circumstances or a search warrant, law enforcement officers can direct a cell service provider to "ping" the GPS unit of a cell handset and report the results back to the officer.

When using this technique, note the potential margin of error. A number of environmental factors can affect the GPS unit's ability to connect to the required number of satellites. As noted, the FCC standard for geolocation of E911 calls is 150 meters. In a dense urban environment, this can lead to a fairly large area where the target may be. While GPS tends to be more accurate than other techniques, such as cell site multilateration/triangulation/trilateration, it can still be off by a significant distance.

Upon request, and with proper legal authority, most cell service companies will provide a law enforcement officer with the latitude, longitude, and margin of error for the reading. Subsequent activations of the GPS unit can tend to the location of the device or the direction of movement. Also, consider the lag time between the query of the GPS unit and the time it takes to relay the information to the investigator. I do not recommend extensive use of GPS "pinging" without surveillance resources in place who are prepared to react to the area quickly.

One law enforcement bulletin from several years ago mentioned a case where the suspect became aware of the investigating officer's use of the GPS "pinging" to locate him. While that may have been accomplished through modification of the phone, the phone's operating system by the suspect, or through some failure of the cell service provider, it remains to be the only published instance I am aware of.

Cellular Geolocation

Cellular geolocation is less precise than the GPS, but it is available to devices that do not have GPS receivers or where the GPS is not available. The technology of geolocation is based on measuring power levels and uses the concept that a cell phone always communicates wirelessly with one of the closest cell sites. This technique's precision varies widely. Precision is highest where a device is within range of at least three cell sites and lowest where only a single cell site can be reached.

Some cell service providers use multilateration to determine the location of certain types of phones. Multilateration, also known as hyperbolic positioning, is the process of locating a cell phone by computing the time difference of arrival of a signal transmitted to three or more receivers. By calculating the differences in the time it takes a signal to reach the receiving cell sites, the provider is able to give a geographic area where the device is. Multilateration is commonly referred to as triangulation/trilateration; however, the processes are different. For purposes of most law enforcement investigations, the terms can be used interchangeably.

If you have specific questions about a cell site or a service provider's operating system, call the provider and ask to speak to an engineer. Don't waste your time with the subpoena compliance people. If you want to try something out of the ordinary, ask to speak to a provider's technical surveillance people.

Per Call Measurement Data (PCMD)

For a long time, many in the law enforcement community have been convinced that cell service providers are not giving all the information or assistance they could with respect to locating a phone. The belief wasn't that the phone companies were deliberately doing so, but that the subpoena compliance people were not aware of the technological capabilities of the underlying system. It shouldn't have been surprising that the legal department's employees didn't have the technical expertise or even the means to extract additional information, which could be relevant for law enforcement investigations. Unfortunately, they are usually the first, and often times the only, points of contact for most law enforcement officers.

Eventually, information began to leak out. There were "hidden" capabilities for recovering information from cell sites. It wasn't as if the cell companies were deliberately keeping the information to themselves. It was more a case of law enforcement officers not knowing the right questions to ask and speaking to the wrong people.

One of the areas of information that has been recently "discovered" is known as Per Call Measurement Data (PCMD). PCMD is an exciting feature built into the equipment of some of the largest cell service providers and has been sitting under the collective noses of law enforcement, and many subpoena compliance departments, for years.

PCMD was originally referred to as Measurement Data in Call Records (MDCR). Both MDCR and PCMD are engineering terms referring to techniques used to measure a cell service's effectiveness in a particular area. PCMD collects select measurement data on every completed call, including information on the nature of terminated, dropped, and normal calls. Another data set collected by PCMD is the time it takes a signal to leave a cellular handset and the return back to the tower. This can reveal previously unavailable information regarding the location of a particular handset.

PCMD was originally designed to be used for quality assurance purposes by gauging the number of dropped calls in a particular area. Based on this information, a cell site engineer could recommend re-orienting a cell tower or other changes to improve the coverage in a particular area. However, someone within Alcatel-Lucent, one of the major manufacturers of cell tower equipment, discovered PCMD is accurate enough to be used for location-based marketing without any upgrades to the existing cellular infrastructure or the customer's equipment. Because the speed of a cell signal is a known constant, it is possible to calculate the distance the handset is from the tower by measuring the time it takes for the signal to make a round trip. When combined with the cell sector information, there is an opportunity to gauge the handset's distance from the cell tower. This might not seem like a revolutionary data set. Knowing that someone is three quarters of a mile away from a cell tower in the middle of downtown Atlanta is not likely to produce any great leads for missing persons or fugitive tracking. However, the data can be much more accurate than it initially appears.

Unlike the cell site and sector information obtained from call detail records, PCMD is captured not only for every phone call, but every text message and data event. Traditionally, most call detail records do not capture the cell site or sector used during text messages or data events, such as voicemail and email notifications. These are not phone calls, so the cell service providers have no motivation to capture the information. PCMD is captured anytime there is a connection or data event between the cell site and the mobile device.

PCMD data has both quality assurance and location-based marketing information applications and is already built into the existing cellular infrastructure. Subsequently, PCMD is able to provide accurate location-based information. Is it accurate as GPS? No way. But it does add a dimension when combined with other information like the cell site and sector. PCMD is not going to magically locate someone's handset but, when used in conjunction with cell site and sector data, it is going to provide investigators clues they might not otherwise have.

Now for the bad news. PCMD is only found in Alcatel-Lucent cell site products on code division multiple access (CDMA) networks. CDMA coverage in the United States is approximately 50 percent of the market, but not all CDMA providers use Alcatel-Lucent equipment. Currently, PCMD is made available to law enforcement pursuant to proper legal demand or exigent circumstances on the Sprint-Nextel and Verizon Networks only. In theory, it should be available from any cell service provider using Alcatel-Lucent equipment. Only those two providers acknowledge the information is available and are able to extract it for law enforcement use.

Here's some more bad news. PCMD is extremely perishable information. It appears the data may only be available for seven to 14 days. This is not something that is going to be routinely saved with the receipt of a preservation letter pursuant to 18 USC 2703(f). If your suspect or victim is using one of the two providers who acknowledge their ability to collect and retrieve PCMD, you must request the information as quickly as possible. I have seen PCMD data that was much older than 14 days, but it was extremely sporadic and would not be helpful in most criminal investigations.

Another limitation of PCMD that falls into the bad news category is that it is not real time information. The data is collected at the switch level and then updated to a call file hourly. Again, this is not done in real time so there will be a delay between when the information is captured and recorded and when it is available for review.

PCMD records a tremendous amount of information. It does not come to you nicely formatted and with a cheat sheet for easy interpretation. Because there is no documentation, the PCMD data could be confusing. You may need to get the assistance of a cell site engineer or an electronic surveillance specialist to make sense of it. Fortunately, Verizon translates this information for you during exigent circumstances requests and will tell you the distance, measured in meters, between the handset and the cell tower.

Verizon

PCMD is routinely provided to law enforcement officials pursuant to exigent circumstances requests, such as missing persons cases, by Verizon. They refer to it as Real Time Data (RTD). However, they do not provide RTD/PCMD during routine requests with court orders and search warrants. This may be because exigent circumstances requests are handled through the electronic surveillance division of Verizon – the people who provision wiretaps and pen registers, and not through the subpoena compliance department. You must specifically ask for RTD/PCMD in your court orders, subpoenas, and search warrants or you will not receive it.

Sprint

Sprint offers its L-Site service to law enforcement officers, which lets an investigator activate the GPS positioning feature of a targeted device on demand. Consequently, investigators forget about the availability of PCMD data from Sprint. As with Verizon, if you do not specifically request PCMD from Sprint, you will not receive the information.

Using Per Call Measurement Data

Before you get too excited about using PCMD, you should know there are some serious challenges working with the data companies provide. In the following example, you can see the data fields available using PCMD. They include the phone number, call start time, duration, switch, cell ID, sector, distance in miles from the cell site, and latitude and longitude.

First, note the problem with the column headers; the cell ID label in this example actually refers to the Cascade ID, which was a blank column to the right of the longitude. Next, look at the longitude. There are a couple of problems here. First, the longitude was reported as the same value for all of the data. The data is not properly formatted and there needs to be a **(-)** symbol in front of the data as well as a decimal point. You might even be tempted to reformat the data represented 121464959 as -121.464959, which would put you right smack dab in the middle of the Pacific Ocean. For reasons that remain unclear, the data sometimes gets corrupted, and in this example the longitude is actually -121.1464959. Somehow, the negative symbol and the first digit were removed from the data. This was only discovered because the investigator knew the longitudes for the area in question. It is unfortunate the longitude was not correctly captured in this case because the specificity of the data would reveal precise, near pinpoint, location data. Another challenge with deciphering PCMD is the distance in miles. PCMD measurements collected the *round trip* data from the cell site to the phone. You need to clarify with the provider if the distance measured is the total round trip distance or the one way distance.

PHONE NUMBER	ESN	CALL START TIME	DURATION (SEC)	SWITCH	CELL ID		SECTOR	DISTANCE (In Miles)	LATITUDE	LONGITUDE
		3/05/10 8:35:07 AM	26	PHX-PHOENIX-MSC_1	PH03XC153	95	2.00	0.625	33.61531542	121464959
		3/05/10 9:43:10 AM	49	PHX-PHOENIX-MSC_1	PH03XC153	95	1.00	0.625	33.62175872	121464959
		3/05/10 9:44:05 AM	77	PHX-PHOENIX-MSC_1	PH03XC153	95	2.00	0.625	33.62131463	121464959
		3/05/10 8:25:16 PM	26	PHX-PHOENIX-MSC_2	PH60XC135	150	1.00	0.890152	33.30296839	121464959
		3/05/10 8:26:58 PM	26	PHX-PHOENIX-MSC_2	PH60XC135	150	1.00	0.890152	33.30898877	121464959
		3/07/10 8:07:08 AM	26	PHX-PHOENIX-MSC_2	PH73XC031	312	1.00	1.297348	33.30907644	121464959
		3/07/10 8:07:44 AM	1	PHX-PHOENIX-MSC_2	PH73XC031	312	1.00	1.297348	33.30990852	121464959
		3/07/10 3:05:31 PM	21	PHX-PHOENIX-MSC_2	PH73XC031	312	1.00	1.297348	33.30907644	121464959
		3/07/10 3:05:56 PM	19	PHX-PHOENIX-MSC_2	PH60XC135	150	1.00	0.92803	33.30915298	121464959

Mapping the Data

PCMD is extremely valuable data for calculating the possible locations for a person's cell phone. The notable applications that come to mind are finding missing people, tracking fugitives, and developing leads for narcotics stash houses.

As of the time of publication there is not any automated way of illustrating PCMD. The manual method can be time consuming and it would take a considerable amount of work hours to plot this data for every instance of recorded PCMD. The case's circumstances will dictate which data you should focus on. For example, in a missing person's investigation you may want to focus on the last calls or those that took place around the time he went missing. For fugitive and narcotics investigations, you could examine high frequency cell sites using a Pivot Table or cell sites that were activated at unusual times or outside the suspect's normal pattern of activity.

To illustrate the PCMD data into something usable, I used Google Earth. First, plot the latitude and longitude of the cell site.

Next, plot the sector's borders using the measurement tool in Google Earth. In this example, the azimuth of sector two is 120 degrees, so the borders are 60 degrees and 180 degrees.

The measurement tool in Google Earth is not as precise as the PCMD data. For example, the first row of data listed above shows the distance to be 0.625 miles. Google's measurement tool is only capable of going to the hundredths' place, so I go with 0.63 miles. It is also very difficult to get to exactly 120 degrees or any other measurement using this tool. Get as close as you can, but for illustration, being off by a couple of hundredths of a degree is not going to make a huge difference.

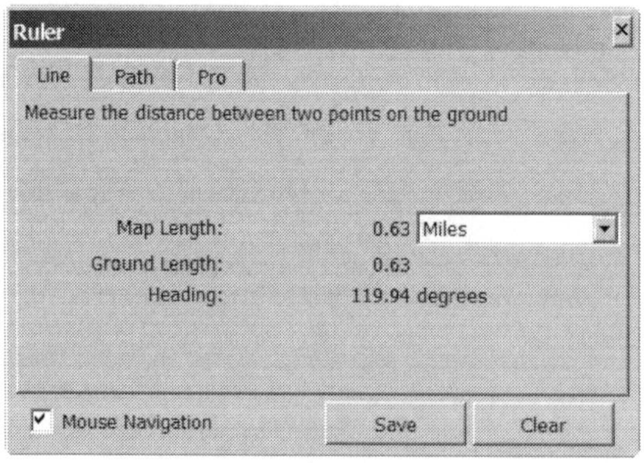

Using the measurement tool, draw a line from the cell site every few degrees. In the illustration below, I went every two degrees, but that can get tedious so you may choose to go every five degrees. Now you have a detailed map of where the handset was when a phone call was made or received.

Timing Advance

The same principles behind PCMD apply to GSM phones, as well. However, in GSM phones the data is known as "timing advance." This measures the length of time a signal takes to reach a cell site from a phone. Radio waves travel at the speed of light, so the exact arrival time of a signal can be calculated and translated into distance. GSM cell service providers do not routinely give timing advance data and it is not something most subpoena compliance personnel are aware of. To access timing advance data, a law enforcement investigator will need to contact the electronic surveillance or engineering departments for the specific provider.

Cell Phone Tracking Using Stingray/Trigger Fish

Before anyone blows a gasket and accuses me of leaking law enforcement sensitive techniques, everything in this book, including this topic, are publicly available.

Cell phone tracking and the use of site emulators are commonly known within our field by the code word "Triggerfish" or "Stingray." The equipment works by emulating a cell tower and querying the particular serial number of a given device and measuring the signal strength to the device. This allows the operators to locate the device and, hopefully, its user. Due to the cost of the equipment, Triggerfish equipment is usually maintained by federal law enforcement (FBI, Secret Service, etc.), state law enforcement, and large municipal law enforcement agencies. Depending on the configuration, the equipment is also capable of capturing the unique serial numbers of all devices in a certain area. This assists investigators to isolate a suspect's previously unknown cell device.

To be used effectively, Triggerfish relies on an operational pen register to locate the cell tower and sector the device most recently used. The Triggerfish team will then go to the area the cell phone sector covers and try to locate the device. In my experience, the equipment is extremely accurate but relies on skilled technicians operating the device, a responsive and competent surveillance element, some degree of intelligence regarding the suspect's associates, and an active cell phone associated with your target.

If your agency requires the equipment, I suggest checking with a larger state investigative agency. Federal law enforcement will use their equipment to assist local law enforcement investigations; however, they typically require "adoption" of the case. This can be a cumbersome process, depending on which federal law enforcement agency you approach. At times they are required to replicate the investigative work already completed to ensure that a federal standard is met.

Is Cell Site Information Valuable Evidence or "Junk Science?"

> *Prosecutor: "Your Honor, I'll submit Agent Shute as an expert in cell site analysis."*
> *Judge: "Okay. Any voir dire?"*
> *Defense: "I wouldn't know how to cross examine him. I'll have to accept him."*
> *Judge: "I will therefore certify that Special Agent Shute is an expert in historical cell cite [sic] analysis."*[9]

It took a while, but defense attorneys are finally coming to grips with cell site location information (CSLI). Defense attorneys do not like CSLI evidence or any evidence that leads to their client's conviction, for that matter. A few of them have taken their experiences and readily shared it with other attorneys. It is helpful to review their literature so you will know what you might be up against when you testify about CSLI. Be aware that the prosecutor probably knows just as little about CSLI as the defense attorney does and take the time to educate them before going on the stand.

Cell site evidence can never be the smoking gun; rather, it is a valuable piece of evidence that complements the entire picture. From reading many defense articles involving cases where CSLI is relevant, it appears the defense was completely unprepared for the introduction of this evidence. Without effective cross examination, judges and juries involved in early CSLI cases were left with the impression that the evidence was ironclad and irrefutable. Unfortunately, not only do the facts not support this, but defense attorneys have found a couple of hired guns who are willing to testify to the contrary.

It's important to learn from the mistakes of others so we do not repeat them. If the defense has success casting doubt on the expertise of a witness because they were unaware if the specific tower was

[9] From the testimony of FBI Special Agent William Shute in US v Sims 06-674 Eastern District of Pennsylvania

undergoing an update or maintenance, you can be sure that they are going to share it with the rest of their community. A defense attorney is going to attempt to instill reasonable doubt into the minds of a jury by pointing out all of the variables that could have affected how a cell site handled a call. Be prepared to address these.

There are several ways for a defense attorney, without the aid of a hired gun expert witness, to cast reasonable doubt on CSLI evidence. Here are a few:

Defense Technique #1:

Subtle belittling of the evidence. The police and prosecution call this "call detail records," but the defense calls it "accounting data and billing records." To subtly lessen the impact of these records – no matter what one calls them – the defense will often use these terms. It ironic that, on one hand, the defense tries to lower the data's integrity by referring to records as mere accounting records while maintaining it would be invasive to require a search warrant. For an example of this contradictory logic, as well as some specific defense objections, see "Defense Against the Disclosure, Admission, and Credibility of Cell Phone Tracking Data by Courtney Montiero" in *Federal Criminal Defense Journal Vol. II*

Defense Technique #2:

It takes information from at least three cell sites to locate a cell device. True. Precise location would require trilateration/triangulation, which can be a proactive technique for locating a phone. The defense knows it is impossible to go back two years and get this information for trial; instead, they imply that any CSLI that does not rely on trilateration/triangulation is inherently false. What is at issue is the handset's exact location. I would never testify that a cellular device was precisely anywhere in the first place. There are too many variables that can affect which cell site a device is in communication with. It is not possible to show that a suspect was at the scene of a crime using the information from one cell site. However, it is possible to say the device was in communication with the same cell site and sector that encompassed the crime scene.

There are a couple of subtle differences in the last sentence. I'm providing testimony regarding the device's communication with the cell service provider's network as documented with the call detail records – not the suspect. I have no idea and can provide no testimony as to whether the suspect was actually using the device at the time. There are no absolutes with regards to CSLI, so the device was likely in communication with a particular cell site. And I'm not saying definitively the suspect or his phone were at the crime scene – just in the same coverage area for the cell site.

Defense Technique #3:

There are so many factors that can affect which cell site handles communication with a mobile device that it's impossible to place the handset within the coverage area of a single cell site. This is true but with some important caveats. What is critical is understanding what the factors are and proactively addressing them. It may be impossible to determine whether a cell site was undergoing a routine upgrade or maintenance during the time frame of the crime. It is doubtful investigators will remember if there were any unusual occurrences in their jurisdiction that could have overwhelmed a particular cell site and caused traffic to shift to another site. Learning what conditions can affect a cell site and documenting them during the initial investigation will help to overcome this objection.

Defense Technique # 4:

"Isn't it possible the mobile device was actually ___ miles/minutes away from the scene of the crime?" Be careful of the language the defense attorney is using. It's common to see them refer to driving time in lieu of distance. Thirty minutes of driving time can take you quite a distance in a rural area with no traffic. Thirty minutes of driving time in Los Angeles during rush hour, however, won't take you very far. Also be aware of the introduction of the "square mile" and "miles squared" language. If you took the 25 miles theoretical maximum range of a GSM cell site and represented it as miles squared, it would be 625 miles. Don't be shocked to see a defense attorney start throwing around these big numbers in an attempt to cause the jurors to believe the suspect's mobile phone could have been anywhere in that area. Theoretical maximums are just that: theory.

Defense Technique # 5:

Your experience and training is insufficient to be able to render an expert opinion on anything related to CSLI. I like to read what the defense has to say. In one case they criticized a detective for only have 80 hours of training in call detail records, GPS, and geo-location. While 80 hours of training does not make you the best in the world, it is significantly more training than most investigators receive. In fact, I think one would be hard pressed to find much more than 80 hours of training on those topics offered nationwide. However, the defense attorneys like to portray their expert, who spent his dubious career as a fingerprint examiner or an electrical engineer, as much more capable of testifying about cell sites as someone who attended training on how to use CSLI to complement investigations. The distinction here is to limit testimony to what is in the call detail records and how that information is relates to the case. Unless you are a cell site engineer and understand the complexities of a particular provider's equipment, don't try to testify like one.

Know who the defense is likely to call as a rebuttal witness. See what they've testified to in the past. There are not a lot of them so it's not very difficult to research them and see how they are going to try to attack your testimony.

A resource available to state and local law enforcement officers for technical assistance and to help prepare for testimony is the Federal Bureau of Investigation's Cell Analysis Survey Team (CAST). These teams are located at most major field offices and are comprised of federal agents who have received special training in using the same techniques and equipment as cell site engineers who work for the cell phone companies. The team members can perform surveys of particular areas and then testify as to the results.

Lastly, don't forget to compel the provider to turn over their propagation maps using a search warrant. Not only do they help focus the investigations but they can also serve as compelling evidence.

Defense Technique #6:

Warrantless searches. This area is going to be much more interesting in light of the United States Supreme Court decision in United States v. Jones regarding the need to obtain a warrant for GPS tracker placement. Previously, defense attorneys would often object to the "warrantless search" of cell phone records. Semantics again. They know full well the records were obtained with a court order pursuant to 18 USC 2703(d), administrative subpoena, or other legal procedure. They are trying to taint the jury's perception of the evidence by implying the law enforcement officer did something underhanded to obtain the records without a warrant. This area is going to be contentious until the Supreme Court rules on it, so let me say this loud and clear. When in doubt get a search warrant. If the case is significant or important get a search warrant instead of using a court order. This allows an investigator to preemptively close the door on a potential defense objection.

Defense Technique #7:

Call detail records cannot be authenticated. States each have different rules regarding the authentication and admissibility of business records into evidence, including call detail records. However, the Federal Rules of Evidence has a universal standard for admissibility. The Federal Rules of Evidence require the records be accompanied by a written certification that they were made near the time of the event documented, it was regular practice of the business to make such records, and the records were kept during the normal course of business. These elements must be testified to by a qualified witness such as the custodian of records, or through certification. It can be difficult for a defense attorney to attack the credibility of this information so they will look for other means to raise doubt and confusion among the judge or jurors.

Defense Technique #8:

There is no way to prove who was using the phone at the time. Some of the interesting techniques are bringing up the issue of cloning and other third party misappropriation of the suspect's phone by other methods including identity theft. Let's examine the technological issues surrounding the feasibility of these methods. Once upon a time, cloning of cell phones was big business among criminals. It was possible using some simple techniques to capture a cell phone's unique serial number and then program it into another phone. This caused the legitimate user to be billed for all of the cloned phone's calls. Cloning is an old school crime which occurred when cell phone minutes were very expensive. Currently, the equipment needed to intercept cell transmissions and then program a mobile phone costs more than an unlimited plan from most service providers.

In one frequently cited article, one defense attorney states, "In fact, 80 percent of cellular phones utilized in the commission of a crime are clones." Really? I'm not sure where she got that information because the URL in her footnotes is dead and I can't find the information anywhere. However, be prepared for a defense attorney who read her article to try to introduce this "fact" in trial. The defense's introduction of cloning as a reason for a call at a crime scene also overlooks the technological innovation the providers created to prevent fraud. These include software designed to thwart multiple activation of the same serial numbers and encryption built into the mobile device itself.

Another potential technological issue the defense could use is the target phone was bluejacked or Bluetooth snarfed. These were techniques utilized to access a victim's phone by pairing with the device as if it were a legitimate Bluetooth device. The security flaw that was exploited has since been patched and this would only work if the mobile device was set to be discoverable by any other device in range. Also, it would require continuous proximity of the hacker to the other phone.

Is it possible someone stole the phone and used it to make the incriminating call? Is it possible someone used the suspect's identity to open an account in his name? It is absolutely possible for either one to happen. It is also pretty easy to show otherwise. A common mistake many investigators overlook is to focus only on a narrow time frame during which the crime occurred. Take a look at the entire record and find the phone calls to the suspect's work, significant other, and family members. It's not likely someone stole the victim's identity and then used the phone to call the victim's mother. The simplest way to overcome this is to ask the suspect if it is his phone. Most of them want the phone back and don't realize the incriminating value of their statement of ownership. Even if they won't admit ownership, documentation of where the phone was located can help establish dominion and control.

Chapter 5: Cell Phone Forensics

Evidentiary examinations of cell phones are a complex, and often frustrating, process. However, the rewards can justify the time and costs associated with the process. This section will provide an introduction to the basics of cell phone forensic examination but is in no way a complete or comprehensive treatise. Cell phone forensics is a constantly changing dynamic field and the available tools do not always keep up with technology.

The National Institute of Standards and Technology (NIST) in their 2007 report *Guidelines of Cell Phone Forensics* states, "Mobile phone forensics is the science of recovering digital evidence from a mobile phone under forensically sound conditions using accepted methods."

Unfortunately, cell phone technology rapidly outpaces the forensic sciences and there are no "forensically sound conditions using accepted methods" available. Cell phone forensics requires an examiner to use techniques that have not been evaluated as an "accepted method." While I always recommend using a forensically sound process or tool, they are not always available. The examination technique involves the use of non-forensic methods such as using a manufacturer's content management software, highly technical programming or repair tools, and third party software designed to modify the operating systems of phones.

There is a wealth of evidence and intelligence to be recovered from even the most primitive cell phones. Combined with the information cell service providers retain, it is possible to glean critical insights into criminal activities and conspiracies which eluded earlier generations of law enforcement officers.

Handset Evidence-Collection of the Device

On a regular basis, an officer will present a cell device for examination by pulling it out of his pocket and dropping it unceremoniously on my desk. Worse than that, is the presentation of what I like to call the "bag o' phones." Following a successful gang sweep, where the officers were directed to seize cell phones to advance a conspiracy investigation, I arrived at my office one day to find a present awaiting me. In a brown paper bag were 12 phones for forensic examination. I approached the lead officer on the case and the following conversation ensued:

> Me: *So, these are the phones from the gang sweep?*
> Lead Officer: *Yep. We did just like we were told. We seized every phone we could get our hands on.*
> Me: *That's great. Did you get all of the phones from one suspect?*
> Lead Officer: *No.*
> Me: *But, which phone belongs to which gangster?*
> Lead Officer: *Ummmmmmmm... I think we got the red phone from Flaco and the silver phone from Gordo. Let me check with the other guys.*

The officers in this case did exactly as they were instructed. They seized every cell device but failed to follow basic evidence handling guidelines. Perhaps it is because cell devices are so commonplace the officers did not think to handle and process the phones as any other piece of evidence.

Without a doubt if they had seized 12 firearms the process would have been completely different. The firearms would have been photographed in place to precisely document the location at which they were found and to establish dominion and control. The firearms would have been carefully handled with gloves to avoid destroying any fingerprints and rendered safe. The make, model, and serial number would have been recorded and they would have been separately placed into specialized packaging materials. The identifying data from the weapons would be checked against national databases to ascertain the registered owner and to determine if they had been stolen. Likely, the weapons would be later submitted for laboratory analysis to see if they had been used in other crimes.

Yet, most officers think nothing of seizing a device, manually scrolling through the phone to see if there is anything of immediate evidentiary value, and then plopping the device down on the desk of an examiner, expecting that evidentiary miracles will be performed. It is imperative to seize, process, and handle a cell device as they would any other piece of evidence, but with additional steps.

If you are going to seize a phone pursuant to a search warrant or incident-to-arrest, seize everything associated with the phone. If you were seizing a firearm you would also seize holsters, ammunition, user manuals, and purchase records. The same concept applies to cellular devices. If you are going to seize the phone you should also seize the user manuals, power or data cables, docking stations, service contracts, SIM card holders, and packaging materials.

Someone once asked me why I recommended seizing the user manuals for the device along with the other items. The first reason is people often write important things about their phones in the user manuals. This can include passwords, email account names, and other information. The second reason is that not all phones are simple to operate and you may actually need the manual in order to find how to access Airplane mode or where the Micro SD card is stored.

Physical Evidence Preservation

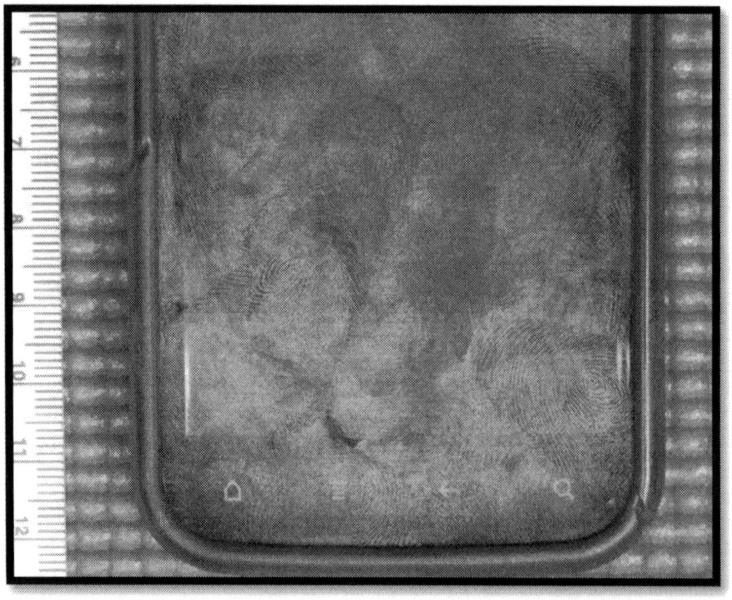

Prior to a forensic examination of a cellular device, it is critical to determine if there is physical evidence that may be of value to the investigator. The first step is to handle it like any other piece of evidence. Care must be taken to preserve any physical evidence that may be contained on the device. If it is going to be examined for fingerprints, it must be handled with gloves and in a manner to avoid destroying existing evidence. Modern touch screen phones such as the iPhone present an ideal surface for recovering latent fingerprints. But do not discount the possibility of recovering partial fingerprints from the buttons on the key pad. While these surfaces are unlikely to retain a full fingerprint due to their size, they may retain a sufficient specimen for comparison with a known suspect.

The battery compartment and external storage media slot are overlooked areas for fingerprint examination. It would be difficult for a suspect to explain the presence of his fingerprints on the battery stored inside the phone. A word of caution: when examining the phone for fingerprints, I would recommend against using those metallic or magnetic powders especially on the interior of the phone. While I am not aware of any documented cases where the use of these mediums resulted in the loss of data, it does not seem like a good idea to introduce them into the interior compartments of the device. I always direct the officer or technician examining for fingerprints to avoid placing fingerprint powder on the electrical charging and data cable connection ports. It is unlikely these areas will retain fingerprints and the powder can prevent the proper connection of the data cables necessary for a digital examination of the device. In fact, on multiple occasions I have had to spend time cleaning the electrical contacts of a mobile phone with cotton swabs and alcohol in an effort to clean the fingerprint power out and establish a connection with a data cable.

DNA evidence should also be considered. Whether the phone is a modern glass faced touch screen or $40 burner phone purchased from the local convenience store they both present an ideal area to recover DNA evidence from. The mouthpiece area in particular is prone to collecting microscopic saliva samples and the glass screen may also contain facial skin cells. Both of these areas are likely to contain sufficient biological samples for DNA comparison.

A phone should be examined for biological evidence prior to digital evidence recovery to prevent cross contamination and preserve any evidence. However, consideration should be given to the facts and circumstances of the case as there might be situations where delaying the digital examination while waiting for the physical evidence recovery could lead to the loss of data. Also, physical evidence recovery could destroy traces of password or pattern unlocks that might exist on the phone screen. Wiping the screen with a dry cotton swab will minimally obscure, and likely completely destroy, trace fingerprints on the screen, as well.

Let's face it. For most criminals, personal hygiene is often a low priority. This can lead to a significant amount of dirt, debris, and other mystery crud in the charging port or data port of the phone. You may find the phone is able to be charged but unable to establish a data connection. If the phone is dirty you are going to need to clean it. Cotton swabs and alcohol or distilled water work well.

Biohazard Evidence

It happens often enough that it deserves address – an evidentiary phone similar to the one pictured below recovered underneath the body of a homicide victim.

In cases similar to the one above, dried blood partially or completely obscures the screen, the power button, the keyboard, and the data port. In order for a forensic examination to be done, the blood will need to be cleaned off prior to examination or sending it to a lab. No forensic examiner is going to risk contaminating his laboratory by investigating this phone at his work station. Swab the blood and preserving the sample before cleaning it. Also remember a lot of blood does not mean a lot of cleaning agent. Any liquids that enter the interior of the device can cause irreparable damage and data loss. So don't pour a bottle of alcohol, distilled water, or bleach over the phone to speed up the process.

Your local cell phone repair shop or repair equipment supplier will probably have an ultrasonic cleaner available. If you suspect the interior of the phone might be contaminated with biohazard material or the phone has been immersed in other liquids, ultrasonic cleaning is highly recommended. The device will be stripped down to the circuit board level and immersed in denatured alcohol during the cleaning process. This does not only kill any potential bacteria or viruses but also displaces any fluids that would otherwise cause the system to short circuit when the phone is turned on. The cleaning must be done prior to attempting to turn the phone on or you risk irretrievably damaging the processor or memory of the phone.

Documenting Characteristics of the Phone

In an earlier example, I noted how, in my experience, law enforcement officers tend to treat cell phones differently from other items of evidence. It is important to maintain an appropriate chain of custody and document – the where, how, when, from whom, and by whom.

Was the device turned on or off? It is important to document whether the device was on and whether it continued to receive incoming calls, text messages, or emails before the device was prevented from communicating with the service provider's network. *Remember remote deletion is a reality!*

It is not always possible to determine the make, model, and serial numbers of a phone unless you remove the battery cover and the battery. When the phone power is back on this could cause the handset security lock to engage. It may have to suffice to describe the device in general terms until it can be examined in a forensic environment.

Usually placed underneath the battery is a sticker applied by the manufacturer, containing important information regarding the device. The sticker can contain the make and model of the device, as well as, the operating system and software version initially installed on the device. You will also find a range of serial numbers including variations of the unique serial number assigned to the device. These serial numbers are unique to the device and vary by the type of phone. They can include the electronic serial number (ESN), mobile equipment identifier (MEID), international mobile equipment identifier (IMEI), and manufacturer's serial numbers as seen in the following example.

These numbers are sometimes represented in two formats: traditional numeric format, which is normally labeled DEC; and hexadecimal format labeled HEX. While hexadecimal formatted numbers appears completely different from their traditional numerical counterparts, they are in fact the exact same numbers.

The number 1234567891011121 as represented in traditional format.

The same number in hexadecimal format would appear as 462D53C984631.

Hexadecimal formatting is used to reduce a number's total length. Most modern calculators, including the ones featured on Windows and smartphones, have a scientific function that will demonstrate the conversion of hexadecimal representations. However, this doesn't work for the hex values associated with cell phones.

You must use a different algorithm to accurately convert hex to decimal such as the one found at the website http://www.meidconverter.com/.

You may recall several years ago hearing or reading about "organized groups of Middle Eastern appearing men" purchasing prepaid phones in bulk from retail outlets. The original belief was that this could be terrorism related and the phones were intended for use as remote improvised explosive device (IED) detonators. After the hysteria died down, it was discovered that these phones were intended for resale after the subjects rewrote the operating system, also known as “flashing” the device.

Prepaid phone services make their money on the phone calls, not the sales of phones. In many cases prepaid phone services actually take a loss when they sell a cell phone handset with the hope of recovering the cost of the device through subscription renewal. Some entrepreneurs will buy pay $25.00 for a $65.00 cell phone, “flash” the device using inexpensive hacker tools and software, and then sell the phone for $40.00 to $50.00 at flea markets or on college campuses. The tools used in “flashing” the device can also rewrite the unique electronic serial numbers, including the ESN, IMEI, and MEID.

These serial numbers are roughly equivalent to a vehicle identification number in that they are assigned by the manufacturer and are difficult, but not impossible to change.

Make sure you document the overall condition of the device at the time it was seized. Is the screen cracked or damaged? Is the battery taped into the compartment? Does the phone have a protective case or screen?

There are countless accessories for cell phones. These can include protective cases that may have an overall innocent appearance but when examined in the greater context of the investigation may lend to a deeper or hidden meaning. For example, I have found gang graffiti on the outside and inside of cell phone protective cases. Also, the case itself may have significance. A red handkerchief pattern on a protective case may appear innocent, but in the context of a Norteno or Blood gang investigation this could add weight to the assertion the owner of the device is a gang member or affiliate.

Several devices allow a user to customize the appearance of the home screen with graphics, image files, or a banner message. These user customized features appear when the phone is on or is screen locked. These images or messages can also provide evidence of gang affiliation and/or narcotics trafficking. If the device is locked, the presence of these images or phrases may assist an investigator with developing probable cause to seek a search warrant.

Preserving Electronic Evidence

As long as the device is on and communicating with the network, the possibility of unintentional or deliberate loss of evidence due to incoming calls or text messages exists. Additionally, there are methods for remote deletion of some or all of the phone's digital contents.

Current recommendations regarding the seizure of a cell phone dictate the phone be turned off in order to prevent the possibility of accidental or intentional deletion of evidence. However, some users will program the device's security settings to activate the handset security lock when the power is turned on. Circumventing or removing handset passcode locks is a field of forensics that is still evolving and it is not always possible to bypass the lock. If a forensic examination can be completed quickly I recommend leaving the device powered on as long as steps are taken to prevent the phone from communicating with the service network. If the examination is delayed turn the device off. If the device is going to be left on, I use one of the following techniques to prevent the device from communicating with the network.

Preventing the Phone from Communicating with the Network

There are several methods for preventing a device from communicating with the service provider's network. Principally these methods rely on using an adjunct device or material to block the cell site's radio signal from establishing a connection with the device. It is important to note that none of these methods have been studied scientifically and it is unlikely any given mechanical signal blocking method will be effective on 100 percent of devices in the market and in every situation. A cell phone with a full battery charge in close proximity to a cell tower may still establish a connection despite the best attempts to mechanically block the signal.

There is another factor to consider when using mechanical methods of signal blocking. A cell phone not actively being used will issue a signal to the cell tower with which it has the strongest connection to every 30 to 60 seconds. This helps a service provider to locate the device in the event of an incoming phone call, text message, or data. When the device senses it does not have a connection with a cell tower, it begins to search for a signal at a much higher rate; in some cases as quickly as once per second. This cellular device's increased activity can drain the battery much faster than if the phone was in regular standby mode. Some of the methods discussed do not allow for a power cable to remain attached to the device while it is being protected from communicating with the network. The end result could be a dead battery when the phone appeared to have a full charge when placed inside the mechanical signal blocking method.

Turn it off

This is the easiest way to secure a device, but it can come with a price. Some users may have enabled a handset security lock that engages when the phone is powered off and then back on again. Prior to turning the device off, the investigator should attempt to preserve the evidence on the screen by using digital photographs or video. Turning the phone off may also result in a loss of information stored in the phone's volatile memory.

Airplane Mode

One of the most effective methods for preventing a device from communicating with the network while maintaining the battery life is to place the phone into "Airplane Mode." This feature allows the device to remain powered on but "disconnects" the antenna of the device. The phone will not make or receive calls, text messages, emails, data events, GPS updates, or program updates while in this mode. The software programs and features of the phone that do not require interaction with the Internet or the cellular network will continue to work. It is possible to complete a forensic examination of the device while the phone is in Airplane Mode.

As seen in the example below, Airplane Mode is found in the device's **Setting** or **Tools** menus and sometimes in the **Phone Settings** sub menu. Unfortunately, this only works on devices that do not have an activated handset security lock. Curiously, not every phone has Airplane Mode as a setting, but the majority of phones currently in use are equipped with this option.

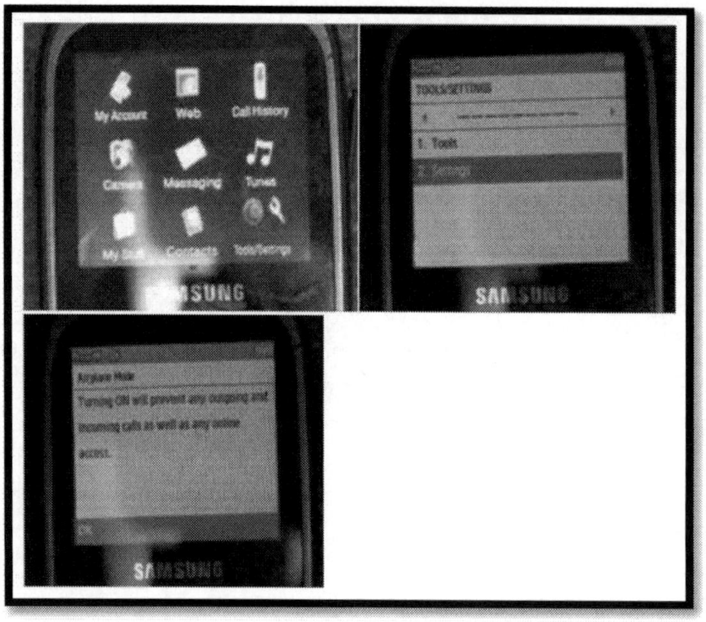

Removing the SIM card

GSM or iDEN phones are equipped with SIM (Subscriber Identity Module) cards. In the following example, removing these cards from the phone prevents it from communicating with the network. A forensic examination may be conducted on the card, which can reveal some information. However, an examination of the device will require the SIM card to be re-inserted. Several forensic companies offer the training and equipment necessary to program "cloned" SIM cards. These cards are manually programmed with the original card's unique identification numbers. The manual programming does not include the encryption key necessary for the device to activate to the cellular network. This tricks the device into allowing it to operate as if the SIM card were present, but does not let the device communicate with the network. A forensic examination can be done without the possibility of evidence deletion caused by communication with the cellular network.

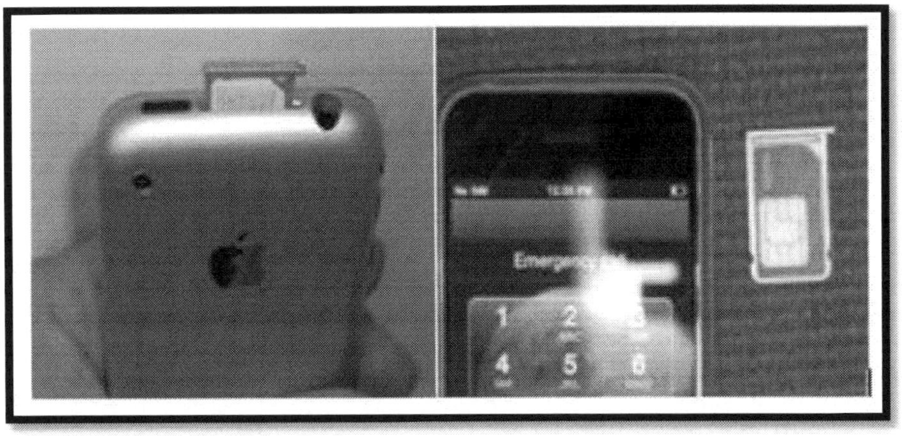

Faraday Bags

These reusable bags consist of several layers of metal mesh such as nickel, copper, and silver plated nylon mesh, which have radio frequency interference properties. The metals inside these bags attempt to block the signal communications between the mobile device's antenna and nearby cell tower. It can get expensive if the bags are to be used on every cell phone seized (approximately $25.00 to $30.00 each). You would not think the use of a bag would require the user to follow complicated instructions, and yet I have seen law enforcement officers consistently fail to use a Faraday bag appropriately.

As noted on the instruction sheet inside the bag, the seized device must be placed into the inner pocket of the bag and not simply placed inside the fold of the bag and then Velcro'd shut. Failure to properly place the device into the inner pocket can result in a false sense of security and lead to data loss. It is possible to secure the Velcro around a power cable and leave the device charging while it is inside the bag. Unfortunately, the bag is opaque and it is impossible to tell if the signal is, in fact, completely blocked as you cannot see the screen while the device is inside the bag.

Metal Mesh

Several cell phone forensic training and equipment companies offer metal mesh for sale with radio frequency interference capabilities. One of the advantages of this material is that it semi-opaque, allowing the user to still view the device's screen and determine whether the signal is blocked. Also, it is possible to wrap the device and attached cable in the material and perform a forensic examination, including manipulating the keys and screen, while shielding the device from the network. The material is less expensive than the Faraday bags and is reusable.

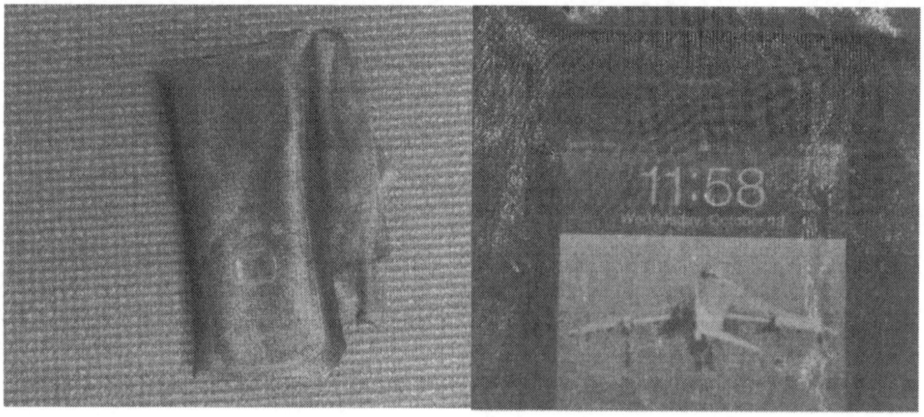

Improvised Signal Blocking

You can also use common metal foil (aka aluminum foil) as an improvised shielding method. It is necessary to use between three and six layers of the foil to prevent signal interaction between the device and the tower.

While inexpensive there are a couple of issues to consider when using this method. The first is the material is completely opaque and it is difficult to do a forensic examination of a device when it is wrapped in six layers of metal foil. As with the Faraday bag, it is impossible to see the screen and determine if the signal is completely blocked. It is also important to consider jury presentation and perception when using any improvised method. It may not appear professional when you are on the stand and present your evidence – a cell phone that now resembles a baked potato. An additional improvised method is using empty paint cans, sometimes referred to as "arson cans." These cans do provide some shielding, but it is not feasible to view the device when it is inside a metal can, nor is it possible to conduct a forensic examination while it is shielded in such a manner.

Shielded Examination Box

Many law enforcement agencies and forensic labs have a shielded box that they use for forensic examination. These boxes resemble those used in biomedical research laboratories when working with hazardous or contagious materials. The box has shielded hand openings for manipulating the device, shielded power and data cable connections, and a viewing port. These devices can be expensive ($1,600 to 2,500 each) and are not intended to be used in the field. With these devices, an examination can take place without danger of losing evidence due to incoming calls or remote deletion. The manufacturers provide some documentation regarding their products' radio frequency shielding, but to my knowledge these claims have not been independently verified.

Signal Disruptors (aka Cell Phone Jammers)

This is a contentious area of signal disruption for phone forensics. The Federal Communications Commission (FCC) enforces the Communications Act of 1934 and FCC rules prohibit the manufacture, importation, marketing, sale or operation of these devices within the United States (See Section 302(b) of the Communications Act, 47 USC § 302a(b) and Section 2.803(a) of the FCC's rules, 47 CFR § 2.803(a)), which prevents the use of any device which can interfere with a wireless signal.)

The statute prevents the use of any such device and threatens possible criminal or civil penalties for violations. Congress recently enacted the Safe Prisons Act, which permits the study and use of these devices inside correctional institutions. However, the contingencies for their use may prevent widespread deployment. It is important to note these devices can block ALL signals in the affected area, including radio transmissions.

Most devices cannot distinguish between which area of the radio spectrum to block and which ones to allow. There are some manufacturers that are able to program their devices to allow transmissions along a certain section of the radio frequency spectrum, such as the radio frequency used by correctional officers within an institution. Correctional institutions in urban environments may have difficulty using these devices as there is no way to completely limit the jamming to the institution and may affect those outside the institution. These devices may also lead to a false sense of security inside a corrections institution as the device may not be uniformly effective in all areas. Prisoners will inevitably discover the areas where the jammers are less effective and they can still use their contraband cell phones. While signal disrupters show promise for limiting the widespread use of contraband cell phones inside correctional institutions, they have not completely solved the problem.

Court Order

It is also possible to direct a service provider to remove a phone from the network by issuing a Court Order or Search Warrant. In effect, you are directing the provider to turn off the service as if it were discontinued due to non-payment or service termination. The downside to this method is that it can take time to draft the order, locate a magistrate, serve the order, and then have the provider comply with the order. This works on all phones, whether or not the handset security lock is enabled or not.

The primary use for this technique would be in tactical situations. For pre-planned tactical operations it is possible to obtain an anticipatory Search Warrant that directs the service provider to remove a phone from the network upon request. The application for this technique may be situations where the suspect's cell phone number is known in advance and there is concern his neighbors, confederates, or lookouts will alert him to law enforcement's approach. Coordinate this with the cell service provider well in advance of a planned tactical operation. Even if the service is turned off, the suspect will still be able to call 911 in the event the operation turns into a barricaded suspect or hostage situation.

Preservation Letters

No matter which method of signal shielding is used, submit a preservation letter to the service provider pursuant to 18 USC 2703(f). It is important to understand that when the phone is prevented from communicating with the network, all of the incoming voicemail messages, text messages, and emails remain stored on the provider's system until the network can update the device. When a signal shielding method is used to preserve the evidence on the phone, those communications are prevented from being delivered to the phone. If the suspect subsequently discontinues service to the device, which is likely once they know law enforcement is interested in the phone, all of the un-delivered items can be deleted from the provider's network. The service provider's reasoning is that if there is no longer service to the device then it does not need to store the undelivered content.

Storing the Phone Evidence

A surprising number of law enforcement agencies store mobile phones in evidence with the battery removed. I've heard theories about why this happens, including one that says the lithium ion battery is a fire hazard and removing it from the phone prevents fires. I don't believe this to be based on reality. Instead what happened is in the early days of cell phones, the property technicians had difficulty in silencing or turning the phones off. Who wants to sit in an evidence storage facility or property room with multiple phones ringing or sounding message notifications? The easiest thing to do was to pull the battery. This is a huge mistake that may result in the potential permanent loss or alteration of data.

A typical cell phone has two batteries: the one you are familiar with under the back cover and an internal battery. The internal battery's principal function is to power the internal clock of the device. The internal battery is small and is charged by the main battery. Due to the internal battery's small size and limited power consumption, it is capable of maintaining a charge for a considerable amount of time by drawing power from the main battery. This occurs even if there is insufficient power in the main battery to operate the phone – it still retains enough power to keep the smaller battery charged.

Removing the main battery causes the internal battery to start to drain. If the smaller battery is completely drained it will reset the clock to what is known as epoch time. Epoch time is used in computer programming as the standard for measuring time and is the number of seconds from Midnight January 1, 1970. Many cell phone operating systems use epoch time to calculate time stamps for data such as incoming calls and text messages. If the internal battery dies, all dates and times will be listed as 1/1/1970. There is no way to get the correct date and time back; there is only a way to prevent it from happening in the first place.

Apple phones are, of course, different. Apple's epoch date is January 1, 1980. If you were unfortunate enough to require Apple's assistance with recovering data from a locked iPhone, you probably have encountered epoch time. Because their epoch time is different from those manufactured by other companies – you cannot use the standard epoch time converters found online. Blackbag Technologies makes a converter for Apple epoch times available on their website for free. Unfortunately, you need an Apple computer to use it.

For some clarification, the internal battery that powers the clock may do so for a long time. I have re-examined phones that had been in evidence for years and while the main battery was drained, the internal battery still reflected the correct time and date. However, I have also re-examined phones that had not been in evidence for very long but were stored with the main battery removed and found the times on the files all reset to epoch time.

Cold Phones

Much in the same way it is possible to recover physical evidence from older or "cold" cases, it is also possible to recover data from phones stored in evidence. These cold phones were either not examined or were examined with the forensic tools available at the time of examination. As cell phone forensic tools evolve, there is a greater possibility of recovering data, included deleted files.

Chapter 6: Digital Evidence

Much of the information that can be retrieved from a device is often not even known or visible to the user. There are two main categories of retrievable digital evidence from a cellular device: logical information and physical information. Logical information refers to those items that are visible to the phone's user and includes the graphical user interface (GUI), menus, applications, and other user features of the device.

An examination of the physical information refers to the stored digital information that is not visible to the device's user. This includes programming information, underlying databases used to compile, store, and display the logical information as well as internal application data. Physical information retrieval is an exciting and evolving area of cellular forensics, as during these examinations it is possible to retrieve information such as the handset security unlock code and deleted data such as contacts, text messages, image, and video files.

Several of the following methods for recovering information from a phone, SIM card, and external storage media involve the use of techniques and software that are not forensically validated. Inability of existing forensic tools to adequately obtain information from a device, the time, effort, and cost associated with the tools, and the failure to get forensic assistance in a timely manner are justifications for not using forensic technique. To mitigate some of the potential problems that may be encountered during court proceedings, it is essential to follow the same steps for verification of other types of evidence.

1. Make certain you possess the requisite search authority. When in doubt, obtain a search warrant. Standard exceptions to the Fourth Amendment such as consent or supervised release may be insufficient legal justification for advanced searching techniques such as deleted file recovery.
2. Establish and maintain an appropriate chain of custody. Thoroughly and completely document the movement of physical and digital evidence.
3. Determine the goal of the examination. Are you interested in digital image files? Do you really need to attempt to recover everything off the device? Make sure you or the case agent really want the complete download files from a 64 gigabyte smartphone.
4. Have an evidence storage plan. As the memory increases with smartphones, where are you going to store multiple examination files containing many gigabytes of data?
5. Verify the integrity of the evidence using established forensically imaging and hashing functions whenever possible.
6. Use validated tools first whenever possible. Use alternate methods after validated tools. The difference between validated tools and non-validated tools and the pros and cons for using each will be discussed in detail later.
7. Document the processes used and the reason certain tools or methods were effective or ineffective.
8. Analyze the evidence and intelligence obtained and prepare a report documenting it. This may need to be done by the case agent or investigating officer if different from the person doing the examination of the device.
9. Ensure the process used is repeatable based on the documentation used. Would another forensic examiner retrieve the same evidence using the techniques you documented using?
10. Prepare to defend the examination techniques based on your training and experience, and the tools and methods available to you at the time of the examination.

Contact Information

One of the most valuable sets of information that can be retrieved from the handset is the electronic phonebook or contact list. This area is often minimized or overlooked in favor of other immediately gratifying information such as text messages and image files. However, depending on the investigation, it is possible to identify a subject's friends, family, and associates based on the contents of his electronic phonebook. In the example below, gang members are designated with a red heart.

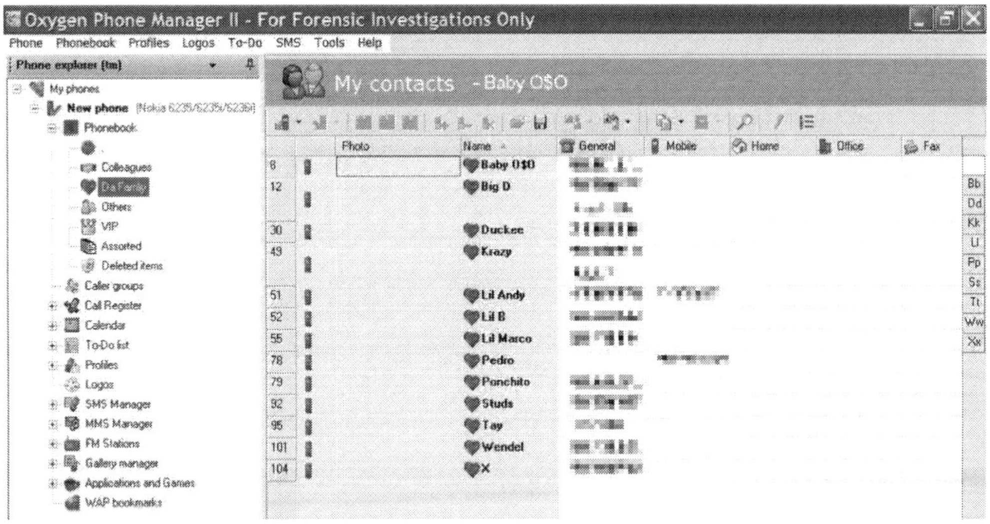

Investigators may presume the phonebook contains precisely what the title describes, the name and phone number of a contact. However, many devices have data fields where a user can input additional data including other contact numbers such as home and work phone numbers, email addresses, UFMI (Universal Fleet Mobile Identifier aka Push to Talk) numbers, and custom photos and ringtones associated with the contact. Also, basic phone owners use the phonebook to store additional alphanumeric information such as passwords, personal identification numbers, and other identifying data.

E	F
Stored Name	Stored Number
none listed	1265
Pin	3306
Baby pw	51692
[illegible]	[illegible]

Researching the names and phone numbers stored in the device can also provide clues and evidence in gang investigations when attempting to identify a person's moniker. In my experience, gang members and narcotics traffickers will identify a contact by his moniker and even establish calling lists, or groups of favorite contacts, for their specific gang or criminal enterprises. Also, I encourage investigators to pay attention to additional alphanumeric characters within a specific contact. I have seen instances where a narcotics trafficker identified customers and their respective drug quantities by the use of extra characters within the contact. For example when the caller identification on the phone indicated the incoming caller was "Joe." the suspect knew he wanted to purchase a small amount of drugs. When an incoming call from "Fred" came in the suspect knew he wanted a much larger quantity, based on the number of periods after the suspect's name.

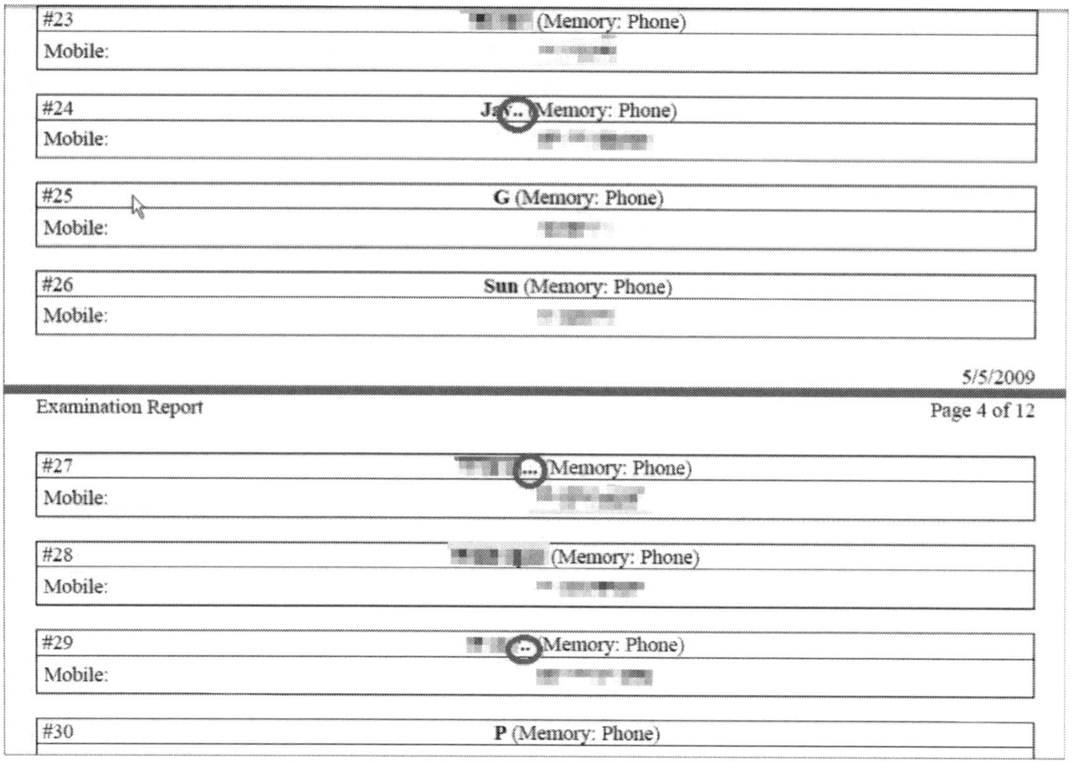

#23	(Memory: Phone)
Mobile:	

#24	J... (Memory: Phone)
Mobile:	

#25	G (Memory: Phone)
Mobile:	

#26	Sun (Memory: Phone)
Mobile:	

5/5/2009

Examination Report Page 4 of 12

#27	... (Memory: Phone)
Mobile:	

#28	(Memory: Phone)
Mobile:	

#29	.. (Memory: Phone)
Mobile:	

#30	P (Memory: Phone)

I have seen phonebook entries with a subject's moniker and his gang listed (e.g. Flaco-A St., indicating Flaco is associated with the A Street gang). Other subtleties may elude the forensic examiner but be apparent to a gang or narcotics investigator. In the area I work it is common for northern Hispanic gang members to cross out the character S to show their disrespect for southern Hispanic gang members. In an electronic format, this can be represented by replacing the character S with the $ dollar sign.

While it may seem obvious, it is possible to identify the previously unknown user of a phone by looking for obvious references to family members in the electronic phonebook. Most users will enter listings of family members such as "Mom" and "Dad." As many older adults retain landline phones in addition to cell numbers, it is much easier to identify them and subsequently their offspring.

An interesting aspect of the electronic phonebook can be the order in which the phone numbers were extracted using a forensic device or software program. When manually viewed, the electronic phonebook will be alphabetized. However, depending on the operating system of the phone, a forensic software tool will retrieve the phonebook in order of data entered into the phone. For example, when viewed manually an electronic phonebook would show the entries as:

Aaron 510-555-1234
Adam 408-555-5678
Alicia 415-555-9101
Angry M 650-555-1112

However, when viewing the results of a forensic examination the contacts may appear in order as:

Flaco 510-555-6789
Big Dave 510-555-5533
Angry M 801-555-5055
Wife 925-555-8929

These results may indicate the order the user stored and entered the phone numbers. This could be significant when investigating a suspect's associates and co-conspirators, as in this example three contacts were entered before the entry "Wife." This could infer a close relationship between the suspect and the first three contacts as evidenced by the order of entry into the phonebook. There is no guarantee this is 100 percent accurate and the order of entry into the electronic phonebook can vary depending on the device's operating system.

Call Logs

Call logs store a fixed number of recent calls the target device has made or received. These logs are divided into Sent, Received, and Missed calls. Most devices have a finite storage capacity for these logs, which may be limited to as few as 10 entries in each one. This information is easily overwritten by incoming calls that will delete older entries. Due to the finite storage capacity it is not possible to retrieve this information from the device during forensic examination. However, it is usually feasible to retrieve the entire call history by requesting the call detail records from the service provider.

When comparing records obtained during forensic examination with the call detail records from the cellular service provider, it is usual for there to be discrepancies in the times of the calls. Primarily the device causes this by altering the time displayed on the device. Many devices automatically correlate the time displayed on the device with the time broadcast by the network. However, the user may alter the time in the settings menu. This is done to grant the chronically late user a degree of leeway when keeping appointments. Make sure to document any differences in the time displayed on the device and the time displayed on the examination equipment.

Calendar

Calendar and alarm functions are other overlooked areas of a cellular device. It is presumed that criminals do not use the calendar functions as they do not have many regular events to keep track of. Most criminals are not likely to enter significant criminal events in their phone such as "2/14/2013 3:00 PM Meet Gordo behind the liquor store and kill him." However, I have found many significant non-criminal events stored in these areas including court dates, meetings with probation and parole officers, and family events. These can provide valuable clues in determining the user of a particular device. This information can also be used to help establish timelines and may be useful during interviews and interrogations to prove or refute alibis. They are not often supported by existing forensic tools and may require the use of an alternate tool or method.

Image Files

These files are usually examined first, particularly when the search occurs out in the field. Most law enforcement officers are looking for immediately actionable information or evidence that may confirm the crime under investigation. These files can contain some of the most valuable and incriminating evidence. But there are some important considerations when examining image files and presenting them as evidence.

Perhaps one of the most important image files is the "self-portrait." Social media and Internet dating sites often require the user to take and send images of himself. These pictures are extremely useful for establishing dominion and control of the phone, especially when it is serviced by a prepaid provider. Due to the lack of identification requirements from most prepaid providers, the self-portrait image may be some of the only direct evidence of who actually used the phone. Fortunately, many phones available for purchase do not have the rearward, user facing camera that forces the user to take the infamous bathroom mirror picture. This is good news for investigators because these images not only depict the user but also capture a nice image of the phone handset as well, making it difficult for them to deny using the device.

The first step is to ensure the image files are recovered in a forensic manner if possible. Recovered images should be saved in their original format before attempting to view or manipulate them. Most image files are captured by the device's built-in camera are designed to be viewed on its screen. Because of that, I strongly recommend that all image files be viewed on a large screen whenever possible. It is possible to see details and items in the background that are not fully visible when viewed on a smaller screen. In the past I have been able to ascertain vehicle license plate numbers, weapons, weapons serial numbers, and details of tattoos that were not easily seen on the device but were when viewed on a 17-inch computer screen. You can enhance image files using software such as Adobe Acrobat or Irfan View and discover or illustrate finer details than what is seen on the small screen.

EXIF Data

Image files also contain what is known as EXIF data or metadata. EXIF or metadata can simply be described as data about other data. This information can include a wide variety of information about the equipment and circumstances surrounding the image, including the model of the device utilized to take the image, whether a flash device was used, or if the device was held upright or on the side. Perhaps of most interest to the investigator is data which may include the date and time the picture was taken. While this information is sometimes decipherable from the image file name (e.g. 091620102233.jpg would indicate the picture was likely taken on 9/16/2010 at 2233 hours). However, if the user changes the image's title this information may not be readily apparent.

Also, many devices equipped with GPS are now recording the latitude and longitude of the handset when the image was taken. It is possible to examine EXIF and metadata information to determine the date/time and location a particular image was taken.

There are a number of free and low cost software programs that will examine an image and create a report of the EXIF or metadata. Opanda iExif is a free software program that lets an operator view the EXIF data of image files. As illustrated in this example, the software was used to examine an image file depicting a dangerous suspect armed with a golf club. The software correctly identified the make and model of the device used to take the image, in this case an older iPhone. The software also notes the date and time the image was taken.

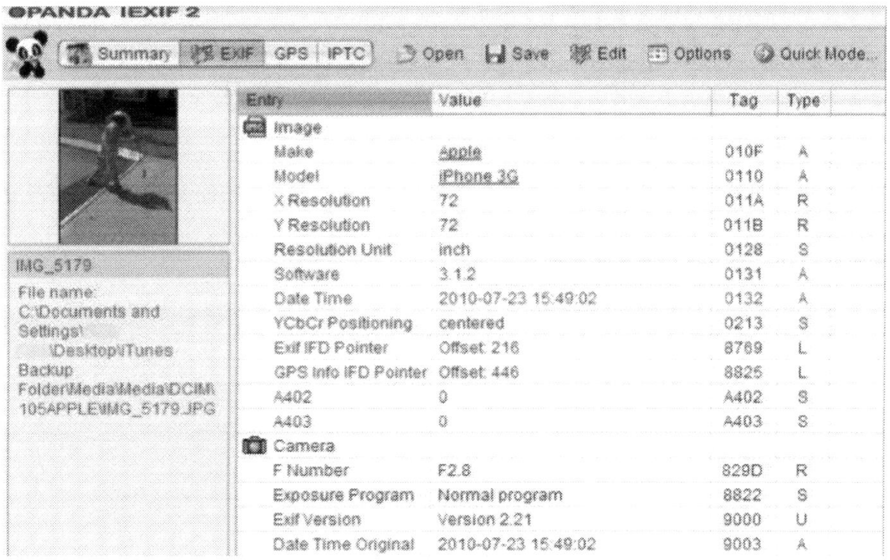

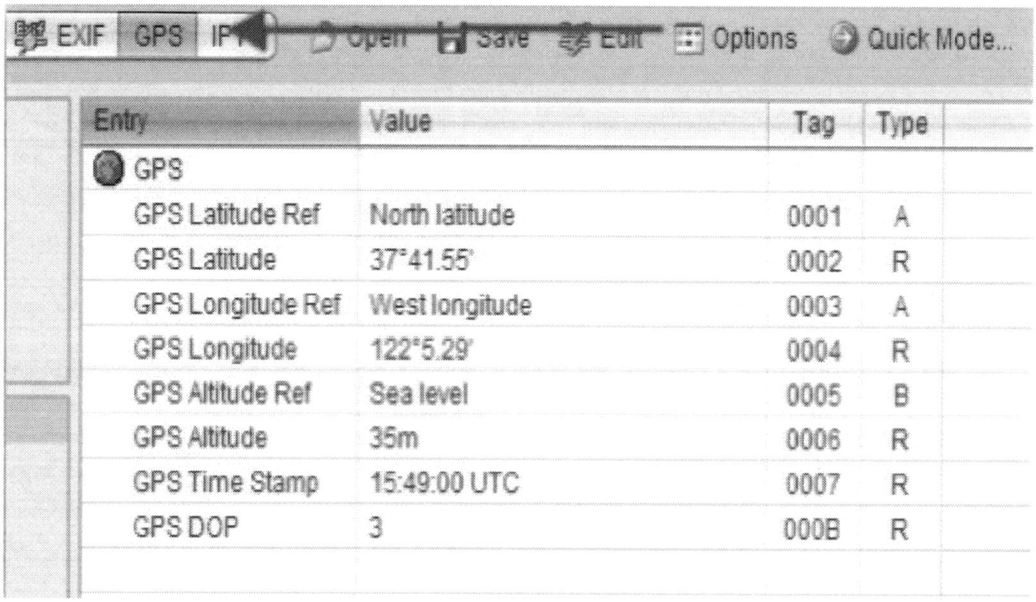

Selecting GPS brings up the latitude and longitude of where the image was taken.

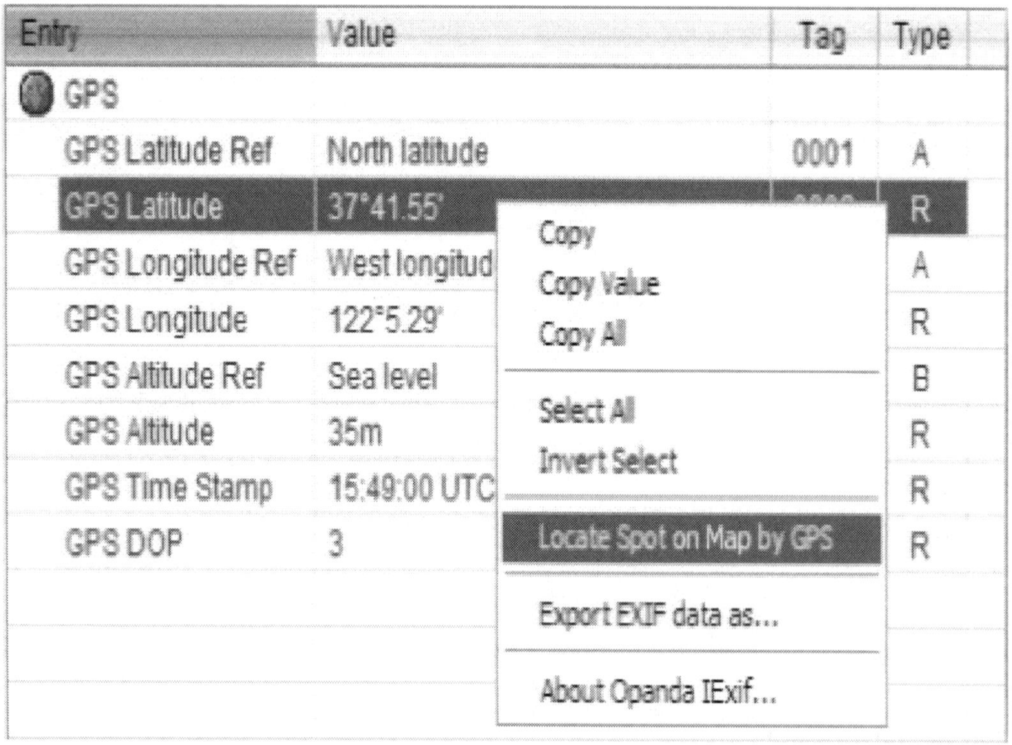

Right mouse click in the data fields to bring up the map interface.

GPS coordinates are automatically plotted on Google Maps in the example abbove. Other free programs available for reading EXIF data in image files are Exif Read, Exif Reader, and ExifPro Image Viewer. None of these programs can be counted on as a forensic utility because they offer the ability to read and write (change) EXIF data.

Sanderson Forensics: Forensic Image Viewer

While Opanda is very simple to use it lacks many of the features and safeguards that would preclude questions regarding the authenticity of the data. A solution to some of these issues is found in a very easy and free program offered by Sanderson Forensics called Forensic Image Viewer.

Forensic Image Viewer (FIV) offers a number of handy features for viewing, manipulating, and logging information related to an image file. The graphical user interface is easy to operate for most users. Simply navigate toward the location the images have been saved and click on the desired image. The EXIF and GPS information are automatically displayed under separate tabs. FIV also incorporates a number of basic tools for image manipulation (rotation, zoom, and crop). FIV differs from other image utilities by creating a log of the user's actions upon opening the software, which can be important for documenting actions taken with an image file. FIV also has a few basic enhancement functions for use on a section of the image, such as the vehicle license plate in the following example.

Hash Verification

One of the inevitable issues when dealing with image files is the validity of the data. In other words, proving the image file recovered from a device has not been tampered with and was collected using sound forensic techniques. Digital images can be easily manipulated and there are a number of paid and free tools available to edit them. Forensic tools may automatically generate what is known as a hash function, also called hash values, hash codes, hash sums, checksums, or hashes. A hash is a mathematical function that evaluates a large amount of a data, such as an image, and produces a fixed length string of numbers and characters. This is used as a security function since even a small change in the underlying data will result in an immediately visible change in the checksum or hash value. Note the changes in this example found on a Wikipedia posting regarding the MD5 hash algorithm:

"The quick brown fox jumps over the lazy dog" reflects a checksum of 9e107d9d372bb6826bd81d3542a419d6

"The quick brown fox jumps over the lazy dog." reflects a checksum of e4d909c290d0fb1ca068ffaddf22cbd0

The simple addition of a period at the end of the sentence causes the checksum's entire value to change. The application for cell phone forensics is that it is possible to take a file, in this case an image file whose authenticity is called into question, and examine the hash value that was automatically generated by the forensic tool using another tool with the same algorithm. This should result in an identical checksum that would refute the possibility of image tampering.

There are several different hash algorithms available for use. One of the earlier and more popular systems was known as MD5. MD5 was widely removed from use in favor of other algorithms such as the Secure Hash Algorithm (SHA). MD5 was shown to have the potential for duplication in large data sets. Early software versions of the Cellebrite Universal Forensic Extraction Device (UFED) used MD5 exclusively, resulting in some concerns about the data's validity. Subsequent versions of the UFED software use both MD5 and SHA.

Secure Hash Algorithms were designed by the National Security Agency and published by the National Institute of Standards and Technology (NIST). The underlying security of hash algorithms rests on the fact it is computationally infeasible to find two different files that produce the same digital representation and that any change in the underlying file will result in a completely different digital representation. A free tool that calculates the hash value of a file, using most of the available variations of algorithms can be accessed at http://www.fileformat.info/tool/hash. Other free tools can be obtained using a simple Internet search for hash calculators.

Pornographic Images

There are pornographic images discovered during the forensic examination of cellular devices. These images can be commercial downloads but are often "homemade." Frequently, juveniles will send or receive images of themselves or others using email or multimedia messages (MMS). These images can be valuable in the criminal prosecution of adults who have received these files even if it did not start as a child pornography investigation. For example, the forensic examination of a cell phone seized during a gang investigation may reveal image files sent by the suspect's underage girlfriend. This can constitute child pornography and could be a valuable tool in prosecuting the gang member. Review the applicable statutes for your jurisdiction and check with your prosecutor.

Child pornography charges can also be a valuable tool in cases involving juvenile blackmailing and harassment investigations. A juvenile may voluntarily submit a naked or semi-naked picture to a known acquaintance who then uses the image to torment, harass, or blackmail the sender. In addition to the underlying criminal charge, this may constitute distribution of child pornography.

One common field expedient method for retrieving image files from a suspect's phone is to send the image from his phone to a law enforcement officer's phone. While this technique allows the officer to retrieve the image, it also provides the suspect with the officer's phone number. If the suspect is aware of, or suspects the actions of the officer, he or she can retrieve the officer's phone number when he examines his bill. In some areas of the country this method also opens up the officer's personal cell phone and call detail records to be examined by the defense attorneys.

Graphic Files

Graphic files differ from image files in that they refer to operating system icons, downloaded animation, and other content. These files are recovered by forensic tools as image files and may provide additional evidence, particularly in gang and narcotics investigations. There are a number of commercially and user-generated graphics files that include particular gang slogans and images.

Video Files

Video files can be compelling evidence. As with image files, video files are designed to be viewed on a handset's screen and subtle detail is discovered when viewed on a larger screen. When the file is viewed on a larger screen, the images often pixelate and the content's clarity and definition can degrade, making it difficult to view. It is possible to enhance these files and produce a better quality file than the original. Unlike image files, there are not as many commonly available tools for video enhancement. One tool that I have had reasonable success with for enhancing cell phone video files is vReveal from a company called Motion DSP. The company originally developed enhancement techniques for videos obtained from unmanned aerial vehicles and translated those techniques into law enforcement video enhancement. They also offer a free trial and low cost full version of easy software for video enhancement, which works well on cell phone videos. Remember that with any enhancement tool, the results rely on the original image's underlying quality. If the video file was shot on a one megapixel camera, don't expect to read a vehicle's license plate that is 50 yards in the background.

Text Messages

These days everyone loves to text message. According to the Cellular Telephone Industry Association, more than 1.56 trillion text messages were sent and received in 2009 (or 4.2 billion per day). There are two main varieties of text messages, short message system (SMS) and multimedia messaging service (MMS). Both types are often used interchangeably as text messages but there are some distinct differences. A SMS is a text communication service in which a maximum of 160 characters may be sent or received. MMS extended the capabilities of SMS by allowing larger messages to be sent, including multimedia files such as image, video, and music files.

The examination of text message content allows for valuable investigative insight into a suspect's communications with others and is often one of the first areas searched by law enforcement officers. As valuable as this information is, care must be taken to preserve existing content and prevent the intentional or accidental deletion. Text messages are one of the most vulnerable areas on a device. Some phones have a potentially limitless capacity to store text messages, but most have a limited storage capacity for maintaining text messages (typically limited to 60 to 90 sent and received messages). A "conversation" using text messages may have multiple sent and received files, which can quickly exceed the phone's capacity and cause older messages to become deleted. Also, several phones have user-defined storage periods that will automatically delete a "read" text message after a set period of time. Once deleted, it can be impossible to retrieve them.

Additionally, use caution when searching these messages. Most cell phone operating systems notate whether the message has been read after it has been received. A forensic examination may show a field search of text messages that indicate the message had been "read" when, in fact, the phone had been in police custody and the suspect had not actually viewed the message.

Unfortunately, several widely used forensic tools available do not capture SMS or MMS text message content, or they capture the content but do not capture the date/time the message was sent or received. Alternate methods for capturing text message content, dates, and times will be discussed later.

The multimedia attachment of a MMS may alternately be stored within the message itself or stored in the system's appropriate media file. Available forensic tools do not always capture the MMS messages' content files.

Memos or Notes

The files are sometimes overlooked during forensic examinations because very few commercial tools capture the information. There is an unfortunate trend in law enforcement which presumes that if a commercial forensic tool did not capture the information from a device there was nothing there to capture in the first place. Most often this is false particularly when it comes to this particular area of examination of a device. It is important to remember that many commercial forensic tools were initially designed to transfer or backup content from one device to another in a retail setting. At the time the most value was found in copying the contacts and media files and not a lot of attention was given to other features such as the Memo or Notes area. In my experience it can be a hit and miss finding relevant evidence in these files. It largely depends on the suspect and whether he or she made extensive use of these features. Due to the fact that many commercial forensic tools do not capture these files it may be necessary to use alternate techniques.

Voice Recordings

Similar to the Memo or Notes feature, mobile phones could be equipped with the ability to make voice recordings. Some people utilize this feature to keep audio notes on their device, particularly when they are unable to use the key pad functions while they drive. I have encountered cases where an informant used his phone as an improvised body wire and recorded incriminating conversations with a suspect. During forensic examination these files may be acquired as audio or music files. This can be confusing, particularly when the device contains a large number of music files. As the forensic acquisition of music files can take a long time, law enforcement officers may skip these files during the examination or review process.

Music files are usually in common formats such as .mp3 or .wav, but voice recording files may be stored as .qcp files. These files are made using a proprietary program from Qualcomm, a leading manufacturer of cell phone components. Most computer software programs are unable to play .qcp files in their native format. There are services, both free and paid, that will convert .qcp files into other common formats, but this not always the case. When Qualcomm licenses its software to cell phone manufacturers, it allows them to make proprietary changes to the software, which can prevent the files from being played outside of the device. While Qualcomm makes their .qcp player available free on its website, it will not always work if the manufacturer has made changes to the underlying software.

Improvised methods for retrieving usable copies of these files include using a digital recorder to make a recording from the speaker of the device. This method can pick up background noise, though. A further refinement is to use a cable converter plugged into the headset jack of the device to record the messages onto a computer. A free software program to record these files is Audacity, which can be downloaded from http://audacity.sourceforge.net/.

Internet Activity

Mobile devices may let a user access the Internet. Depending on the investigation this information can be critically important, but it can be very difficult to retrieve. Additionally, this information may not be retrievable if the device has been prevented from communicating with the network. Numerous commercial forensic tools are not equipped to capture this information or will only capture portions of it. It is easier and sounder, from an evidence standpoint, to obtain the information from the cell service provider. However, this information can be extremely perishable and a preservation letter should be submitted to the provider as soon as possible.

Maps

Mapping features are also unnoticed on some advanced devices. Smartphones, such as those using the Apple or Android operating systems, will usually retain the last queried search in their memory. Using some advanced techniques and/or software, it may be possible to recover previous search queries, directions, and map tiles.

In addition to the native mapping software there are a number of location-based service applications such as Foursquare, which leverage the existing GPS functions of the device. Application specific forensics is an evolving field and examining software programs that interface with the device's native GPS may need to be done manually.

Chapter 7: Types of Examinations

There are generally two types of examinations for cell devices: manual and assisted. A manual examination is exactly what it sounds like. The low-tech, "fat fingered" technique of navigating through the phone and identifying items of intelligence or evidentiary value.

Items of interest may then be written down or photographed. Identifying evidence or intelligence during manual examinations is quick, simple, and can be completed anywhere. However, recording the evidence found during a manual examination can be a labor-intensive as is the case when trying to hand write several hundred phone book entries. Unfortunately, the very real possibility exists for accidentally altering or deleting data during a manual examination. I have, on more than one occasion, accidentally activated certain functions of a device that I later had to explain in my examination report.

A manual examination is best used when there are no other options for evidence preservation or it is imperative to return the device to the user before he becomes aware of law enforcement's interest in it. It is very difficult to validate the findings of a manual examination conducted out in the field as the only record, which will likely exist in the report or notes generated by the investigator conducting the exam. While I conclude most forensic examinations with a manual examination to validate the findings of the forensic tool, I usually don't start with this technique unless there is no alternative.

An assisted examination relies on using a software or hardware tool to automatically duplicate or recover files from a phone and translates them into something presentable. This is the preferred method for examining devices, but there are downsides, as well. Many forensic tools are not designed to be portable or easily used in the field. Some software and hardware programs are designed to be used in an office or laboratory environment and not in the front seat of a patrol car.

These tools can be quite expensive, which puts them out of reach of law enforcement agencies. They need frequent updating as the software or hardware manufacturers provide updates to encompass new devices on the market.

Most of these tools require some basic level of training and a familiarity and comfort with computers and technology. There is no one tool or software program that captures every piece of information from each device. Manufacturers and sales representatives might tout that their tool works on "over 5,000 devices" but fail to mention it will only read the phone book contact list for 3,000 of them. As with all toolkits it is crucial to have multiple tools available, presenting an ongoing financial burden to law enforcement agencies.

Basic Examination of Cellular Devices: SIM Cards

There are two major types of cell service, Code Division Multiple Access (CDMA) and Global Standards for Mobile/Global Systems for Mobile (GSM). An important subset of GSM devices known as Integrated Digital Enhanced Network (iDEN), and commonly referred to as "push to talk" phones, also exist.

GSM is currently used on over 80 percent of the world's mobile networks in over 210 countries.[10] Its technology is clearly the worldwide favorite but does not have the same dominance in the United States. Major GSM carriers in the United States are T-Mobile and AT&T.

A major difference between a CDMA phone and a GSM or iDEN phone is the presence of a Subscriber Identification Module (SIM) card. A SIM card is a smart card that is inserted into a cell phone and primarily

10 http://www.gsmworld.com/newsroom/market-data/market_data_summary.htm

used for identification, security verification as well as data and speech encryption. SIM cards can also store data such as user information, phonebook entries, dialed numbers, text messages, and mobile phone settings. A SIM card can be examined exclusively off the phone handset and can contain important information that is not normally visible to the user.

SIM cards, and their contents, are much sturdier than the phone that houses them. Researchers in France exposed SIM cards to high temperatures and found they were still able to recover data from them.[11] SIM cards also may have a greater ability to resist water damage than a phone. This can be good news if the suspect attempts to destroy the phone.

Examining a SIM card requires using a SIM card reader. There are plenty of SIM card readers on the market, but the majority of them are poorly made. Investigators who use commercial SIM card readers could become quickly frustrated with the reader's inability to acquire data from the card. The problem with the SIM card reader is that it was designed to read second generation SIM cards (and not the current 3G cards) or they are not compatible with the cards used by phones operating on the iDen network. Be sure to check compatibility with current SIM cards before purchasing a card reader.

Each SIM is internationally identified by its Integrated Circuit Card ID (ICC-ID). ICC-IDs are stored in the SIM card and can be engraved or printed on the SIM card's body during a process called personalization. The number is up to 20 digits long with an addition of a single "check digit" for error detection.

A typical SIM (19 digits) example, **89 91 10 1200 00 320451 0**, provide the following details:

- The first two digits (89) refer to the service provider.
- The next two digits (91) refer to the country code (91-India).
- The next two digits (10) refer to the network code.
- The next four digits (1200) refer to the month and year of manufacturing.
- The next two digits (00) refer to the switch configuration code.
- The next six digits (320451) refer to the SIM number.
- The last digit, which is separated from the rest, is called the "check digit."

These digits can be further grouped for additional information:

- The first three to four digits represent the Mobile Country Code (MCC) – (Some cards only have three digits to represent the Telecom ID and country code.)
- The next two digits represent the Mobile Network Code (MNC, AKA the mobile operator).
- The next 12 digits represent the Home Location Register.
- And mentioned above, the "check digit."

The website http://www.numberingplans.com/ offers tools in the Number Analysis section for deciphering the meaning of the ICCID as well as cellular telephone numbers and identifying numbers. The results from a SIM card reader program can reveal information aside from the obvious (deleted text messages, etc.). These additional fields are:

- **International Mobile Subscriber Identity (IMSI)**: A unique 15-digit numeric identifier that pinpoints the phone/subscription to the GSM network. The structure of the IMSI is somewhat similar to the ICCID, consisting of a Mobile Country Code (MCC), a Mobile Network Code (MNC), and a Mobile Subscriber Identity Number (MSIN). The MCC is three digits, the MNC may be either

[11] http://www.fsijournal.org/article/S0379-0738(07)00793-1/abstract

two or three digits, and the MSIN assigned by the carrier takes up the remainder.
Mobile Country Code (MCC): A three-digit code that represents the SIM card's country of origin

- **Mobile Network Code (MNC)**: A two-digit code that represents the SIM card's home network
- **Mobile Subscriber Identification Number (MSIN)**: A unique 10-digit identifying number that identifies the specific subscriber to the GSM network
- **Mobile Subscriber International ISDN Number (MSISDN)**: A number that identifies the phone number used by the headset
- **Abbreviated Dialing Numbers (ADN)**: Telephone numbers stored in memory of the SIM card
- **Short Message Service (SMS)**: The Short Message Service (SMS), commonly referred to as a text message, file contains text and associated parameters for sending and receiving messages from the network. SMS files have text and header information, the time an incoming message was recorded by the mobile phone network, and the sender's phone number. The status of a SMS can be a received message to be read, a received message that has been read, an outgoing message that has been sent, or an outgoing message waiting to be sent. Deleted messages are usually listed as free space and the content is left unchanged until overwritten by newer data.
- **Public Land Mobile Network (PLMN) selector**: Network service providers the mobile phone communicated with.
- **Forbidden PLMNs**
- **Location Information (LOCI)**: General information regarding the geographic location of the mobile phone. The LOCI field contains the Location Area Information (LAI) for voice communications. The LAI is composed of the MCC and MNC of the location area and the Location Area Code (LAC), an identifier for a collection of cells. The LAC can provide general information about the handset's location.
- **General Packet Radio Service (GPRS) location**: contains the Routing Area Information (RAI) for data communications over the General Packet Radio Service (GPRS). The RAI is composed of the MCC and MNC of the routing area and the LAC, as well as a Routing Area Code (RAC), an identifier of the routing area within the LAC.
- **Service Provider Name (SPN):** An optional field that contains the service provider's name
- **Abbreviated Dialing Numbers (AND):** User entered names and numbers
- **Last Number Dialed:** This field may contain one or more of the last outgoing numbers dialed by the handset
- **Language Preference (LP)**
- **Emergency Call Code**
- **Fixed Dialing Numbers (FDN)**
- **Own Dialing Number**
- **Temporary Mobile Subscriber Identity (TMSI)**: A number assigned when the user is roaming outside the home network
- **Depersonalization Key**[12]

---

[12] http://www.forensicswiki.org/wiki/SIM_Cards

SIM Card Adapters

SIM cards used to be an extremely important part of forensic examination because they were accessible even when the phone was protected with a passcode. Certain data elements were stored on the SIM card itself and not in the memory of the phone. Stored data included the phone book, the last 30 call logs of incoming, outgoing, and missed calls as well as some deleted text messages. As increased memory on phones became less expensive and subsequently more prevalent, though, fewer data is currently stored on SIM cards. While still an important piece of evidence to examine, the likelihood of recovering useful data from the SIM card has steadily decreased. Currently, most GMS mobile phones use the SIM card for network authentication and little else.

The reduction of the SIM card's importance as a storage medium has led to its smaller size. An older, larger SIM card can take up a lot of valuable real estate inside the phone, so manufacturers have developed smaller SIM cards like the micro SIM and now the nano SIM. The micro and nano SIM cards usually require special adapters to configure them for use in standard size SIM card readers.

SIM Card Security

Some users enable a security feature on their SIM cards that requires a personal identification number (PIN) to unlock. After three attempts to unlock the card using the PIN number the card defaults to a higher security code requiring a Personal Unlocking/Unblocking Key (PUK). The PUK is an eight-digit code that will unlock the card. After 10 unsuccessful attempts at the PUK code the card will permanently lock. PUK codes are available from the service provider with a Court Order or Search Warrant.

SIM Card Switching

The presence of multiple SIM cards in a suspect's possession can be an indication of SIM card switching. As noted earlier, the majority of cell phones in use worldwide operate on the GSM network. One of the convenient features of GSM technology for international travelers is the ability to switch a SIM card from their home country's service provider with one from another country's service provider. When the user switches the SIM card from his home network to another he is given a local phone number. This prevents the user from incurring international roaming charges while he travels. Some prolific travelers will have multiple SIM cards, one corresponding to each of the countries they frequently visit.

Sophisticated criminals sometimes use the same technique to cover their tracks. By changing their SIM cards frequently the suspects are able to change their phone numbers. They can also retain their phone handsets and do not have to replace them, or their content, every time they need to change their number. What most of the public, including SIM card switching criminals, are unaware of is the data is recorded whenever a call is placed. As noted earlier, all cell phones have a unique serial number assigned to the device. When a call is placed with a GSM phone the unique serial number of the SIM card and the unique serial number of the phone handset are broadcasts to the network. This way the service provider can verify the status of the device and the billing information. So no matter how many times suspects switch their SIM card, the phone's serial number remains the same. It is possible to track a suspect who switches his SIM cards by tracking the phone handset's serial number.

The investigative units of several correctional institutions in California have reported it is common for there to be one phone in a housing unit that inmates share or rent. In many instances the inmates have their own SIM cards that they insert into the phone when it is their turn to use it. Finding the phone or even one of the SIM cards can reveal all of the other SIM cards used with the device and their associated phone records. Even though each SIM card has its own unique identifier, the unique identifier of the phone it is inserted in is also broadcast. Through a court order or search warrant you can identify the serial number of the phone the SIM card was used in and then identify all of the other SIM cards which use the same serial number of the phone.

SIM Card Software

There are a number of commercial and freeware software programs designed to recover information from a SIM card. One of these programs, SIM CON, was designed for law enforcement and distributed freely until the company was bought by a corporate interest who now charges for the software. It is still possible to find copies of the software program in law enforcement circles. However, the original software has not been updated and may not work with all SIM cards. Ares of interest on a SIM card are the assigned phone number and carrier, the phone book contact list, and SMS text messages, including the possibility of recovering deleted text messages.
Vendors for SIM card software, both forensic and non-forensic are:

- SIM Card Seizure by Paraben Corporation (http://www.paraben.com)
- SIM Explorer by Dekart (http://www.dekart.com)
- CPA SIM Analyzer by BK Forensics (http://www.bkforensics.com/)

External Storage Media

Cell phone manufacturer industry estimates are that approximately 85 percent of cell phones will ship with some sort of memory card.[13] This external storage media is in the form of Micro SD cards and to a lesser extent M2 cards (exclusively used with Sony Ericsson phones). Calling these cards external storage media refers to the fact that they are not part of the internal memory. However, they are still inserted into the phone. Frequently storage media is inserted into a slot on the side of the mobile phone but can also be found under the battery.

Micro SD cards are the most common form of external storage media. SD stands for Secure Digital and is a format for removable flash memory. Micro SD cards utilize non-volatile memory, which means they retain the data contained on them even when the device does not have power. While most commonly associated with cell phones, regular sized SD cards and Micro SD cards can also be found in handheld GPS devices, portable media players, expandable USB flash memory, digital cameras, and some handheld gaming systems.

Micro SD cards are the smallest commercially available memory card, approximately the size of a fingernail. They require an adapter in some devices and card readers. Other card readers are not compatible with a newer format of Micro SD cards, Micro SDHC (High Capacity). Micro SDHC cards have a storage capacity of up to 32 gigabytes of data, but a new generation of Micro SDXC cards can store up to 64 gigabytes of data. As an example of how much data these cards can store, one gigabyte (GB) is roughly equivalent to:

- Over 890,000 pages of plain text
- Almost 4,500 books
- Approximately 340 digital pictures

---

[13] http://www.sdcard.org/developers/tech/sdcard#Micro SD

- 250 MP3 audio files

Some devices require a Micro SD card to operate. Frequently, the phone user dictates where certain user generated files, such as images, are stored. Other operating systems automatically write certain information to the Micro SD card unbeknownst to the user.

Recovering Deleted Files

In addition to recovering visible files from a Micro SD card it is also possible to recover deleted files as well. Before discussing the recovery of evidence from a Micro SD card, you should address the topic of proper forensic procedures. In all cases it is preferable to have an experienced and trained forensic technician recover digital evidence. While some larger departments have dedicated sworn or civilian forensic examiners, other agencies do not have the financial resources to support those programs in-house. There are federally funded Regional Computer Forensic Laboratories and regional high technology crime task forces located throughout the United States that can assist with forensic examinations. These laboratories and task forces triage incoming cases to adequately allocate their resources to the highest priority crimes. Unfortunately, this can often result in lengthy delays in getting your evidence processed. Your high priority cases may not meet the laboratory's criteria for prompt examination and getting forensic assistance on weekends or after normal business hours can be extremely difficult.

The keys to performing your own forensic examination are training and proper documentation. Before undertaking any forensic examination of evidence an investigator should have received thorough training in forensic recovery techniques. The key piece of evidence in your investigation should not be the first time you have tried a particular tool or technique. Thorough documentation of the examination's circumstances and a forensic resource's availability are critical in laying a foundation for admitting evidence.

To recover all files it is important to understand the structure of these cards, how files are stored, how they are deleted, and how recovery programs work.

Micro SD cards are a form of flash memory, which is a non-volatile storage chip that does not require power to store data. The chip can be electrically programmed, erased, and then reprogrammed.

Files stored on a Micro SD will stay there until they are deleted. Flash memory is resistant to water, pressure, and heat and it may be possible to recover data from a memory card that has been exposed to any of those elements.

Micro SD cards are pre-formatted with a file system, known as File Allocation Table (FAT) 16 or FAT32. With this formatting, the cards are accessed on virtually any device with a SD card reader. However, since the card can appear as a removable hard drive to the host operating system, the card is susceptible to being overwritten or reformatted to any file system the operating system supports.

All flash memory media have individually erasable segments. These segments have a limited number of erase and re-write cycles before they become unreliable. An unreliable sector or segment can cause the entire device to fail and become unreadable. To counteract this, device manufacturers created a process known as wear leveling. Wear leveling prohibits the chip from reusing a block of memory until all of the other unused sections have been utilized. This evens out the wear on the chip and lengthens the amount of time the flash memory device can be used before a segment fails. Similar to computer memory, when a file is erased from a flash memory device, the underlying data remains on the device but is no longer visible. What is deleted is the visible link to the file, not the actual data. The space the file occupied becomes what is known as unallocated or free space. This area of the flash memory device is available to be re-written. This is good news when it comes to recovering deleted files because the data still remains on the flash memory. Newer files will be written to other unused sections of the device first and will not necessarily overwrite the recently deleted file.

When attempting to recover data from a Micro SD card, or other flash memory device, it is important to prevent new information from being written to it. This can contaminate the evidence as well as result in potential data loss. To counteract this, forensic examiners use what are known as write blockers to prevent the computer used by the examiner from writing any new data to the flash memory. Write blockers allow read commands to pass through to the target of the examination but prohibit any write commands or features. A write blocker should be used any time an examination of a MicroSD card is going to be made. It is not forensically sound to simply plug the card into any card reader without taking steps to prevent the data on the card from being altered.

When using a write blocker, you should know whether the product works or not. The National Institute of Standards and Technology (NIST) established a Computer Forensic Tool Testing website where they publish results of their independent evaluation of commercially available forensic tools. The test results can be useful when evaluating a product for purchase.

Write blockers come in two flavors: software based and hardware based. They advertise the same benefits of preventing the underlying evidence from being altered. The various devices have respective pros and cons, which are not always applicable to forensic examinations involving flash memory devices. For example, software write blockers are regarded as having faster acquisition speeds than hardware write blockers. This could be a valuable consideration for a computer forensic examiner who is attempting to access a one terabyte (TB) hard drive. However, this may not be when examining a 4GB Micro SD card. A downside to using a software write blocker is that the NIST has only evaluated a few programs. In fact, the NIST has evaluated far fewer software write blockers compared to hardware write blockers.

Hardware write blockers have some distinct advantages such as portability. Hardware write blockers are also more accepted within the forensic community. Whichever method an investigator decides to use, he should have a basic understanding of the underlying software or technology and be prepared to explain it to a judge and jury. There are no NIST-evaluated tools designed exclusively for flash memory.

I use a hardware write blocker called the Ultrablock Forensic Card Reader made by a company called Digital Intelligence. The product costs approximately $80.00, making it affordable for almost every law enforcement agency. I like that it is portable, easy to use, and has a manual configuration that ensures the device is in write block mode. To examine a Micro SD using this device it will still be necessary to obtain a SD card adapter.

Imaging the Card

In addition to preventing new information from being written to the Micro SD card you should avoid working with the original media. In other words, simply putting a Micro SD card into a hardware or software write blocker and examining it is not the best practice. Standard forensic process is to make an exact duplicate of the device to be examined and to use forensic tools to examine the copy, not the original. This exact copy includes not only the visible files but the entire unallocated space where deleted files are found.

The first step in obtaining evidence from a Micro SD card is to make a forensic copy of the card. There are tools available for making a forensic copy. One well documented, forensically accepted, and free software program available for imaging a Micro SD card is called Forensic Tool Kit (FTK) Imager provided by Access Data. FTK is one of the main programs used in computer forensics. Access Data makes the imaging component of its software available free but charges for its full suite of tools, recovery and reporting. The software can be downloaded from the company's website and is included with several free and open source forensic packages.

There are two versions of the software available: FTK Imager and FTK Imager Lite. The Lite version of the software is designed to be run from a portable USB drive, which is helpful for law enforcement agencies that lock down their computers and prevent new software from being installed.

To get started, insert the Micro SD card in the card reader and attach it to the workstation. After installing and downloading either FTK Imager or FTK Lite, launch the program. The main interface screen for the program is shown below.

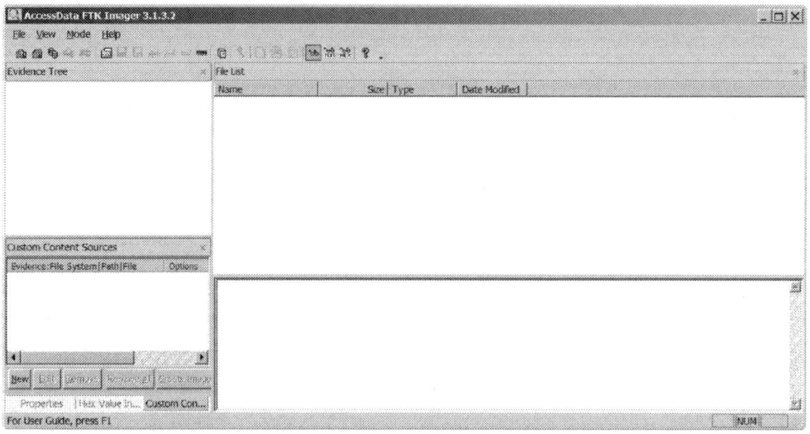

From the main screen select **File**>**Create Disk Image**>**Select Source** (from the drop down menu) >**Physical Drive**>**Next**

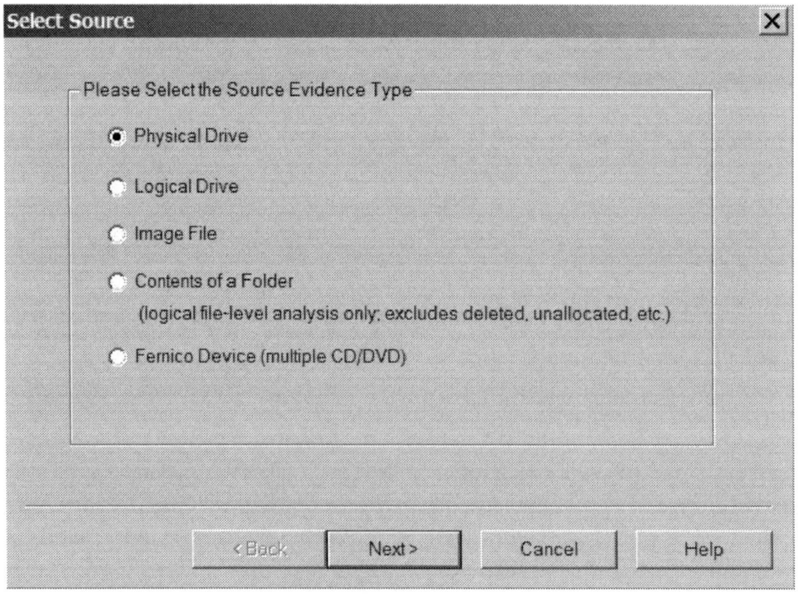

From the Create Image dialogue box select **Add Image**.

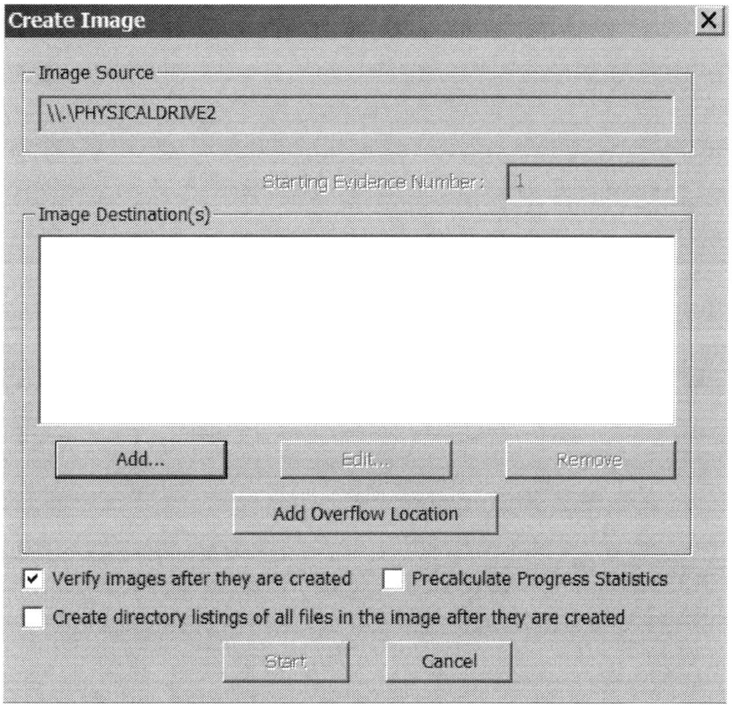

Select **Image Type>Raw (dd)>Next**

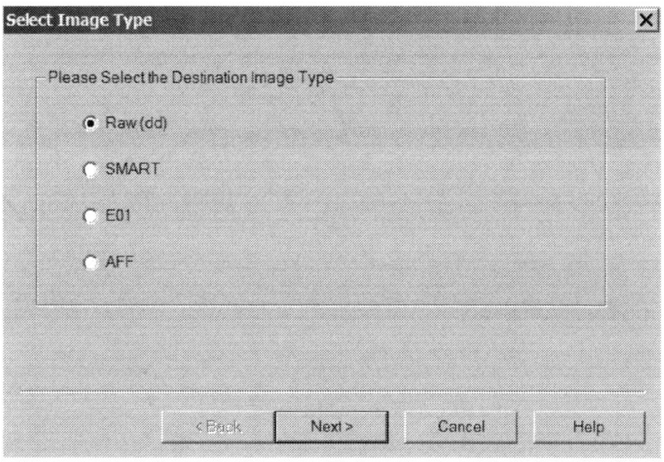

The next dialogue box brings up the **Evidence Item Information** screen. Although this information is not required to make an image of the Micro SD card it is good practice to include it. The information you enter is these fields will be incorporated into an automatically generated report.

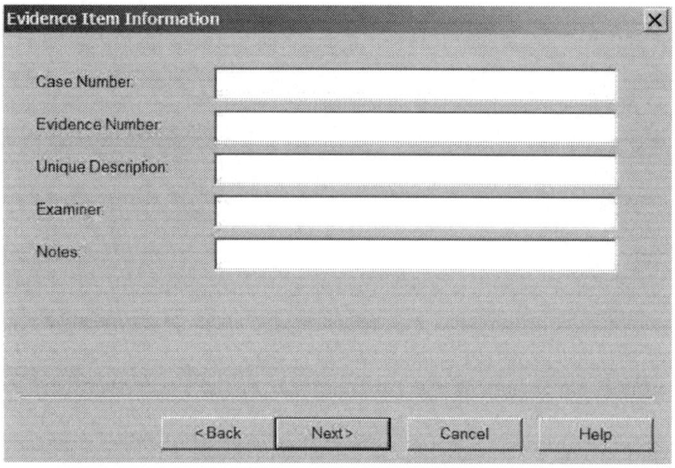

The **Select Image Destination** dialogue box asks where you want the image of the card to be written to. This should be your storage media. FTK Imager gives you options for where and how to save the image of the Micro SD card. There are options for storing the image on CD or DVD, but this can be problematic, especially when creating an image of a 16 GB Micro SD card. Use a single destination for the resulting image file such as a USB drive. If using a USB drive, you will need to have storage media with greater capacity than the target. Most USB drives do not have a significant amount of storage space allocated for the operating system of the device. If you have ever inserted a brand new 16 GB drive into a computer and found you only have 14.7 GB of available space you know what I'm talking about.

By default the Image Fragment Size is set to 1500 MB. Change the setting to 0. Once you select **Finish**, the source and destinations will be loaded.

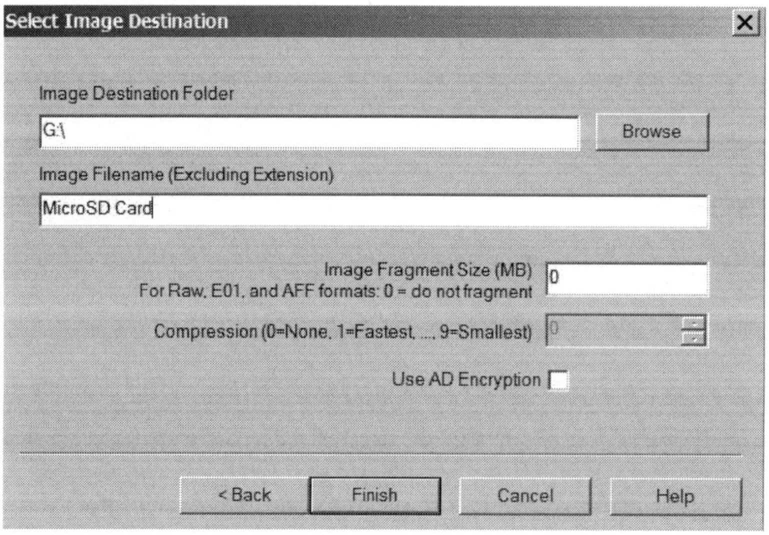

From the **Create Image** screen, select **Start** to begin the process.

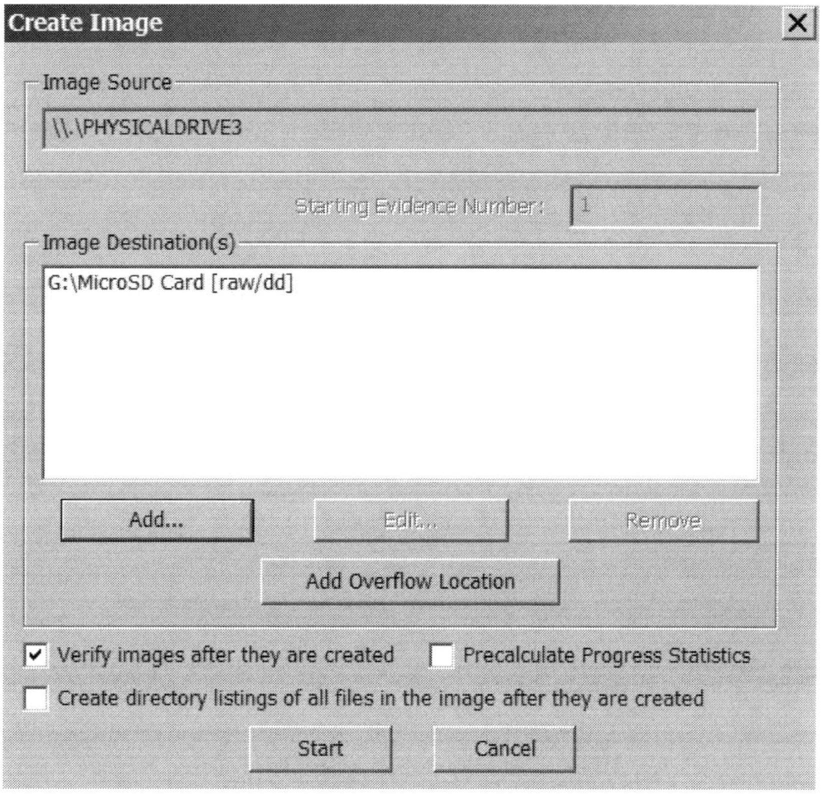

As the software creates the image you will see a status bar depicting the process. This can take a significant amount of time depending on the storage capacity of the target device. For this illustration it took slightly over nine minutes to image a four gigabyte Micro SD card.

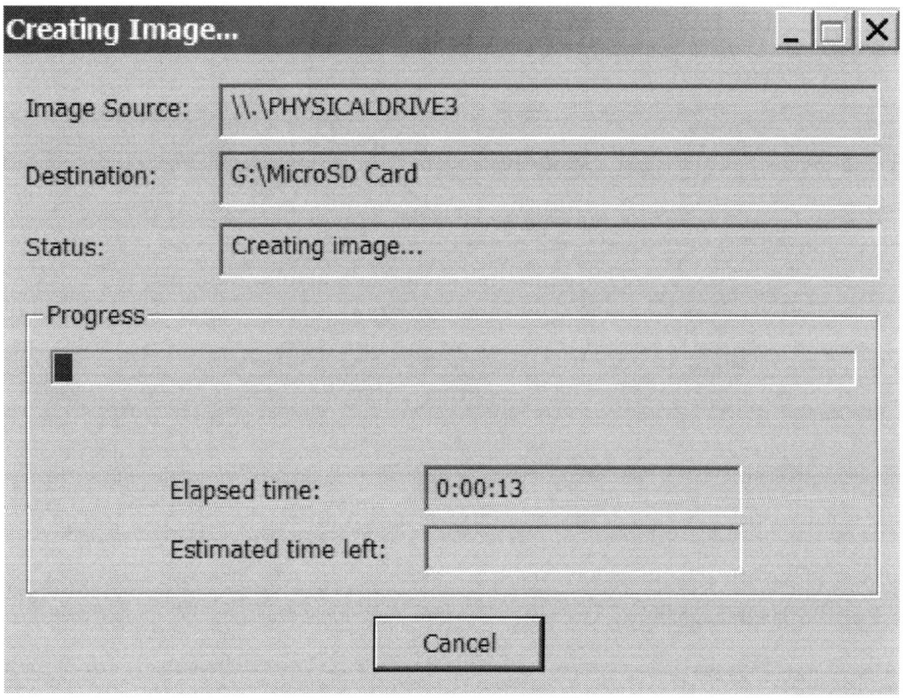

At the completion of the imaging process the files will be verified for their integrity.

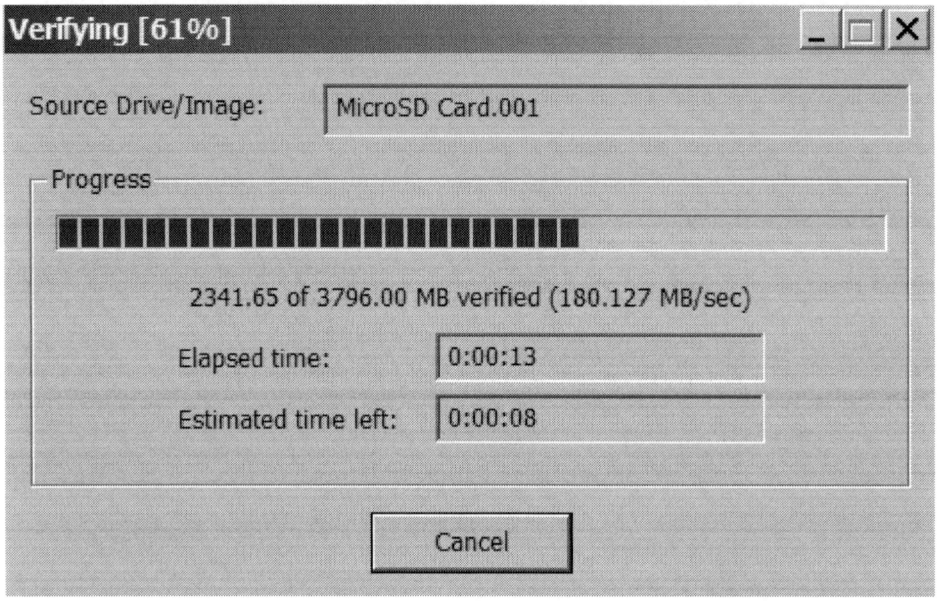

The software also gives you a status of the completion of the imaging process and the amount of time it took to complete.

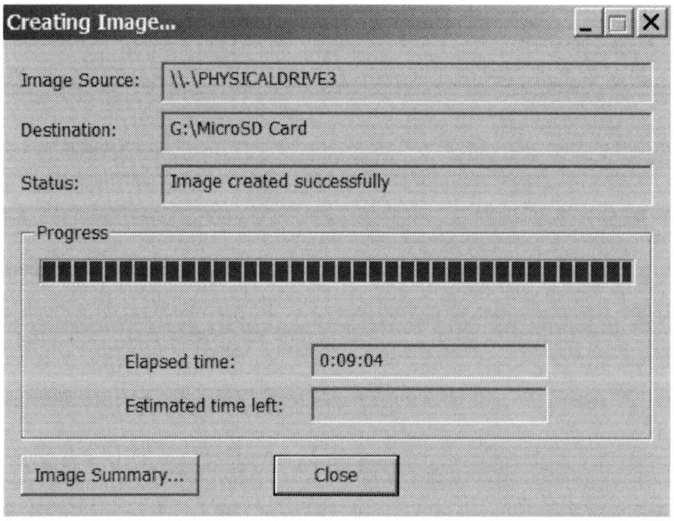

Additionally, FTK Imager displays the hash values of the image. These values are included in an automatically generated report for inclusion in your documentation.

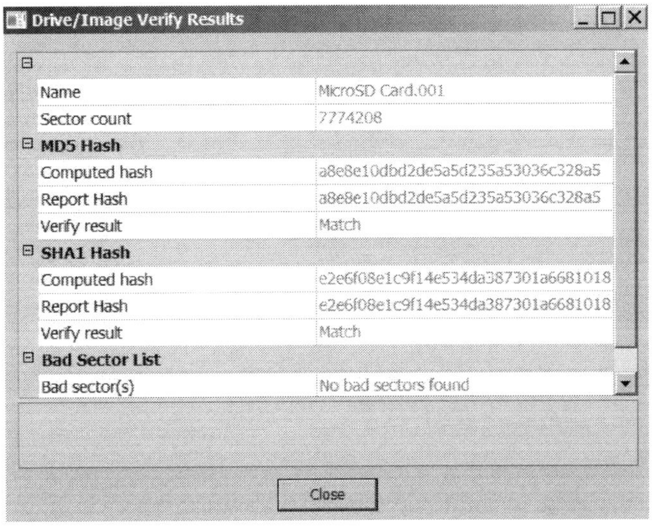

On the media where the evidence image is stored there should be two resulting files. The first is the image of the Micro SD card.

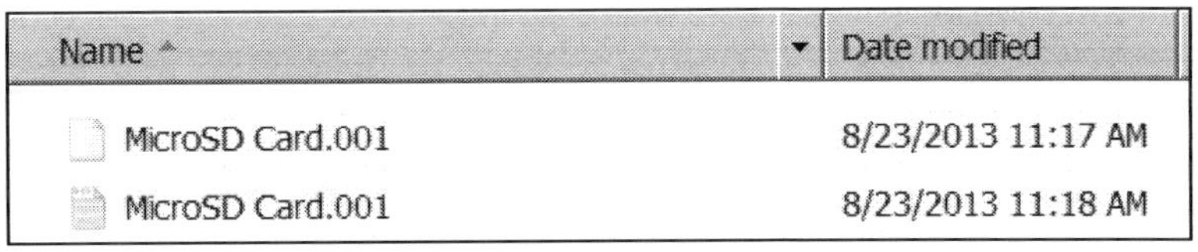

The second file is a text file containing information about the Micro SD card, the time it took the process to complete, and the hash values of the file.

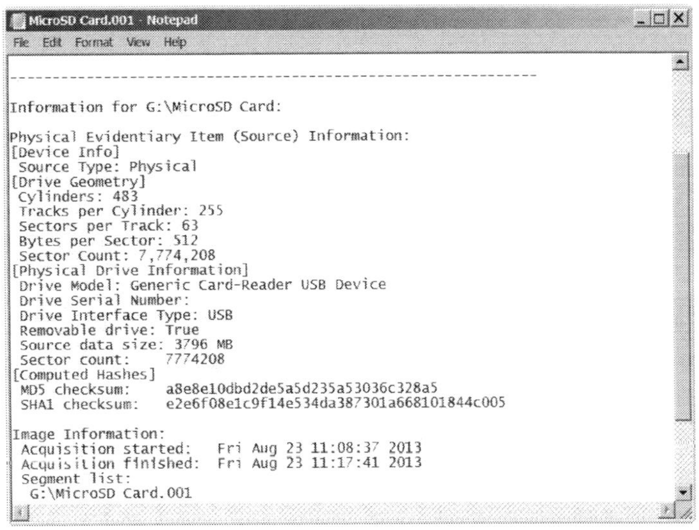

The results of the imaging process create the evidentiary file. The original duplicate image should be booked into evidence. The deleted file recovery process should be performed against a copy of the file.

The copy of the file must be mounted to the computer. This process causes the computer to view the image file as a disk drive and allows the recovery software to work against it. From the main screen of FTK Imager select **File>Image Mounting**. Navigate to the copy of the image file you created, highlight then select **Mount**.

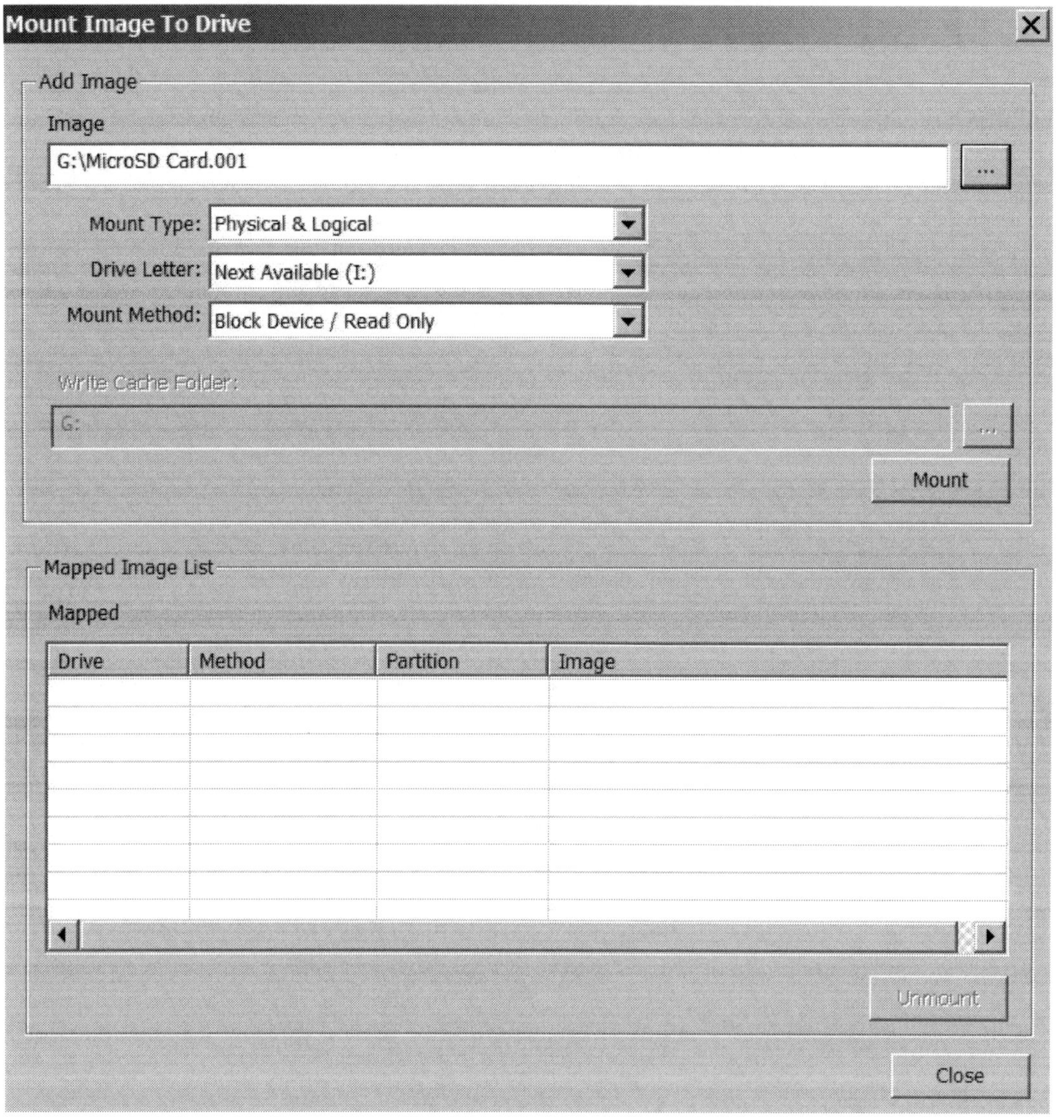

Once the image of the Micro SD card has been successfully mounted it will appear as a disk drive in the device manager.

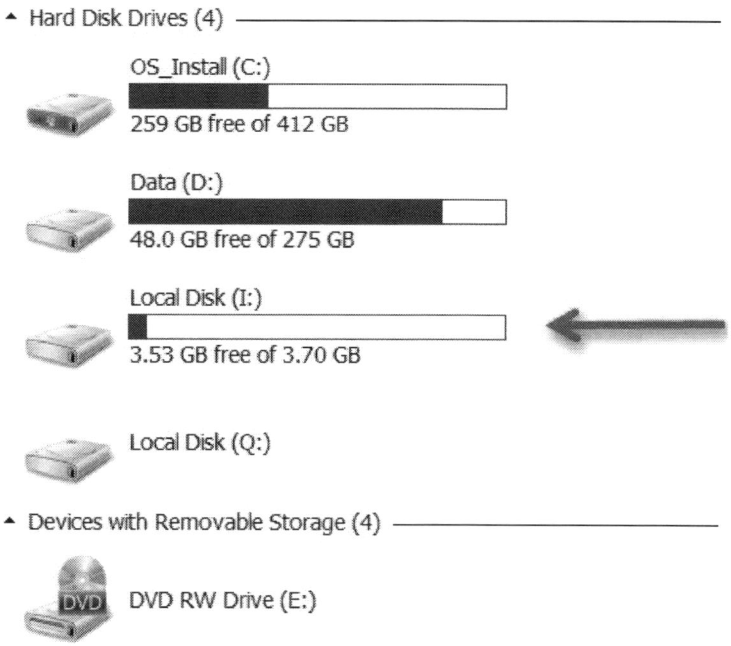

At this point you are ready to begin recovery of deleted files from the image of the Micro SD card. An important point to remember is at the conclusion of your examination you must "unmount" the image of the Micro SD card so it no longer appears as a drive to the computer. From the main screen of FTK Imager select **File**>**Image Mounting**> Select the mounted drive >**Unmount**. Also make sure to properly eject the storage media at the conclusion of the examination. Failing to properly eject the storage media can result in file corruption.

Recovery Tools

Once you have obtained a forensic image it is possible to examine the image just like any other attached media. The mounted drive will be viewed as a drive to the computer and you can now use a variety of tools to examine the image and forensically recover files, including deleted content. There are free and paid versions of software for recovering files. Not all of the tools may be able to stand a competent defense attorney's scrutiny in the courtroom. Commercial software developers are more likely to respond to requests for assistance from law enforcement. No one wants to use or purchase a tool that has been challenged in court.

One of the better tools for recovering files and deleted content is Photorec from CG Security. Photorec is a companion of the larger suite of free tools called Test Disk used for recovering information from damaged computer hard drives. These programs are distributed with several packages of forensic software tools. Photorec works with all major operating systems and is not limited to recovering image and video files like some other programs. It is capable of recovering over 320 different types of archive files, many types of multimedia files, office and document files, and other proprietary and system files. It is a read only program, so it technically could be used with a write blocker to examine a Micro SD card directly. In fact, Photorec is capable of working directly with an image file without mounting the drive. However, these are not the best forensic practices.

Photorec can appear a little intimidating to use if you are not comfortable working with a command line interface. Fortunately, the developers have made the software easy to use by defaulting to the most common commands. Photorec can be downloaded at: http://www.cgsecurity.org/wiki/TestDisk_Download.

PhotoRec is cross platform, meaning means it works on all versions of Windows operating systems, Linux, and Apple OS X (finally something for you Mac users.) It also works on computer hard disk drives, most major flash memory cards, USB thumb drives, and even iPods.

Even though PhotoRec sounds like it only recovers pictures, it actually can recover nearly 400 different file extensions such as compressed archive files, document files created by all of the major software applications, a variety of proprietary or specialized software formats, and a whole slew of multimedia files including many specialized formats used by mobile phones. In fact, if you are attempting to recover files from a Micro SD card that had been inserted into a mobile phone, this is the software you want to use. Some mobile phones store video in formats that are not detected by other recovery tools and are missed during the automated searches. Also, among the many file types supported by PhotoRec are Nokia's .vmg text message format. This means if the text messages were saved or stored on the external media inserted in the phone, they may be recoverable. An early note about unusual recovered file types: just because you recover them doesn't mean you can open or read them. You might need to download and install specialized reader or converter software to view many specialized file formats.

Downloading and Installing

The developers of PhotoRec, CG Security, had two main products. PhotoRec was one of them and was developed as a standalone software application. In 2011, CG Security stopped making it a separate program and incorporated it with its other tool, TestDisk. Both TestDisk and PhotoRec are are in the same download package and you need to be really careful to only run PhotoRec. TestDisk has a number of advanced recovery options that can permanently and irretrievably damage your computer if you use them incorrectly.

If I haven't scared you off, go to www.cgsecurity.org and download the last stable version of the software, which at the time of writing this is TestDisk 6.13. You'll see there is a version 6.14, but that is a test version so grab the one that is stable. If you are using a computer with any operating system other than Windows select it from the download option.

As you start the download you will notice the software comes in a compressed .zip file so you are going to need additional software to unpack it. If you do not already have an appropriate software program, download the 7 zip file compressor and extractor. It's free and can handle this .zip file and most other compressed files.

As with all deleted recovery programs, do not install this on the same hard disk drive or media you are attempting to recover data from. TestDisk is capable of being run from a USB thumb drive so you can download and install it there if you need to.

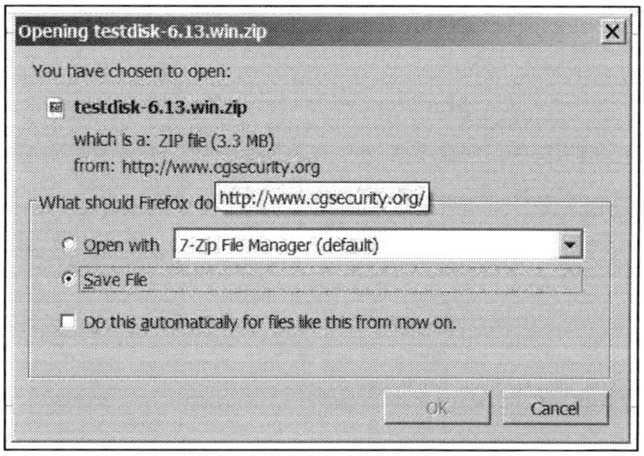

Once you have successfully unzipped and installed the program you should have a resulting file folder that says TestDisk 6.13. Click on the file to open it and view the contents. Remember, to run the program you want to click on the PhotoRec application and *not* the TestDisk application. But before you start the program there is an important step you need to take first.

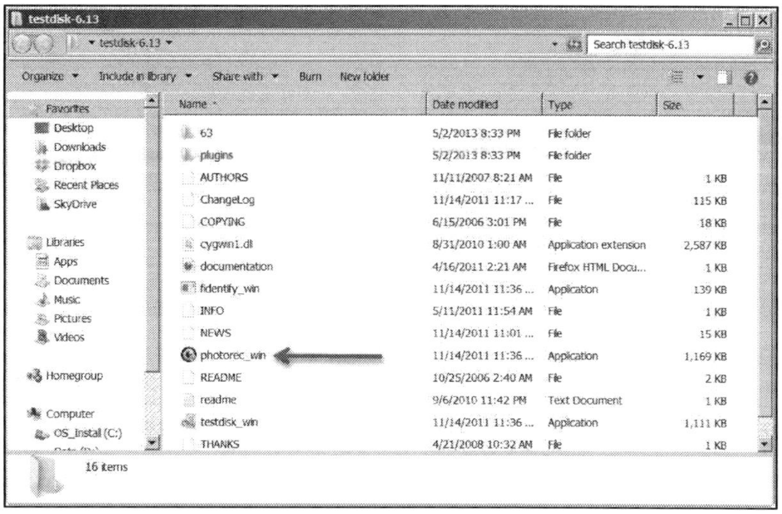

Before You Start

You see in the next couple of pages that PhotoRec can be a little challenging to navigate. You will also find PhotoRec needs a place to put the recovered files. To make this as easy as possible you need to create additional folders within the main TestDisk folder to house the data. To create these new folders **right mouse click** inside the TestDisk folder and select **New**. Scroll down to **Folder** and **left mouse click** or hit **Enter**. You need to name the file something you will remember. I use the obvious and call it "Recovered Deleted" but you can name it anything you want – just something you'll remember for a few minutes.

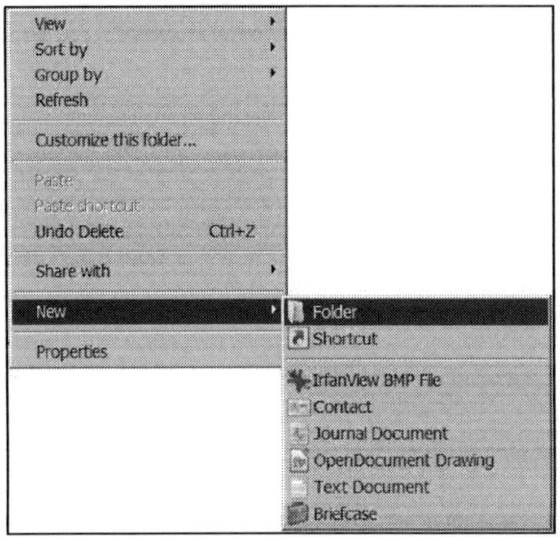

Using the Program

Now that you have created the file folder, to store the recovered data click on the PhotoRec application icon, the main TestDisk folder. Remember: PhotoRec, *not* TestDisk. Once you have launched the program you should see a screen similar to the one below. Your view may change depending on what you have plugged into the computer. If you are attempting to recover files from portable media like a Micro SD card or USB thumb drive you should see those listed on this screen. If you do not see them listed you will need to close PhotoRec, safely eject the media, re-insert it, and then restart PhotoRec. One important thing to note in the example below: all of the other recovery utilities initially recognized on My Computer's hard disk drive as having two partitions, C: and D:. In this screen PhotoRec just looks at it like one big hard drive. Even though this screen appears, this intimidating program is actually simple to use. You cannot navigate using your mouse. You must use the **up or down arrow keys** on your keyboard. Press **Enter** once you have selected the target disk drive or media.

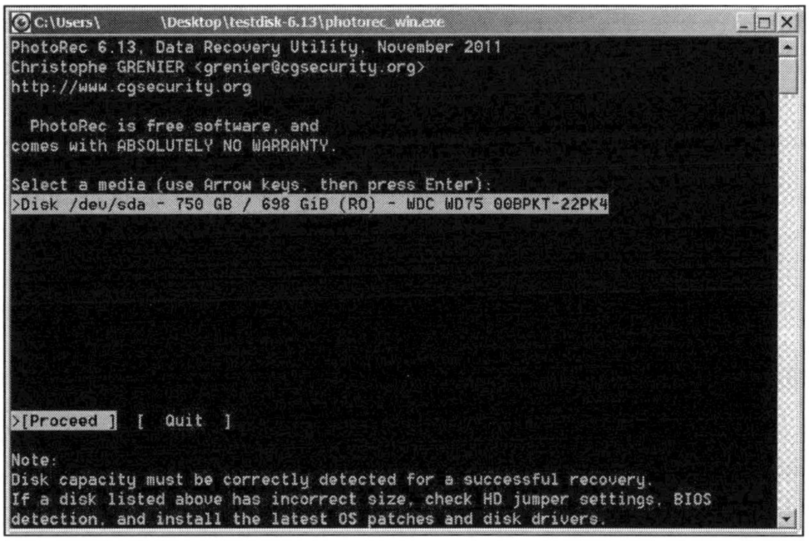

At this point PhotoRec has discovered the partitions on the hard disk drive, but by their names not by their location. From earlier examples you might remember drive C: contains the operating system (OS Install) and drive D: contains the bulk of the data (DATA.) This screen is asking you where you want to search for deleted files from. Make the appropriate selection using the **up or down arrow keys** and then press **Enter**.

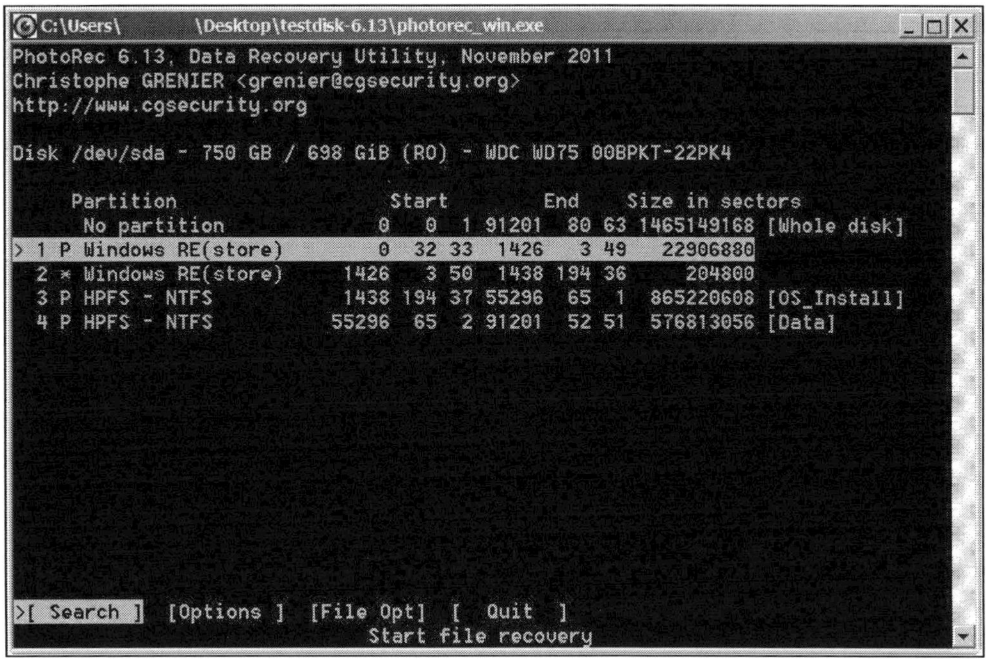

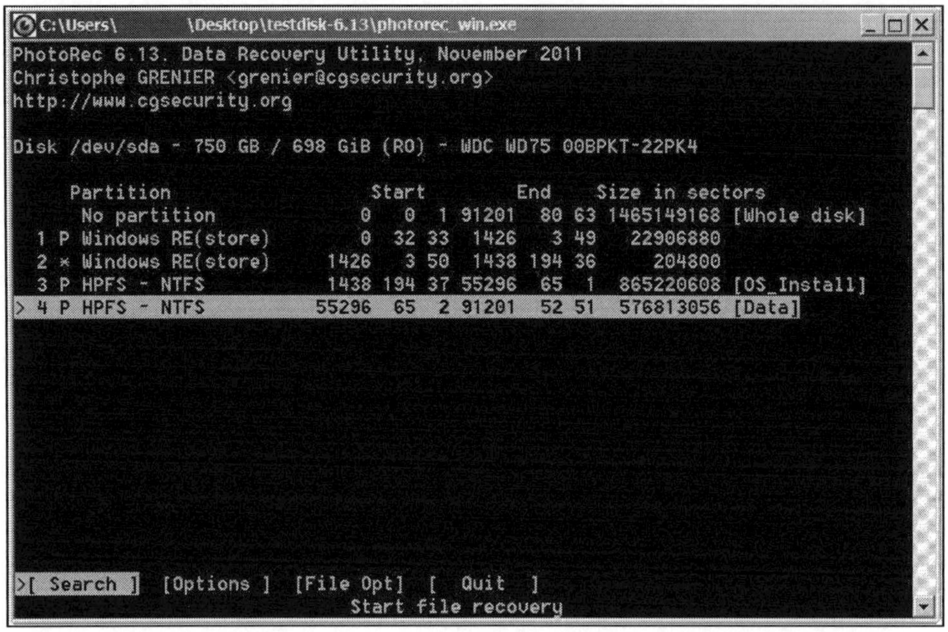

This next screen might concern you, but don't let it. By default PhotoRec should have selected **Other**. Press **Enter**.

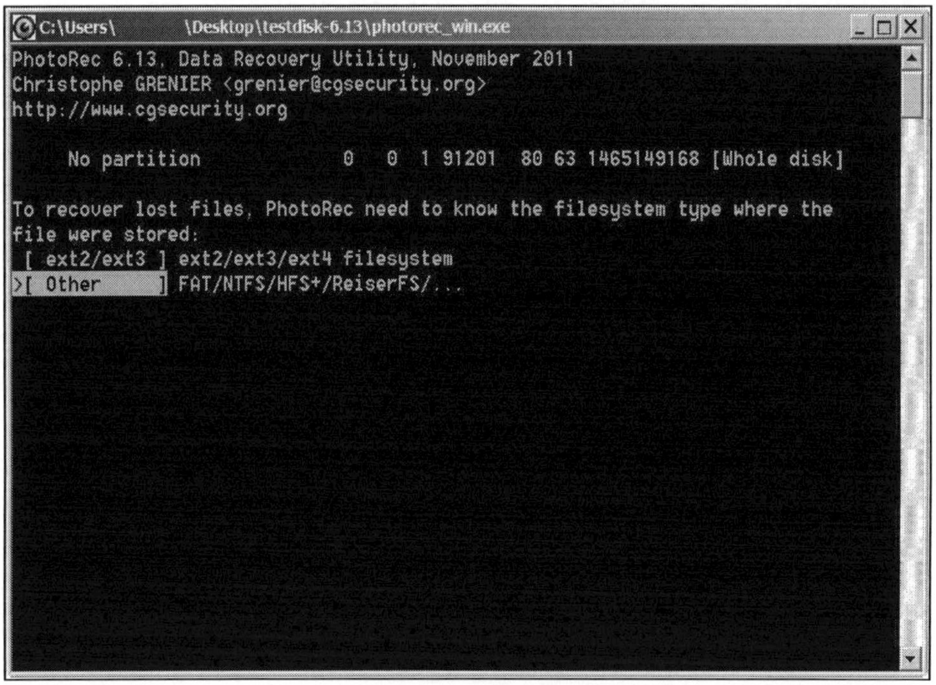

In this screen PhotoRec is asking you where you want to place the recovered files. This is why I strongly encouraged you to remember the file name you created. This interface is not great for browsing around and there is no cute little button to change the destination location. Navigate down to the file you created using the **up and down arrow keys** on your keyboard and then press the **C button** to confirm your choice.

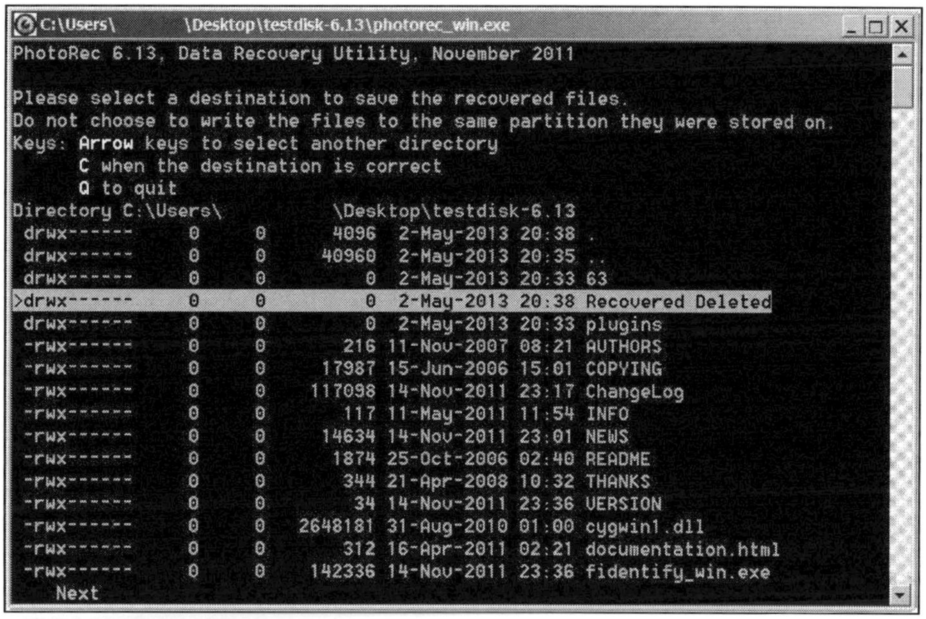

On the next screen simply press the **C button** to confirm.

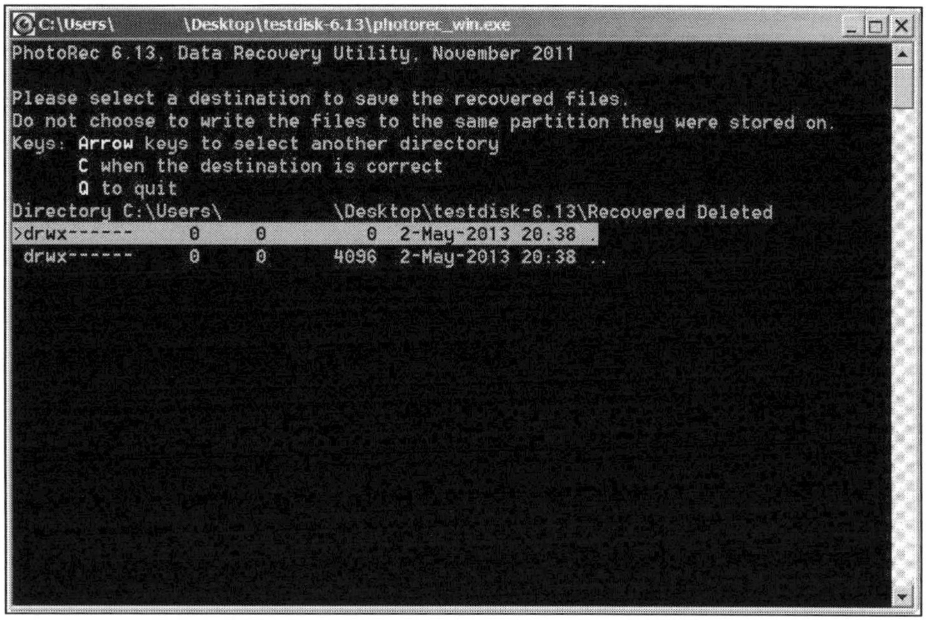

You're almost done. The last choice you have to make is if you want to recover every file from the hard disk drive or removable media or only those files that have been deleted. I'm pretty sure you don't want to recover hundreds of gigabytes of data from the hard drive, so make sure to select **Free** to scan for files in the unallocated space. Press **Enter** to start the search.

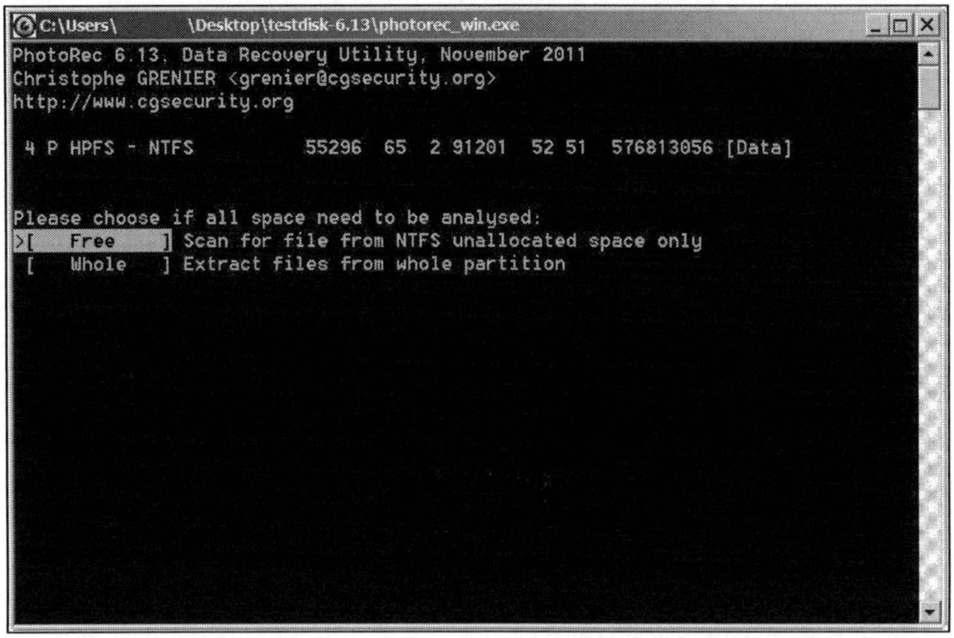

PhotoRec will now scan all of the unallocated sectors to recover deleted files. As the program progresses you'll be able to see a running count of what types of files have been recovered and how many. It will also give you an estimate for how long the process is going to take. Remember, it is only an estimate so make sure you have allotted enough time to allow PhotoRec to finish. This process is going to take a while, so if you start seeing a lot of file extensions you are unfamiliar with, now is a good time to start researching them. The easiest way to do that is to use your favorite search engine. Make sure to put a period before the extension to reduce the amount of unrelated results you have to go through on the Internet.

Reviewing the Results

As PhotoRec recovers files, it places them inside the folder you designated. Depending on how much data is recovered, PhotoRec will create sequentially numbered sub-folders containing the recovered data recup_dir.1, recup_dir.2, recup_dir.3, and so on. As the program works you can click on the file folders to see what kind of files have been recovered.

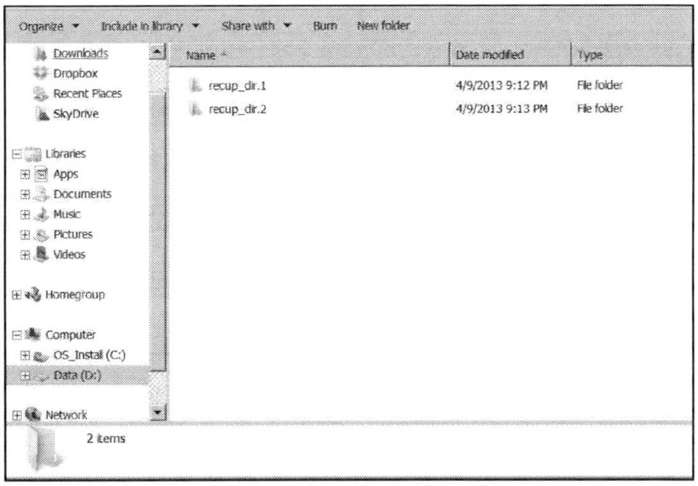

PhotoRec does not have a nice user interface like some other paid recovery programs. You will only be able to see the files' details from this screen. You will also not be able to see some of the other data, which might be displayed in some of the other programs including the file name and file path.

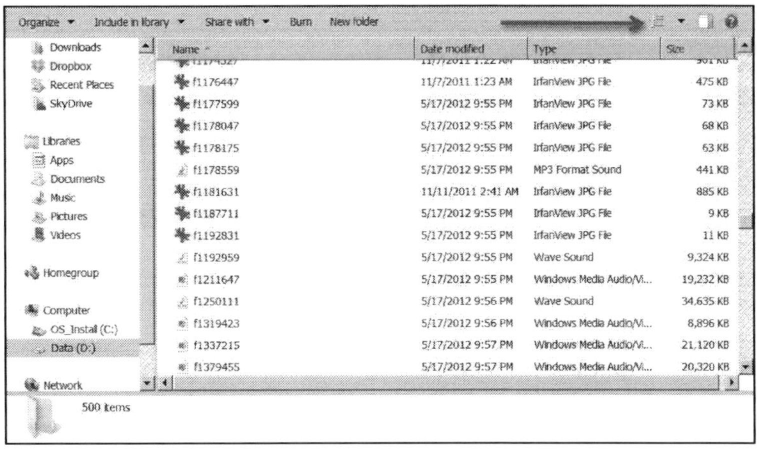

To quickly review the resulting files you might want to change the viewing option settings. Select the icon illustrated below. It may be different on various versions of operating systems, but what we are trying to get to is an icon view.

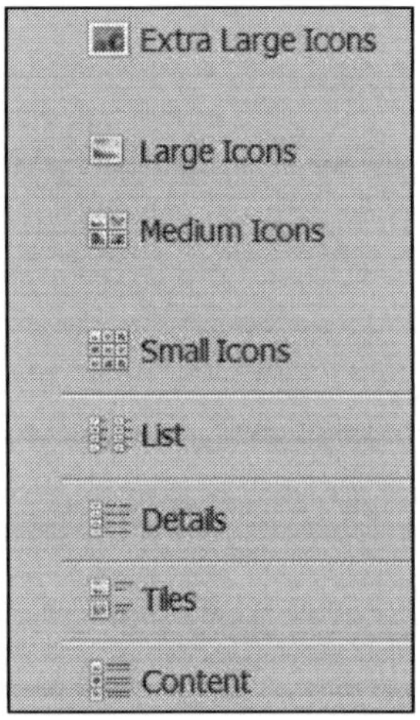

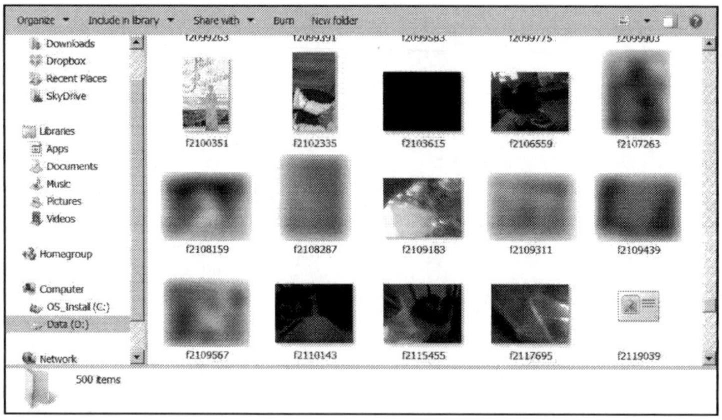

TestDisk is an extremely powerful deleted recovery utility but it does not have a lot of the (useful) bells and whistles as some of the other software. However, given the nearly 400 different types of files it can recover, and its usefulness in examining memory cards from phones, PhotoRec is definitely a program every law enforcement officer should keep in his arsenal.

Chapter 8: Using Cell Phone Forensics

There are approximately 20 different commercial tools and a variety of freeware programs used to examine mobile phones. Some tools work better than others. It is important to examine forensic software carefully before purchase. Forensic software vendors place great emphasis on the amount of devices their particular product works with. An examination of many manufacturers claims show that although a particular mobile phone may be listed as "supported," the product is capable of only retrieving the electronic phone book from the device. While this may be a useful and time saving feature, it does not help you if you are interested in evidence found in text messages or image files.

You should also be wary that vendors are international companies and support phones in other countries. Unless you are encountering a lot of foreign suspects the ability to examine phones found exclusively in Europe or the Middle East does little for you. Almost all of the major vendors have a spreadsheet that shows what data is available from a particular model of phone. This spreadsheet is not always included with the sales literature nor is it easily found on some of the companies' websites. You want to see that spreadsheet. If the vendor says it supports 6,000 models of phones, but half of their manufacturers are unknown to you, the product is not what was it is advertised to be.

It is also important to know what kind of phones you intend to examine. Departments may make a purchase of a tool that heavily supports GSM or CDMA phones without knowing what types of phones are common to their community. A software tool with strong support for GSM phones in an area with a large saturation of CDMA phones will lead to frustration when the product does not work as described.

Most mobile phone forensic vendors will allow a prospective buyer to evaluate a full or limited feature version of their software. Ask to have a trial version of the software before making a purchase. It also helps to have a sample of phones ready to test the product on. I recommend using the same test phones and evaluating a number of forensic tools before making a purchase.

Cameras

A mobile phone forensic tool is similar to any traditional hand tool in that it makes an otherwise difficult task much easier. Unfortunately, many of the currently available forensic tools require significant time and money in order to use them effectively. The digital camera is the most widely available and simplest tool to operate.

Nearly every law enforcement agency in the United States possesses digital photography capability and it requires minimal training to use. As one vendor states, taking digital images of phones is guaranteed to work on 100 percent of (non-password protected) mobile phones. There are some important considerations in using a digital camera to record evidence. Some law enforcement officers are content to record the images they believe are relevant to their case, but you should take photographs of everything in the phone. You do not always know what is going to be important to the investigation down the road.

If you only take pictures of what you believe is important, you may miss other critical evidence. Now imagine how difficult this can be when there are several hundred text messages on a mobile phone, each requiring a separate image. Additionally, recording the images may not be enough in all cases. To use the underlying data contained on the images of, for instance, a cell phone's contact list, it needs to be transcribed. Taking the pictures and transcribing the data could last hours.

Manually viewing the contents of a mobile phone is a common method to verify a forensic tool has captured all of the visible information. The step is usually one of the last in the examination process. It is possible to inadvertently manipulate or alter data on the phone, so extreme care should be taken. In cases where you need to record an unread incoming text message, this isn't possible. Opening the file to take a picture alters the underlying data, which should be addressed via thorough documentation.

Sometimes digital photography is the only way to capture the evidence. I advise using a table top vice or other tool to hold the phone in the same position while manipulating the keys. Using the vice with a small table top tripod can make the process easier and also results in consistent images of the evidence.

There are some vendors that have combined a hardware and software package to simplify the process of gathering evidence using a digital camera. One of the tools is so poorly manufactured and difficult to use I won't even mention it by name. Unfortunately it attracts some business from law enforcement due to its low price.

A better, but more expensive, alternative that I have used with success is the ZRT Mobile Device Screen Capture Tool by Fernico. Fernico actually offers two tools: the ZRT for still images and the ZRTV, which records audio and video from a device. I have not used the ZRTV so I'll limit my comments to the ZRT still capture model. ZRT consists of a high megapixel or HD camera on an articulating arm, which can be easily adjusted. The beauty of this package is in the camera system's software interface. The software warrants custom report generation and documentation of the device's characteristics.

Adjusting the camera's settings and taking photographs with the system is accomplished through the software interface. It is not necessary to manually adjust the camera. Another of the systems strong points involve the capture of the image files themselves. When taking a picture using ZRT, the photo is transferred directly to the computer and is not stored on the camera itself. Also, the software comes with an automated CD/DVD copying function. These features minimize the number of times files are moved or copied from one file to another. ZRT differs from other off the shelf digital cameras by applying two hash algorithms to the captured images and including the hash values on the automatically generated report. This can help a law enforcement officer properly document and testify to the integrity of the digital images and show they were not altered.

Cables

In the early days of mobile phone forensics, cables and software drivers were the bane of many mobile forensic examiners' existence. Every manufacturer had a different cable for its devices and those cables could change among different models even though they were made by the same company. Today, most phones have standardized the Micro USB cable, save for some notable holdouts like Apple. Apple has two different data cable models, the 30 pin and the lightning cable. The cables necessary for forensic examination are usually included with most forensic tools. I always find it helpful to have extra cables handy and the standardization of Micro USB has greatly decreased variety you used to need.

A low cost option for obtaining cell phone data cables and their drivers is to purchase Data Pilot by Susteen. Data Pilot was originally marketed to consumers to let them transfer their content from one device to another, easily edit or enter information on a phone using a computer interface, or create a backup of the phone's content. What's nice about their products is that they include an abundance of commonly used data cables and the necessary software drivers. Susteen's tools have also been utilzed to obtain data from cell phone handsets, but it is not recommended to use the consumer version for law enforcement investigations. You can use Susteen tools in conjunction with other software tools to conduct an examination and recover evidence. Its products are available on the Internet and at many consumer electronics stores. Susteen also has a law enforcement version of its software for forensic examination.

Improvised (Free) Tools

I know I am going to get some professional criticism for proposing the use of non-forensic tools and methods for obtaining data from a cell phone. Most of this criticism will be generated from computer forensic examiners who are accustomed to standardized acquisition protocol using validated tools or they are new to the phone forensics field. Before the advent of $10,000 software and hardware combinations, forensic examinations were done with whatever tool worked because there were no alternatives. The dependence on commercial tools has, in my opinion, led to a loss of the ingenuity, which was required in the fledgling stages of phone forensics. My rebuttal to those who take issue with the mere discussion of using something other than a validated forensic tool is: "What would you rather have: a phone with evidence on it that cannot be recovered using the available tools or the evidence from the phone?"

Some of the earliest tools used to assist law enforcement investigations were free and paid content management software. Cell phone manufacturers developed and distributed content management software to provide their customers with a simple interface to transfer data to and from their device. However, content management software can also recover data when the expensive forensic tools don't support a device, do not support the recovery of the evidentiary data, or fail to work at all.

Not too long ago my agency was flooded with phones taken from suspects – the least expensive phone offered by MetroPCS was the Nokia 1006. The fact the company was a prepaid provider, did not require any identifying information to open an account, and took several months to respond to court orders and search warrant, made it even more imperative to obtain intelligence and evidence from the device. Unfortunately, the major forensic software and hardware equipment makers only supported recovering the electronic phonebook and nothing else. Contacting multiple manufacturers to request support for the device resulted in exactly one company rising to the task but only after several months had passed and the provider no longer offered the model. To get the call logs, text messages, and images from the phone required content management software.

One of the best free content management tools for recovering data from a phone when other methods have failed are the content management software distributed by Nokia. Nokia supports two software programs, Nokia Ovi Suite and Nokia PC Suite. With Nokia Ovi, consumers can download applications, share reviews and maps, and manage their phone's content. Nokia PC Suite lets consumers manage their phone's information, edit and exchange files between the phone and a computer, and create backup files of the phone. As the manufacturers authorized these features, they are capable of recovering information from phones that are not supported by any of the major forensic tools. In all cases a forensic tool should be used first, but these tools are a viable option when other tools fail.

There are two main options for using Ovi Suite and Nokia PC Suite to recover evidence from a phone. With these tools it is possible to directly backup the electronic phonebook, text messages, calendar, and media files directly from the device. The second option is to create a backup file of the contents of the device and then use other software to recover those items from the backup file.

Nokia PC Suite

In the world of improvised forensic tools there are few better than Nokia PC Suite. Like other content management programs, Nokia PC Suite is designed for users to transfer content between their phone and computer. It's a shame Nokia is losing so much market share because not only does its content management program work well for doing an improvised logical extraction of data, but there is also a companion program that works extremely well.

As a reminder, content management tools are improvised tools and should not be used as a first line forensic examination tool. However, you might find the $10,000 piece of hardware or software your agency purchased is only capable of recovering a device's phone book and what you really need are the image files or text messages. In that situation make sure to appropriately document the facts and circumstances surrounding your decision to use an improvised tool.

Downloading and Installing

Locating Nokia PC Suite is as easy as entering the terms in your favorite search engine. There are other versions hosted on other file sharing websites, but I would only download from Nokia's webpage. Some of the other versions of the file sharing websites are outdated or virus infested. Installation is easy, just follow the prompts to install the program.

[NOTE: Nokia PC Suite will not work on a locked device. Once the software has been downloaded and installed double click the icon to bring up the connection wizard.]

Connection and Connectivity Issues

For most users the cable connection will suffice and it is also the preferred connection option. Nokia manufactured phone models with the infrared connection capability, but that has pretty much gone the way of the dinosaur. Bluetooth connectivity is an option, but be warned. Recovering data via Bluetooth can take a long, long time even for a download without a significant amount of data. Once you have selected the connection option Nokia PC Suite will prompt you to connect the device.

If there are any challenge using Nokia PC Suite it will be at this stage. The software comes with the drivers necessary to enable communication between the workstation and the phone for most of its current models. If you are using this software with an older model Nokia you may run into some connection issues. For this tutorial I deliberately chose a discontinued model to illustrate the driver challenge and show you how to overcome it. In this example, Nokia PC Suite is not able to connect to the device.

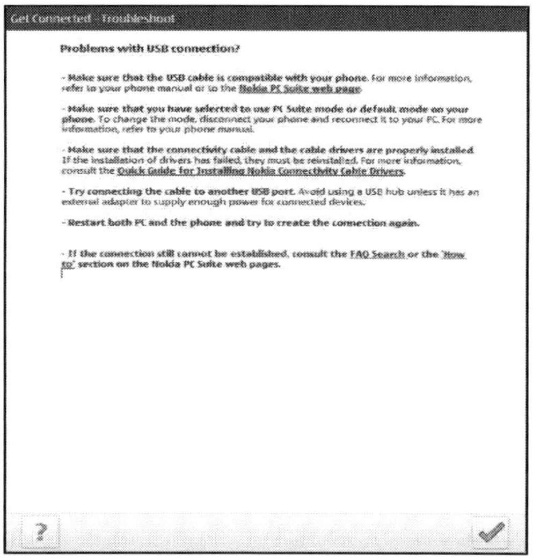
Get Connected - Troubleshoot
Problems with USB connection?
- Make sure that the USB cable is compatible with your phone. For more information, refer to your phone manual or to the Nokia PC Suite web page.
- Make sure that you have selected to use PC Suite mode or default mode on your phone. To change the mode, disconnect your phone and reconnect it to your PC. For more information, refer to your phone manual.
- Make sure that the connectivity cable and the cable drivers are properly installed. If the installation of drivers has failed, they must be reinstalled. For more information, consult the Quick Guide for Installing Nokia Connectivity Cable Drivers.
- Try connecting the cable to another USB port. Avoid using a USB hub unless it has an external adapter to supply enough power for connected devices.
- Restart both PC and the phone and try to create the connection again.
- If the connection still cannot be established, consult the FAQ Search or the 'How to' section on the Nokia PC Suite web pages.
?

Clicking on the question mark will bring up a dialog box on how to troubleshoot the connection. In this case none of the easy solutions Nokia PC Suite offered worked. The solution was to download the Nokia Connectivity Cable Drivers. It is an additional standalone download designed to enable connection with Nokia devices when the drivers are not included in the standard Nokia PC Suite download. You will know you have a good connection with the device when, after connecting it via cable to the workstation, you see the Windows dialog box below. You should also see the phone connected to the computer by checking the hardware connected to the device.

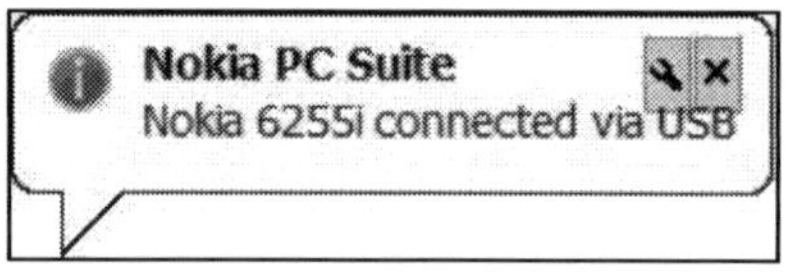

Using the Software

Once a connection has been established with the device, you should see the Nokia PC Suite interface. Hover over an icon with your mouse to see what each function is. For the most part, investigators are going to be interested in the Phonebook, Messages, Pictures, and Video files.

Phonebook

Selecting the phonebook icon will bring up an alphabetical list of the names, phone numbers, email addresses, and any other data stored in the electronic contacts file. There is an option to select and copy the entries from this screen, but I advise against it. By default the files are stored in .vcf or v card format. This enables the entries to be read and imported into the contact lists of popular email clients like Microsoft outlook. It also makes reading them limited to using one of those email programs. To avoid the potential for importing a suspect's contacts into your email client, I would recommend an alternative method for copying the contact list.

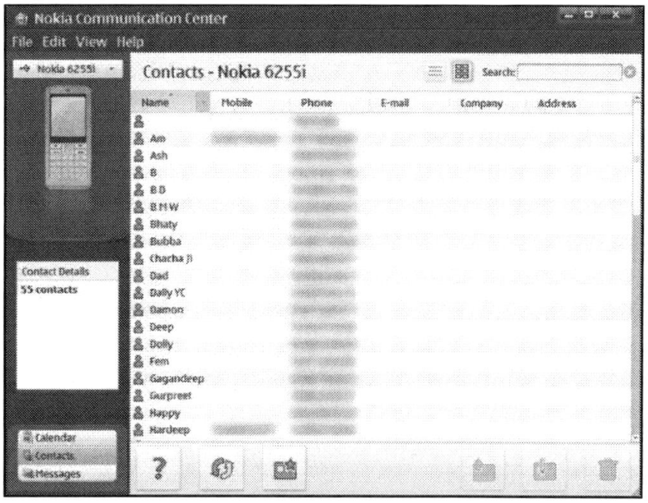

In the upper left corner of the screen is the **File** tab. Select that and scroll down to **Export**. This export function will copy all of the data in a comma separated value (.csv) file capable of being read and sorted in Microsoft Excel. Similarly, text messages can be read, copied, and exported using the same method after clicking the messages tab.

Images

Images can also be copied from the device, but there are a couple of added steps and areas to be cautious with. Select the images icon and a dialogue box will appear. To copy the image files, PC Suite needs you to designate a home for them. You can create a file somewhere easily accessible to transfer the images to. Be careful here. By default PC Suite will create a folder for each image. These folders will be created with the date of transfer and not the date they were taken. This could make sorting through the resulting image files a challenge, so change the settings on this screen to **use shooting date**.

If the setting is not changed PC Suite will erase the images from the phone once they are transferred. You don't want that, so make sure the appropriate box is checked.

Backup Files

There is additional information that can be recovered from Nokia phones using PC Suite, but to retrieve that data you need to create a backup file of the phone and then use another software program. The first step is to create the backup file using PC Suite. Click the safe icon in the upper left corner of the screen.

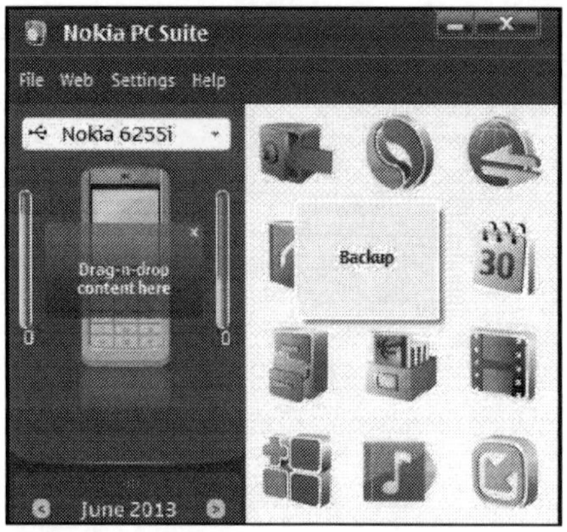

PC Suite will list the data that is going to be stored in the backup file. This data will vary by device. Once the process has finished you should be left with a file with the extension **.nbu**.

At the end of the extraction process you should delete the Nokia PC Suite from your workstation. The reason for this is pretty straightforward. The software remembers the data extracted from the device. The next time you hook up another phone, PC Suite is going to give you the option to write the older data to the new device. To prevent accidental deletion or contamination of data, and to remove possible challenges to the admissibility of evidence obtained using this tool, it is wise to start with a clean copy. Nokia PC Suite wants to run in the background unless you properly close it. The software is a resource hog and can slow down some older computers or those with minimal amounts of memory (such as those purchased by most law enforcement agencies).

NBU Explorer

Now that you have created a backup file using PC Suite it is time to examine it using another software program called NbuExplorer. The software is free and hosted on the reputable Source Forge website. Navigate there using your search engine, download, and install the software. Once you have installed the software open it, select **File>Open** and then point the program to where you stored the .nbu file created using PC Suite.

Once you have loaded the .nbu file you should see a tree view version of the file system loaded in the left side of the screen.

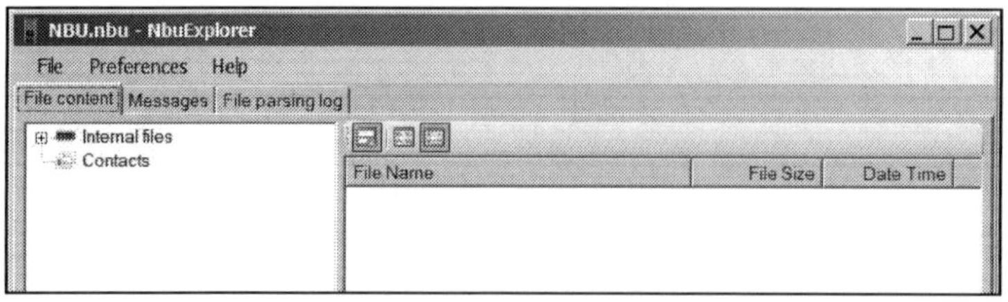

Double clicking any of the items will populate the data in the screens on the right. In this case there does not appear to be a lot of data contained in the backup file, but that is not the case.

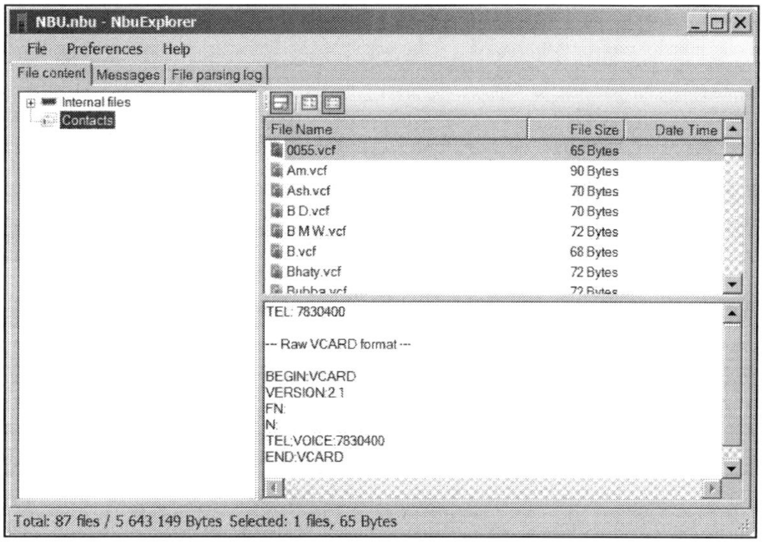

Selecting the **Messages** tab at the top of the screen will bring up another window showing the **Sent**, **Received**, and other (draft) SMS messages stored on the device. You might have noticed during the creation of the backup file that text messages were not included as a data set being copies. In this case the suspect deleted all of his text messages prior to his arrest so there was no data to be backed up. However, were there messages they would be retrievable from this screen, as well.

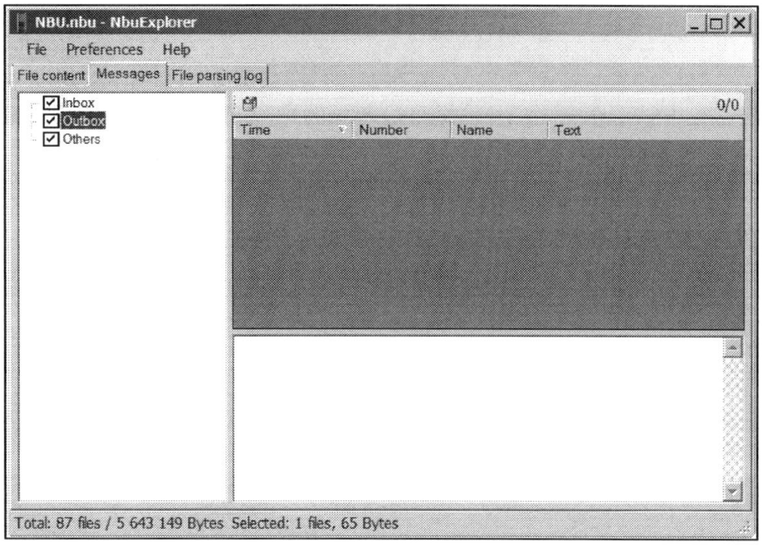

If you select **File** at the top of the screen you can direct NbuExplorer to **Export all** files contained in the backup by parsing the data. This process captures additional data not viewable using PC Suite. You can also save the parsing log to show what files the software processed and when. This can be useful information to include with your examination file to document the process.

Samsung PC Studio 7

Samsung PC Studio 7 looks and operates exactly like Nokia PC Suite – so much so it appears to same company created both programs. However, instead of being downloadable from Samsung's website it appears to only be available from third party websites, including some sketchy ones. I would recommend downloading it from softpedia.com.

Free Android Forensics Tool for Law Enforcement

In the early days of phone forensics there were not a lot of tools available to examine a mobile phone. Law enforcement officers had to use free content management programs to capture the information from a device. Over the last several years, the availability of free tools has diminished as large, well-funded companies move into the field with products costing thousands of dollars. Via Forensics went against the grain and made their entry level product available to law enforcement for free. AF Logical is a tool for collecting information from an Android device using software installed onto a Micro SD card and then inserted into the target phone. This software is capable of grabbing a tremendous amount of information. However, it is a pain to use and the user directions provided by Via Forensics leave some things to be desired. Fortunately, its product developers and technical support personnel aggressively support its free distribution.

Recoverable Data

Before we get started you should know what you are going to recover using this software program. As the name suggests, you will obtain logical information, which is what the user can see. A logical examination differs from a physical examination. Physical examinations are aimed at recovering data the user cannot see such as the file system and deleted data.

In this case, the software is capable of retrieving:

Browser Bookmarks	Contacts Settings	Internal Videos
Browser Searches	External Image Media	Maps-Friends
Calendar Attendees	External Image Thumb Media	Maps-Friends contacts
Calendar Events	External Media	Maps-Friends extra
Calendar Extended Properties	External Videos	MMS
Calendar Reminders	IM Account	MMS Parts Provider
Calendars	IM Accounts	Notes
Call Log	IM Chats	People
Calls	IM Contacts Provider	People Deleted
Contacts	IM Invitations	Phone Storage (HTC Incredible)
Contact Methods	IM Messages	Search History
Contacts Extensions	IM Providers	SMS
Contacts Groups	IM Provider Settings	Social Contracts Activities
Contacts Organizations	Internal Image Media	Internal Videos
Contacts Phones	Internal Image Thumb Media	Maps-Friends

I want you to keep this list in mind during the process. The use of this software is not always easy and you might have to extend yourself outside your comfort zone. However, the benefits are well worth the effort.

Required Equipment and Software

To get started you are going to need several things:

1. **A non-password protected device**: I would love it if this worked on a phone protected with a pattern lock or a PIN, but it doesn't. Neither does anything else for that matter, so don't get upset that a free tool won't work on a password protected phone.
2. **USB to micro USB data cable**: If you have existing forensic software you likely have the required cable. If not, go to your local consumer electronics shop and purchase one for $10.00.
3. **A "clean" micro SD card**: To make this work, we are going to install AF Logical to a Micro SD card and put it in the target phone. Preferably, the card would be new and only used for one examination. Micro SD cards can be purchased inexpensively from local retailers and online sources such as Amazon.com. You do not need high capacity cards. For the examples in this tutorial I used a four gigabyte card with no problem so you should be safe with eight gigabytes. Make sure you also have access to a Micro SD card adapter and a SD card reader.

The results of the examination are evidence so you should also probably consider using a forensic card reader to view the results. I suggest using the Ultrablock card reader from Digital Intelligence available for $80.00 at http://www.digitalintelligence.com/products/forensic_card_reader/. If you are going to recycle Micro SD cards you should consider preparing the card to receive evidence by making sure any existing files are securely deleted and the card is forensically prepared.

1. **Administrator access to your forensic workstation**: In my experience I have found this to be the most difficult thing to get as the information technology personnel from most law enforcement agencies freak out about giving lowly investigators admin access to a departmental computer. You are going to need it, so work with your agency's IT personnel to get you the access.
2. **Software drivers**: Forensic software programs may have the appropriate drivers for a phone to communicate with a computer. Depending on the model, you might need to update or add drivers. This is pretty easy to do. Use a search engine to find drivers for the make and model of your target device. I'd suggest downloading them from the manufacturer's or service provider's websites and avoid those that come from third party web sites.
3. **AF Logical**: To get this go to Via Forensic's website at https://viaforensics.com/resources/tools/android-forensics-tool/. Request access to the law enforcement version of the software by registering. Once you have access, download the file AndroindForensics.apk.
4. **The Android Software Developer's Kit (SDK)**: To communicate with the phone from the computer you will need a tool, and associated files, from the SDK. Download the entire package at http://developer.android.com/tools/index.html.

Getting Started

Once you have all of the tools assembled, it is time to begin.

1. Install the Android SDK. It would be best to install the software to the root of your C: drive. You'll see why later, but you do not want to install it to your desktop or any other location. You will also want to rename the containing file folder something you will remember. I recommend you rename it something so it will be the first file in the list of file folders. Again, this will make more sense later, but trust me for now. In the example below I labelled it ADK to keep it at the top of the alphabetized list.
2. While not expressly told to do so by the guys at Via Forensics I also placed a copy of the AF Logical file in the root of the C: drive. This was an issue when I was testing the software and it appears the file needs to be there, as well.

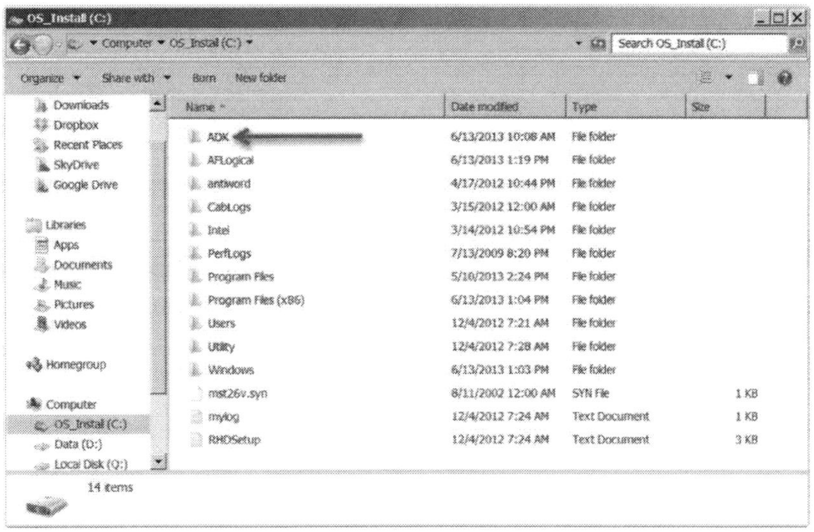

3. Make sure the Android device has USB debugging turned on. To determine this, navigate to:

Settings>Applications>Development>USB Debugging

By default the box next to this is left unchecked. You will need to check the box to proceed. You will probably get the following warning message:

Select **OK** to continue.

4. Open the **Command Prompt**. For those who are not used to this it is located in the **Accessories** folder. The **Command Prompt** sometimes freaks people out. Don't worry. Even though there are no easy buttons to push the output of the AF Logical software makes it well worthwhile.

5. Once you've opened the command prompt is will look something similar to this:

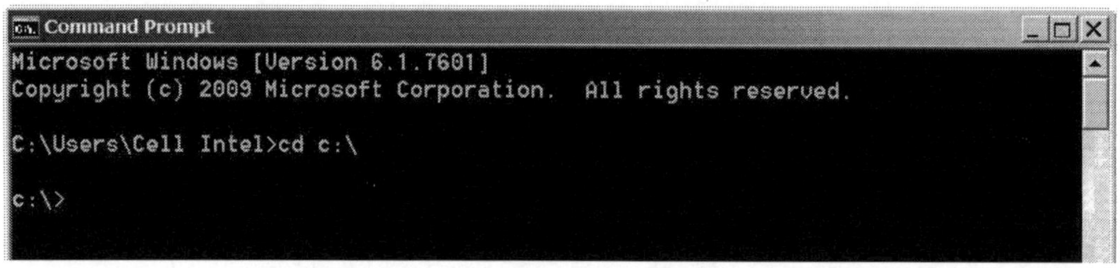

6. You need to change into the directory to where you extracted the Android SDK folder. If you followed my suggestion it should be in the root of your C: drive. If you placed it somewhere else I would suggest reinstalling the SDK at this point to the C: drive. To change the directory type the following command:

cd c:

We need to get to the folder where the files are stored. To do this you could type:

C:\ADT\adt-bundle-windows-x86-20130522\sdk\platform-tools

I find this to be a huge inconvenience because if I screw up anywhere along the line by adding a space, missing a space, or making a typo, the process will not work. I like using an alternative process that, while slower, takes out the possibility of any user error on my part.

I prefer to use the **Tab** button from my keyboard to scroll through the available options. To make use of this shortcut type **cd** and then press the **Tab** button. Once you have found your target folder just keep typing **cd** and press the tab button until the path to the file looks like the command above. Again, this might take a few extra seconds, but it beats retyping the string above after you've made a mistake.

7. Remove the target Micro SD card from the phone. Before you do so, make sure to unmount the card. If you remove the Micro SD card from the phone you can corrupt the card, making the data irretrievable without some serious forensic assistance. To properly and safely remove the Micro SD card from the device go to:

Settings>Storage>Unmount SD card

[NOTE: It may be at the bottom of the screen, so you will have to scroll down to find it. On some versions the Format SD card command is perilously close to the Unmount SD card selection. You do not want to delete all of the data on the card so don't actually fat finger the wrong button.]

8. Insert the forensic Micro SD card with the Android Forensics.apk file on it.
9. Connect the target device to your computer. Give the computer a minute or so to recognize the device attached to it.
10. Launch adb.exe. From the command prompt scroll using the tab function until you find the file and press **Enter** to launch it.

You should see several pages of text rapidly scroll by. At the end your screen should appear similar to this:

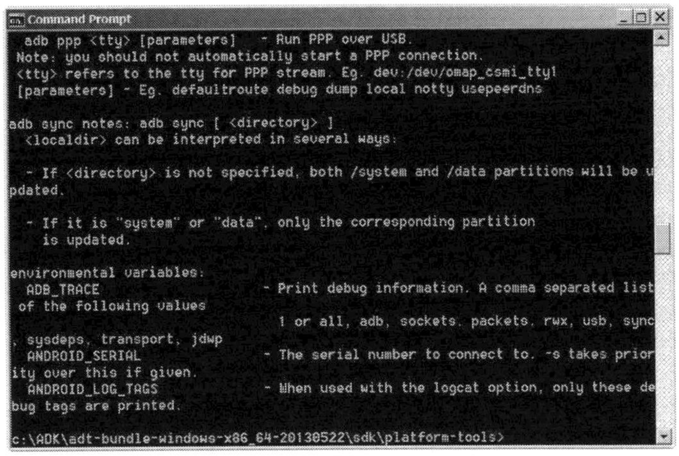

11. From the command prompt type **adb devices.** If you do not see anything under the list of devices attached there's a problem. It is likely a driver issue. Refer to the troubleshooting section at the end of this chapter.

12. From the command prompt type: **adb install AndroidForensics.apk.**

This was where I ran into one of the problems with using the software. If it wasn't installed in the same directory as the Android SDK I got an error message stating the file was not found. When I moved the AndroidForensic.apk file into the C: directory, it worked. It is probably an operator error, but I'm trying to remove all of the possible barriers to make this work for you.

13. From the command prompt type:

adb shell am start –n com.viaforensics.android/com.viaforensics.android.ForensicsActivity

[NOTE: The string here is a little confusing because of how Word formats the text. It is all one command. There is a space after the –nso. It is not two separate lines.]

14. On the device AF Logical should be running. If it is selected, **Capture** in the lower right corner of the screen. If the program is not automatically running you will have to navigate to the Micro SD card through the **Settings** menu and install it manually.

Selecting the AndroidForensics.apk file will bring up the dialogue box below. Select **Package installer** and then **Install**.

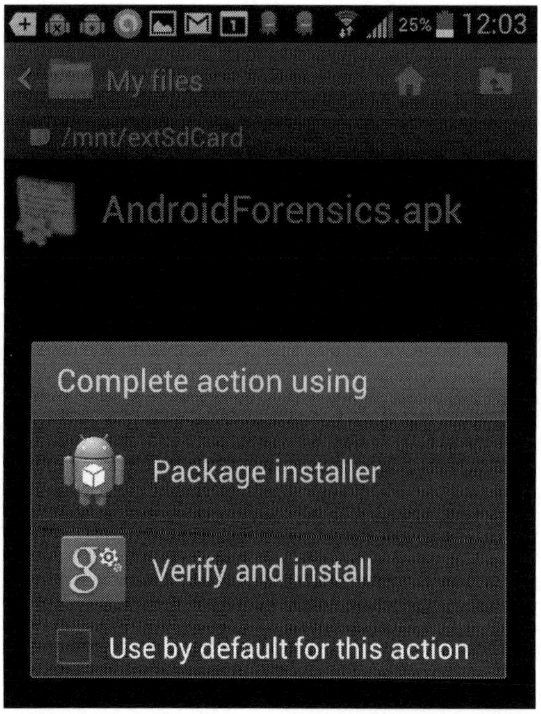

Once the application has finished installing, select **Open** at the bottom of the screen.

15. Once the process is complete go back to the command prompt and type:

adb uninstall com.viaforensics.android

16. Before you disconnect the device, make sure the files were captured on the forensic card. They should be located in a new file folder on the card labeled forensics. The files will be contained in a subfolder labeled with the date and time of the acquisition, such as 201306141204. However, in the tests I ran the files were not located on the removable Micro SD card. Some devices create a virtual Micro SD for file storage and that was the case with my test phone. You will know you are dealing with a virtual storage card when you look in the My Files icon. In this case, the **sdcard** is actually the internal storage for the device and the **extSdCard** is the external Micro SD card with AF Logical installed on it.

17. Unmount the card, remove it, and replace the original Micro SD card in the phone. You may need to selected Mount SD card to get it to work.

The Results

The examination's results were a mixed bag but overall impressive, especially compared to the results of similar examinations using forensic tools costing thousands of dollars. All of the files were placed in the folder created by the AF Logical software. If you were expecting a nice clean user interface you are going to be disappointed. The files are created in a folder and appear similar to this:

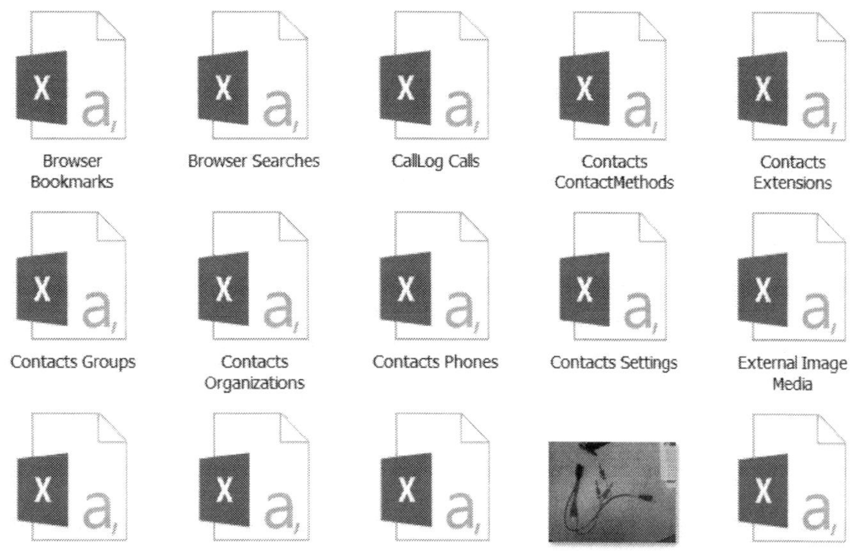

- **Images:** AF Logical recovered a fraction of the images stored on the phone. Based on the results, it appears the recovered images were those sent or received using the multimedia message service (MMS) and not any of the stored images taken with the camera of the phone. Examining the images revealed there was no EXIF or metadata associated with them in the image. The only potentially useful information was the file name, which contained some information about the date and time the picture was taken, but there was nothing else of value in the images themselves.
- **Browser Bookmarks:** Only one browser bookmark was recovered by AF Logical. The other 83 bookmarks were missed.
- **Browser Searches:** AF Logical recovered three browser searches, but there is no accompanying information about when the search was made. There were several hundred browser searches that were not found by the software.
- **Call Log Calls:** AF Logical recovered 500 calls. The data included the other number, the date, duration, type, and associated name from the phonebook. It was a little challenging to work with. The times are stored as UNIX times, so you will need to convert them using another tool. There are a variety of tools for converting UNIX time, but there's the added challenge of changing them from Coordinated Universal Time to your local time zone. I found a handy formula for doing the conversion in Excel at http://spreadsheetpage.com/index.php/tip/converting_unix_timestamps/.
- **Contact's Contact Methods:** This turned out to be contact email addresses and AF Logical recovered 231 of them.
- **Contact's Extensions:** No data contained in this file.
- **Contact's Groups:** Calling groups such as Friends, Family, Coworkers, and Google + circles. As none were enabled on the device it was not surprising no information was recovered.
- **Contact's Organizations:** This data appears to correlate to position title and company for data imported from or linked to Outlook contacts and other data sets where this information is recorded.
- **Contact's Phones:** This is probably closest to the traditional electronic phonebook information you might be used to seeing from other forensic tools. It contains the names and phone numbers stored in the contacts list. One interesting item was the times contacted field, which appears to be a summary of the records from the call logs.
- **Contact Settings:** No data was contained in this field.
- **External Image Media:** This file contained a list of 320 images stored on the external media card of the device. However, in this case what is listed as the external media card is actually the virtual media card inside the device. So these are actually the images stored in the device, not on the Micro SD card. There is a tremendous amount of data contained in this file, including the date added, date modified, date taken, and where the image is stored on the phone. The date information is also stored in UNIX time and will need to be converted, but I found a lot of useful investigative information in this file.
- **External Image Thumb Media:** A list of the image thumb information.
- **External Media:** This resulted in a list of all of the audio files stored on the device, such as ringtones. I was also able to locate downloaded voicemail files and determine their storage location using this report:

13151	/mnt/sdca	1claxon2.mp3	51879	audio/mpe	1.35E+09	0	1.35E+09
13743	/mnt/sdca	voicemail.mp3	5120	audio/mpe	1.36E+09	0	1.36E+09
13744	/mnt/sdca	voicemail-1.mp3	5120	audio/mpe	1.36E+09	0	1.36E+09
60813	/mnt/sdca	voicemail-2.mp3	56216	audio/mpe	1.37E+09	0	1.37E+09

- **External Videos:** The report listed the video files stored on the device's internal media.

_id	_data	_display_n	_size	mime_typ	date_added	date_modified
59875	/mnt/sdcard/DCIM/Camera/20130316_1	20130316_	35194945	video/mp4	1363462266	1363462266
59876	/mnt/sdcard/DCIM/Camera/20130316_1	20130316_	37456506	video/mp4	1363462390	1363462390
59953	/mnt/sdcard/DCIM/Camera/20130323_1	20130323_	17602710	video/mp4	1364065376	1364065376
59954	/mnt/sdcard/DCIM/Camera/20130323_1	20130323_	9599353	video/mp4	1364065391	1364065391
59955	/mnt/sdcard/DCIM/Camera/20130323_1	20130323_	23556117	video/mp4	1364065412	1364065412
59956	/mnt/sdcard/DCIM/Camera/20130323_1	20130323_	39365632	video/mp4	1364067100	1364067100

In this case they were all taken with the device's camera.

- **IM Accounts:** In the test device there was only one instant messaging account and it was located and recovered by AF Logical.
- **IM Chats:** There was transactional information related to instant messaging contained in this report, but I was not able to decipher what the values represented in the fields.
- **IM Contacts:** Instant messaging contacts were contained in this report.
- **IM Invitations:** As there was only minimal instant messaging activity on this device, there was no data recovered from this file.
- **IM Messages:** For what instant messaging traffic there was on the phone, the content of those messages was located in this file.
- **Internal Image Media:** The report from this file reflected some internal images from a gaming application on the test device.
- **Internal Image Thumb Media:** No data was displayed in this report.
- **Internal Videos:** As with the Internal Image Media this file reflected internal videos from gaming applications.
- **MMS:** This file contained the transactional and storage information relating to multimedia messages. It also included any text that was sent with the file.

12	24	1.34E+09	0	1	1	047_2f1737bbeb6232f9_2#L1#6664#NoSpid			appl
8	24	1.34E+09	0	1	1	034_141b368996f76437_2#L1#6664#NoSpid			appl
6	18	1.34E+09	0	1	1	028_3268	I'm innocent!!! I SWEAR!!!	106	appl

- **MMS Parts:** Other text message content associated with multimedia messages was also recovered in this file by AF Logical. The information in this file was different than the data discovered in the MMS file.
- **People:** This file was another combined list of contacts and email addresses but did not include their phone numbers.

- **SMS:** Gold. Sort of. AF Logical recovered 2,525 text messages from the target device.
- However, translating them into something useful is going to take some work. The data includes the message ID, the thread ID. This is so you can sort all messages by the conversation in which they took place, the other party's phone number, the ID number of the person associated with that phone number, the date and time in UNIX format again, the date sent, message status (read/unread), and message traffic. This feature by itself makes the challenge of using AF Logical worth it. Especially if you have ever had to take hundreds of pictures of a phone to recover the contents of text messages.
- **Social Contracts Activities:** Yes that's the file name and not a misspelling on my part. It does say "Contracts." There was no data in the report; not even column headers, so it's difficult to know what this was supposed to capture.

Troubleshooting

I had a lot of issues getting AF Logical to work. The two main ones were making sure the proper drivers were installed and then locating the data once the program ran. Locating the data was solved by checking the device and noting AF Logical wrote the recovered data to the virtual Micro SD card in the device and not physical one. To get the files I copied them from the **sdcard** file to the **extSdCard** file, removed the file, and read them from the computer.

I have a lot of software drivers installed on my workstation for phones to communicate with the computer and erroneously believed that I had the necessary drivers installed on my workstation to get AF Logical to work properly. That's what I get for assuming. In this case the test phone was a Samsung Galaxy S3, also known as the Samsung SGH-I747. When I initially connected the phone I heard the distinct sound of Windows recognizing a new device followed by the autoplay screen. However, this connection was insufficient to enable AF Logical. I had to install Samsung Kies to obtain the necessary drivers. For those who don't know, Samsung Kies was the company's attempt to come up with an iTunes-like interface for the Galaxy and other Android operating systems; it was universally proclaimed a failure after it was launched. The suite does contain the necessary drivers for the AF Logical program to work. Once Kies was installed I discovered that in addition to the indicators I find when attaching a device to a Windows computer the following notifications appeared:

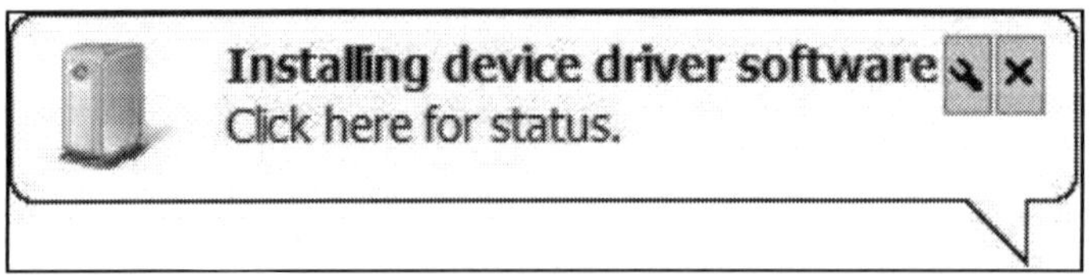

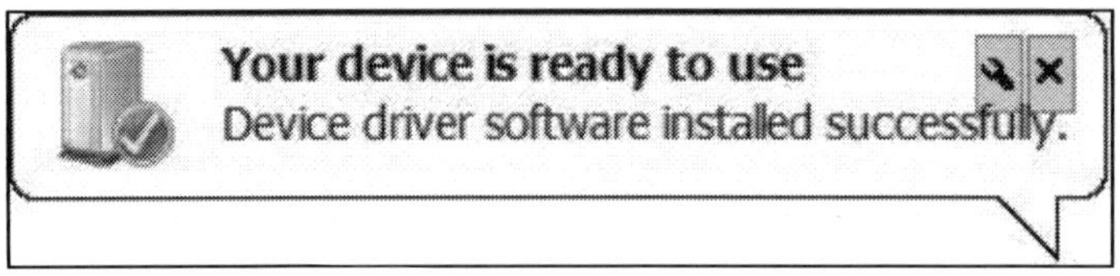

If you do not see those notifications pop-up when attaching the device you probably are not going to have any success with the software. Make sure you install drivers specific to the make and model of the phone you are going to attempt to use AF Logical on.

Issues Using the Software

There's a problem with where AF Logical writes the data. By design, the software writes to what it thinks is the removable Micro SD card, but it actually is writing the recovered files to the internal memory. This could cause some issues by overwriting deleted data and raises the questions of how forensic it is. If the software is writing new files to the phone's storage area it is altering the data. For that reason I don't believe it qualifies as a forensic tool. However, it does recover a lot of data. AF Logical works – but it does so in a non-forensic manner. This might be a problem for some, but for agencies that can't afford tens of thousands of dollars for forensic tools, this may be an option for data recovery. As long as you understand the limitations of recovering logical data only and that it may write data to the memory of some devices, AF Logical still has a place in the law enforcement toolbox.

Additional Resources

Andrew Hoog at Via Forensics literally wrote THE book on Android forensic examination techniques. You can review other methods and tools for acquiring data from an Android device by downloading his free .pdf file at https://viaforensics.com/resources/reports/android-forensics/.

Research in Motion's BlackBerry was adopted early as one of the premier devices in corporate and government environments. This was likely due, in part, to the security features included with the device. In addition to a handset security feature that encrypted the contents of the device, the company offered encrypted email and messaging services. The combination of the encrypted handset and communications made lawful intercept and forensic examination by government law enforcement and intelligence services difficult, or in some cases, impossible. In fact, because of government's inability to wiretap the encrypted communication, some countries had threatened to shut down RIM's service. The governments of India, the United Arab Emirates, and Saudi Arabia were prepared to force the closure of all RIM services if they were not allowed access to the company's servers and encryption programs.

BlackBerry mobile devices have been challenging to forensic examiners for a long time, primarily due to the encryption feature of the handset. At this time, the only methods for bypassing the handset security feature of BlackBerry devices is limited to certain national and international law enforcement and intelligence agencies.

A BlackBerry device, without the handset lock enabled, is easily examined with an improvised technique and some free or inexpensive tools. The technique uses RIM's own backup software to acquire the information.

For older BlackBerry models:

1. Download and install BlackBerry's Desktop Manager

2. Connect the device
3. At the top of the screen select the **Device** tab and then select **Back up**. Select a backup file location you will be able to find. The resulting file should have the file extension .ipd.

For newer BlackBerry devices such as the Q10 and Z10:

1. Download and install BlackBerry's Link software
2. Connect the device
3. At the top of the screen select the **Device** tab and then select **Back up**. Select a backup file location you will be able to find. The resulting file should have the file extension .ipd.

Once a backup file has been created it is possible to use several different methods to parse the backup file into readable and extractable information.

BlackBerry backups contain essential information stored in the device. User data such as email, SMS and MMS messages, Web browsing history and cache, call logs, pictures and photos, contacts, calendars, and appointments.

You might find various freeware downloads available, but they are all older, funky UI trial versions of ABC.

Magic Berry by Mena Step Innovative Solutions

Magic Berry IPD Reader is bare bones and you get what you pay for. There is not much of a user interface or any real explanation of what data is recovered. Any field with a + symbol next to the entry in the tree contains data, so you will have to scroll down the list to find the various available data.

Auto Text: One of my favorite places to look in any smartphone file system. This is the user generated custom dictionary that is created when you type something in a message field and the operating system doesn't recognize it. If you have ever tried to type a swear word in any smartphone you will most likely have encountered this. This is also a place to look for custom and slang words. There is never any contextual information, so you can't tell how it was used in a sentence, but what you can get an idea of when the word was used. You will note from the list it is in chronological order and not alphabetical. Again, there is no context or determination of when the word was used but you can get a general idea of what your target might have written to someone else by examining the **Auto Text** file.

In this .ipd backup, email addresses associated with or accessed from the device were stored in the **Service Book**. This is a great source for follow-up investigation for further subpoenas or search warrants for email providers and any messages stored on their servers.

Memo: A surprising number of criminal suspects use a smartphone's memo function in place of the old pen and paper. In this case the suspect used a memo to create a record entitled **Debts** to document how many people owed her money for drugs. This is the digital version of the old "pay/owe" sheets.

Contact information is stored in the **Address Book** function. The interface is not clean because the phone number, address, and email addresses are all stored in different tabs.

However, there is an easy way to export this into an Excel readable .csv file that will be discussed at the end of this section.

Browser bookmarks are just that. Bookmarks created for a user to easily return to a Web page without retyping the address. This is another valuable source because it can contain information about the suspect's social media sites, banking data, and other evidence and intelligence. In this example, the suspect has bookmarked the homepage of Global Tel Link, the provider of correctional institutions inmate payphone services. It would be possible to take her mobile phone number and ask Global Tel Link to run it through their systems to see which inmate or inmates from which correctional institutions were in communication with her.

Call Logs: This is standard call log information found in the phone. It is difficult to view in Magic Berry IPD Reader but it is easy to export the information and view it in Excel.

SMS Text messages can also be viewed in the .ipd backup. Double click on a message to view the direction, folder, date, time, number, and the message's content.

Viewing files in Magic Berry IPD Reader is a pain. The easiest thing to do is to select the **Manipulate** tab at the top of the screen and then scroll down to **Export**.

To use the export feature you must be in an area where there is data and click on any data field. Once you have selected a field you can choose to export all of the data in that section or all of the available data. Use the **Export All** function and Export to **CSV file**. All of the information stored in the .ipd file will be converted into an Excel readable .csv file. When viewing the .csv file be sure to scroll at the way to the right of the spreadsheet. There is often data contained in these columns that is not immediately apparent.

Magic Berry IPD Reader also gives you an option to export all of the media files contained in the .ipd file. Sort of. For this example I extracted all of the .png graphics files because there were no stored .jpg image files. They were all stored on the Micro SD card. However, even though the software said it was extracting the graphics and gave me the file path where it was extracting the files to I was unable to find them. I repeated the process several times and used a variety of search techniques, but it does not appear the software actually copies the image files out of the .ipd backup file. Magic Berry IPD Reader is a great free tool for investigating BlackBerry backup files, but be forewarned. You get what you pay for.

Berrygrab.com

Berrygrab.com is an online service that will parse your .ipd files but there are a couple of issues you should be aware of before using this service. First, I'm always hesitant to put my case information out on the Internet where I have no control over who gets it or how it's used. Second, the service is slow. Instead of immediately parsing out the data I was forced to wait for several hours for their servers to process the file. I'm not sure why it took so long because any other .ipd file parser does the job in about four seconds. Third, once you are able to download the files you have to do so one at a time. Each separate field of data has to be downloaded as it's own .csv file instead of downloading them all at once.

I was excited to see Berrygrab.com would parse the email files, but when I opened the resulting .csv file all I found was the header information and the importance level with no sender information, dates, times, message content, or attachments. Additionally, the contact's file did not contain any information other than the name of the contact and failed to provide any phone numbers, addresses, email addresses, or any other additional information. Based on my experiences I'd suggest skipping this one.

Amber BlackBerry Converter/Elcomsoft BlackBerry Extractor

Once upon a time there was a program used by forensic examiners to recover data from BlackBerry devices known as Amber BlackBerry Converter (ABC). The software was an improvised forensic tool and was extremely useful for parsing out .ipd files after connecting a device and creating the backup file. In fact, ABC recovered more information, and still does, than most of the $10,000 forensic software devices and programs out there. You can still find older versions of the program, but they are trial versions and only recover 10 items per data field.

At some point the program, or rights to the program, were purchased by Elcomsoft. If you haven't heard of them they are a software development company based out of the Russian Federation. Despite the fact any software out of the Russian Federation conjures images of virus-laden hacker tools, I'm still a huge fan of their products. While most of their software is designed to decrypt passwords they also have a number of really useful software for forensic recovery and they have done right by ABC – now known as Elcomsoft BlackBerry Extractor. Even though the software used to cost $19.95 and now costs $49,199, depending on the version, it is definitely worth it if you are doing any work with BlackBerry devices. Elcomsoft kept the original ABC format but enhanced it. Here's a screenshot showing the available data recovered from a test .ipd file:

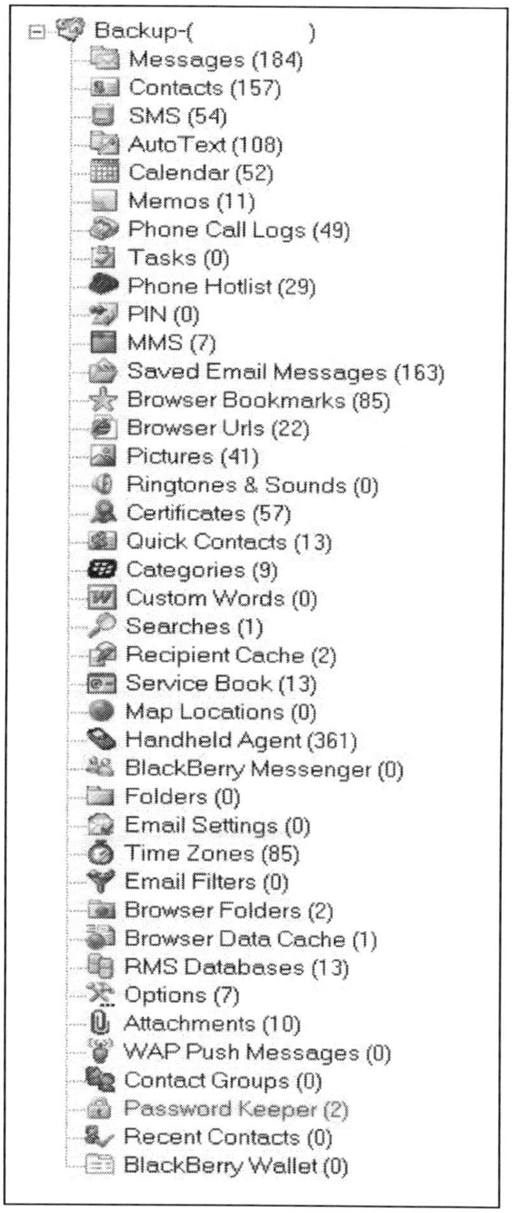

You can see that text messages, contacts, emails, calendar entries, memos and call logs are available. But additionally, multimedia messages with attachments, call groups, and Web browsing activity are recoverable depending on the sophistication of the user. As a bonus Elcomsoft BlackBerry Extractor also recovered image files from the .ipd file that were stored in the phone memory and not on the Micro SD card. There is plenty of information recovered using this product compared to any of the other available tools.

An observant reader would see the **Password Keeper** field located near the bottom of the list of recoverable data. Passwords are one of the Holy Grails of mobile forensic examinations because most people recycle theirs from one account or website to the next. Elcomsoft can recover these passwords, but you have to pay extra for the feature. Password recovery from the .ipd file using Elcomsoft BlackBerry extractor requires the purchase of the add-on Phone Password Breaker for another $199.00. Normally, I object to tiered purchases where my department is forced to pay to get the really good stuff. However, for Elcomsoft, I'll make an exception because not only does its password breaker work on BlackBerry backup files but it also works on encrypted Apple back files. You might never encounter one, but if you do, having this software in your arsenal of tools will definitely be a game changer.

One of the features that is particularly beneficial in Elcomsoft BlackBerry Extractor is the ability to export and save the data in an array of file formats including .pdf and .csv, if you elect to pay for the software. You cannot export any of the data unless you purchase a license for the software.

I need to warn you about one thing before you attempt to purchase this software. Depsite Elcomsoft's law enforcement friendly software it still operates out of the Russian Federation. Should you decide to purchase the software, or any other Elcomsoft product for that matter, you might want to warn the finance or accounts payable department. I have never seen more people freak out when a purchase appeared from Russian on an agency credit card. But Elcomsoft BlackBerry Extractor is the best tool of its kind on the market today and the hassle is well worth it.

iOS Research

There are only a couple of people out there who think for the good of the order. Jonathan Zdziarski is one of them. He was one of the first people to hack the iPhone and has made his years of research into forensics and evidence recovery freely available to the law enforcement community. I'm not even going to touch his material, but if you want a complete set of free forensics tools for Apple devices or you'd like to read his book for free you can either go to his website at iosresearch.org. His book is downloadable in pdf format at http://www.zdziarski.com/blog/wp-content/uploads/2013/05/iOS.

You are going to need a Mac to use his tools. This makes sense – they are designed to work on Apple's operating systems. But this presents a huge challenge for many law enforcement agencies who have been convinced by their IT departments they can, and must, only use Windows. The easiest way to get a Mac is to go down to your property room and find one that is pending disposal and re-purpose it for the agency's use. You will probably find the computer is several years old, but you can update the operating system for about $35.00. I've also heard of different law enforcement agencies having varying levels of success approaching their local Apple store and asking them to donate a machine for the betterment of society. The second thing you need to be aware of is Zdziarski's methods are not for the computer novice. If the extent of your cell phone forensic experience is pushing buttons on a Cellebrite, his methods are probably not for you. Let me put it another way. One of the greatest minds I know in phone forensics, a practitioner for many years and pioneer of many of the law enforcement methods in use today, called Zdziarski's class one of the toughest he had ever taken. He also pointed out that there is no room for error. Make a mistake and you've turned an iPhone into very expensive paperwork. Speaking of training, Zdziarski conducts a training class to teach people how to use his methods and tools. The class is not cheap, but it really should be considered a prerequisite for using the free software and tools. If you are comfortable with iOS and read and understand his book, you still might want to practice on a couple of phones before trying his methods.

Paid Forensic Tools

I have an issue with some manufacturers and developers of cell phone forensic equipment and software. There is misleading advertising about the number of cell phones their product supports. If the product is only capable of recovering the electronic phone book from a device and nothing else, I don't really consider that to be supported.

Unfortunately, most manufacturers consider this as "supported," but it doesn't do me any good if I need image files or text messages. My other contention with many of the commercial tools is the pricing. I understand the profits of a particular product allow for further research and development and that benefits the entire law enforcement community. At the same time I find some of the tools designated as forensic or for law enforcement are much more expensive than the consumer version of the same product. It's like SWAT equipment. Take a pair of bolt cutters, paint it black, call it a tactical lock bypass tool, and quadruple the price. Is the black paint really that expensive? On top of that you need to take a special tactical lock bypass tool class to be certified to properly use the tool, but the class' cost is not included in the inflated price. I digress.

The market for forensic tools is evolving as rapidly as cell phones and there is not one tool that works on everything. I noted at the beginning of this section the adage that if you only have a hammer, everything looks like a nail. But you need to have multiple tools to obtain evidence from mobile phones. In addition to the tools, you must receive training. Some of the tools seem attractive because they appear supervisor-proof because of limited options or buttons. While I'm not aware of any case where evidence obtained from a cell phone was disallowed because the equipment's operator had insufficient training, I believe the time will come has little or no training.

I don't know of any law enforcement agency that allows its officers to administer a preliminary alcohol screening (PAS) device test to a suspected intoxicated driver without training. How hard is it to use a PAS? Put the mouthpiece in and blow, right? However, some agencies think nothing of purchasing a mobile phone forensic tool and putting their officers out onto the streets without so much as an operating manual. Don't be the officer who gets made an example of on the stand. There are a number of free, low cost, and/or really expensive training programs out there. Get trained.

Another significant issue that should be evaluated closely along with the software is the renewal fees. The developers of forensic software programs continually update their products as new phones reach the consumer market. Software updates are usually included for the first year. However, subsequent years can require significant fees in order to continue receiving the updates. Make sure to factor the annual subscription fees into your evaluation.

The following is a survey of some of the available tools and my experiences using them. This should not be considered an all-inclusive list.

Logical Versus Physical Forensic Examination Tools

A logical forensic examination of a mobile phone results in recovery of information that is visible to the handset user. In other words if the file is visible from the handset, a logical tool may be able to acquire it. However, a logical examination does not recover deleted content from a device. Forensic tools may only perform a logical examination of a mobile phone. There are tools and techniques to perform a physical examination of a mobile phone that attempts to recover deleted files. Several vendors have developed hybrid tools that perform both logical and physical examinations.

NIST Evaluation

A law enforcement agency will evaluate a forensic tool based on the number of phones it claims to support and its price point. The amount of supported devices is not a good measurement of a forensic tool because several factors can affect this number. For example, phones that are listed as being supported but only recover one data field like the phonebook and other countries' supported models are unlikely to be examined by the investigating agency. A factor that should be considered before purchasing any tool is whether it has been evaluated by the National Institute for Standards and Technology (NIST). NIST performs neutral scientific evaluations of a variety of law enforcement related equipment, processes, and procedures.

NIST evaluations should be included in the selection procedure prior to any purchase but should not be the sole determiner in the process. In an effort to thoroughly evaluate the various forensic tools, NIST can take a long time to finish the process and publish its results. Often times a developer will have to wait months for NIST to finish an evaluation. Issues identified by NIST have already been addressed by the time its final report is complete.

NIST does not make its results easy to find, but you can access all of the reports for cell phone forensic tool evaluations at http://www.cftt.nist.gov/mobile_devices.htm.

MOBILEdit! Forensic by Compelson Labs

Compelson Labs is a European company that produces commercial content management software and has a forensic version of the program available for law enforcement. The software is capable of communicating with a mobile phone via Bluetooth, infrared (IrDA), and cable connection. Recoverable data includes the phonebook, last dialed numbers, missed calls, received calls, text messages, multimedia messages, photos, other files, detailed information of the phone, calendar entries, user generated notes, and tasks. MOBILedit! Forensic also includes SIM card analysis and can read deleted text messages from the card. Like other tools available MOBILedit! also automatically generates a report of any findings and incorporates hash verification into it. Users can export data from the software in a variety of formats for entry into databases or intelligence analysis software.

As of the writing of this manual, MOBILedit! Forensic supports almost 3,300 devices. The phones supported by MOBILEedit! Include a large number of GSM phones and phones manufactured for the international market. A review of their supported phone lists reveals that many common devices in the United States are not listed as being supported by the product. The software has a General Model acquisition feature which will try to examine a phone which is not expressly listed as being supported.

MOBILedit! is an attractive option for some law enforcement agencies due to its price: $599.00 for the software, $299.00 annual update fees, and $799.00 for the software with a cable kit, which includes USB cables, a SIM card reader, and infrared connection. It also offers a combination package of the software and cables for $1,299. I recommend careful evaluation of the types of mobile phones your agency encounters to make sure they are supported by MOBILedit!

Oxygen Forensic Suite

Oxygen Forensic is another European based software developer that strongly supports a variety of phones. Oxygen Forensic's strength is its support for phones using the Symbian operating system, Nokia S60 operating system, Sony Ericsson UIQ operating system, Windows Mobile 5/6 operating systems, and BlackBerry, Android, and Apple smartphones. Some of the types of supported phones are not as commonly found in markets in the United States, but Oxygen Forensic has unique features that make it an attractive candidate for use in law enforcement toward advanced smartphones such as the Apple iPhone series, BlackBerry, and Android operating system. Oxygen Forensic has tools for processing these devices, which are packaged in one software package. So you do not have to resort to purchasing multiple tools to analyze smartphones.

Oxygen Forensic is capable of acquiring basic phone information, contact lists with corresponding mobile, landline, fax numbers, stored address information, contact photos, calling groups, missed, incoming, and outgoing call lists, photo and video files, calendar, to-do, and memo information, and voice records.

Additionally Oxygen Forensic is capable of recovering multimedia messages with attachments, email messages with attachments, wireless data traffic and session logs, and Web browser cache and bookmark data. The software can also recover data from the SIM card and all files from any attached external storage media.

Oxygen Forensic automatically generates a report of the examination's findings. The integrity of the data is protected with a number of different hash algorithm options. It can retrieve deleted text message information from some Nokia phones using the Symbian operating system. This feature is different from recovering text messages from the SIM card as phones using the Symbian operating system do not store text message information on the SIM card. Instead text message information is found on the internal memory.

Another unique feature found in Oxygen Forensic software is its support for reading information found in iPhone backup files. Whenever an iPhone is connected to an authorized computer a backup file is created that contains logical and application information. This feature can be particularly useful when an examiner has access to the backup file but is not able to access the iPhone. Oxygen Forensic can pull data from an unencrypted backup file as if it were the phone itself.

Some of the more interesting features of Oxygen Forensic are the ability to root Android phones and recover passwords. Rooting gives an examiner access to a device's complete physical file system and leads to additional recoverable data. Oxygen Forensic recovers passwords stored in third-party applications and decrypts the password storing keychain in Apple devices.

The software supports approximately 6,800 devices. It is important to note Oxygen Forensic software does not work with password protected devices. Oxygen Forensics pricing model can be a little confusing. The basic software is $1,299 with an annual subscription cost of $599. They offer enhanced versions of the software that has many of the features listed above. Getting all of the enhanced features can drive the price up to $2,899.

Oxygen Forensic deserves careful consideration as a forensic tool for smartphone analysis. Check if you need all of the features available in the enhanced versions of the software.

Device Seizure

Device Seizure by Paraben supports over 3,000 phones, including those using Windows Mobile, BlackBerry, Symbian, Apple (iOS), and Android operating systems. Along with cell phones and personal data assistants, Device Seizure can also acquire information from GPS devices. The software works with both USB and serial cables. Depending on the device and the operating system it is able to recover the following data: phonebooks stored on the memory of the phone and/or the SIM card, call history including received, dialed, and missed calls with call dates and duration, electronic calendar and to-do lists, image files, and email.

This software has some advanced features for examining mobile phones. Device Seizure is able to examine the entire file system by performing a physical memory dump of the device. This permits an examiner to view system files, multimedia images, Java files, and deleted data.

Device Seizure automatically generates a report and file integrity is verified with two hash algorithms. Files may be exported from the device for data analysis. Paraben integrates its Point 2 Point software with Device Seizure. This feature converts GPS data points into Google Earth for easy visualization.

It is priced at $1095 for the software with a $360.00 annual subscription fee. If you opt for the cable set and radio frequency isolation bag, the total cost is $1,795. The forensic tool kit includes a SIM card reader, 16 cables, and three charger attachments. A trial version of the software can be downloaded for evaluation prior to purchase.

There have been some anecdotal reporting from the forensic community that the software or the software drivers can conflict with other forensic programs. This may require Device Seizure to be installed on a separate computer, isolated from other forensic tools.

Device Seizure by Paraben was one of the first major forensic tools in the United States. As such there were some growing pains and lingering resentment exists among some forensic examiners due to past performance and customer support issues. However, Device Seizure is a major player in the field and deserves consideration for inclusion in any forensic tool kit.

Secure View for Forensics 3 by Susteen

Susteen started marketing content management software and data cables to consumers. As law enforcement investigators began using their products to obtain information from devices, Susteen introduced a forensic version of the software. Secure View can obtain device information, call logs, contact lists, calendar information, to-do lists, pictures and wallpaper, ringtones and music as well as video and music.

Susteen provides an IrDA adapter, Bluetooth USB adapter, and data cables for use with its software. These hardware devices can be used with other forensic tools, as well.

Secure View for Forensics 3 supports more than 2,200 North American and European models from most of the major manufacturers. Smartphone support is currently limited to BlackBerry devices.

Forensic examination integrity is verified with hash verification and an audit trail that keeps track of the operator's actions with time stamping.

Secure View for Forensics 3 is also capable of acquiring data from a memory card attached to the cell phone using USB data cables. This is a nice feature because the operator does not have to resort to using an additional card reader. However, recovering deleted content may still require the user to remove the card to perform a more comprehensive examination of it.

Susteen has also forged a new trail by offering additional software modules for specific functions. SV Probe is a data analysis tool included with Secure View for Forensics 2. SV Probe is able to link data from Secure View for Forensics 3 with data acquired from other sources. The software automatically generates a timeline of call and event data, creating a link chart illustrating associations and contact frequency, searching files for key words and phone numbers, bookmarking items of interest, and identifying high frequency contact numbers.

SV Loader is another add-on module that can combine data and reports from other forensic products such as XRY by Microsystemation, the Universal Forensic Extraction Device (UFED) by Cellebrite, and Paraben's Device Seizure. With this, multiple reports are combined into one for analysis and presentation.

It may be necessary to use multiple tools to obtain a full examination of a mobile phone. This can be confusing when information is contained in the automatically generated report from one tool but not in any other report. SV Loader is one choice but there is a cheaper option. Mobile Forensics Inc. (makers of the free BitPim Cleaner and the Port Scrubber) has also developed a report compiler called DNCompiler. This software combines the exported data from different forensic and non-forensic tools into a single report. There is a free trial version of the software and it is economically priced at $99.00.

Perhaps one of the most intriguing add-on modules from Susteen is svPIN. It is an unlock tool for reading the handset security code from CDMA handsets. Susteen advertises svPIN works on all CDMA phones, not just those supported by their forensic tools. svPIN is available as an add-on module or as a stand-alone utility. There are a variety of other techniques for retrieving, removing, or bypassing the handset security locks on mobile phones, but this may be an option for a law enforcement agency looking for an easy tool.

Susteen will allow an interested party to download a trial version of the software from its website, but you need to obtain the hardware cables from an alternative source.

Secure View for Forensics 3 costs are $2120 for a one-year license and $2715 for a two-year license. Annual renewal is $800.00 without cables and $950.00 with cable support. The additional modules cost $500.00 for svProbe and $350.00 for svPIN, if you have the main software program installed, or $550.00 as a stand-alone.

Universal Forensic Extraction Device (UFED) by Cellebrite

Cellebrite's Universal Forensic Extraction Device (UFED) is one of the most popular tools on the market today. The UFED started life as the Universal Memory Exchanger 36 (UME-36) and was designed for mobile phone retailers to transfer a customer's content from an older phone to a newly purchased one. Law enforcement quickly adopted the tool for use in recovering data from a suspect's handsets.

One of Cellebrite's claims to fame was its support for devices prior to the phones being sold at retailers. As Cellebrite's products were used in the commercial environment it was able to establish relationships with cell phone manufacturers and service providers. The manufacturers and service providers, facilitating customers upgrading to newer products, allowed Cellebrite's advanced information on the devices and operating systems, which allowed them to configure their tool to work with the newer phones. Cellebrite advertises it works with over 140 different manufacturers and service providers including Verizon Wireless, AT&T, Sprint/Nextel, T-Mobile, Rogers Wireless (Canada), Orange (France) and Telstra

(Australia). Other forensic tool developers are forced to reverse engineer a device or operating system to effectively program their products to work with new devices.

The early UME-36, and later the UFED, had several features that made it extremely attractive to use. The first is the design of the unit itself. The UFED is a standalone device, making it portable and able to be used in the field. Extracted data can be stored to a SD card or USB drive for later evaluation. This differs from other software based tools that require installation on a laptop or desktop computer. The design and user interface are simple. The software menus are maneuverable and let the user select which fields he wants to examine.

The UFED supports serial and USB cables, infrared, and Bluetooth. The device also has an integrated SIM card reader. One of the major benefits of Cellebrites product is it comes with a complete cable kit consisting of the listed device interfaces and over 100 individual cables. Many of these cables support older or international versions of phones, but it saves money and time to purchase all of these cables at once. Not all of the cables are capable of charging the device they work with, though. While it is possible to charge phones from the Cellebrite data cables, you still may be forced to obtain a charging cable for the device. This is not as much of an issue with newer phones using the microUSB interface but can be an issue with older phones.

Cellebrite's cables can be used with other software tools. Many of the cables have a RJ45 modular connection, which appears similar to a landline phone cable connector. It is possible to use a RJ45 to USB connector to use these cables with a computer and other forensic tools.

Cellebrite says its product "extracts vital information from 95 percent of all cellular phones on the market today" and supports close to 6,000 models of phones. The exact content of this "vital information" changes from model to model and may not always have the information the investigator desires. Depending on the phone's model and service provider, the UFED can extract the following data fields: device information, contacts, text messages, deleted text messages from some SIM cards, call logs with received, dialed, and missed calls, audio files, video files, and picture and image files.

Cellebrite has developed the Physical Analyzer, designed to recover file system data such as deleted content. The Physical Analyzer is also capable of retrieving chat files, email, Internet bookmarks, Internet browsing history, Bluetooth pairings, GPS information, and other files. Physical Analyzer has advanced support for the iPhone series and obtains logical information along with location based data, Skype contacts, calls, chats, and Facebook contacts. The Physical Analyzer can retrieve information from unencrypted iPhone backup files, as well.

Cellebrite has introduced its ruggedized tablet-like UFED Touch. Anecdotal reporting from some forensic practitioners question the utility of the device – given the price point of over $10,000. Not many law enforcement agencies are going to take their expensive tool out into the field no matter how ruggedized it is.

Not only is the price an entry barrier for many law enforcement agencies, but the required level of financial commitment can lead an agency to put all of its eggs in one basket. I may not make a lot of friends at Cellebrite for this, but I have found the law enforcement community relies too much on its product. Don't get me wrong, Cellebrite's product is one of the best on the market. In fact, I liked it so much I purchased a unit with my own money. That being said, law enforcement agencies might believe that once they obtain a UFED all of their problems are over. This reliance on a single tool can cause an agency to be too confident in its ability to examine devices. Despite all of its wonderful features, the UFED is not the only tool you will ever need to do complete mobile device forensics.

There is also a lack of training on the proper use of the device. While the device is intuitive, not all of the features of the UFED are obvious or self-explanatory. To effectively use this piece of equipment and get the maximum value, obtain training prior to deploying it.

Profileforcing

While Cellebrite advertises support for nearly 6,000 devices, the desired information may not always be supported for the examined device. It is possible to "trick" the machine into acquiring information from a device, though.

Once you have acquired the information available from the supported module, check the surrounding model numbers to see if the any additional information is available. It is common to find the supported model does not allow for obtaining text messages or image files but a similar model has support for those files. If it is, attempt another acquisition using the other model number. The key to making this work is to only select the other information and to avoid selecting data fields that were already acquired. You will also have to explain the process with thorough documentation as the automatically generated report will reflect the information was acquired from a different model number.

The UFED includes a report management software where user can enter some limited case information and the remainder of the report is automatically generated. Hash verification is calculated for acquired information. Much of the data is exportable in common formats for analysis and database entry.

Cellebrite does not currently send out evaluation units but will refer interested parties to other users for references. Cellebrite's tools tend to run toward the higher end of the cost spectrum and you will have to contact Cellebrite for exact pricing information.

XRY by Micro Systemation

XRY by Microsystemation was once a less well known developer in the mobile forensics field but has made some significant software improvements that make it an upcoming player. XRY has three components: a standalone logical acquisition model, a physical acquisition mode, and a combination package that includes both. The software was designed for logical acquisitions but also provide the tools needed to perform more advanced physical data acquisitions.

XRY Logical supports over 3,000 different models, including a large number of GSM phones. Microsystemation has included some recent updates with stronger support for CDMA phones. XRY Logical is able to recover device information, electronic contact information, sent, received, and archives text messages, image files, calendar information, sound files, and SIM card data.

XRY comes with a distinct communications hub where operators connect cables, hardware dongles, and access infrared and Bluetooth connectors from one location. This is helpful as all of these options are accessed from one central location that does not tie up all of the computer's ports. XRY also has approximately 50 data cables, covering the most common phone models.

With the software, you can customize the logotype and agency information of the report template. The automatically generated version of the report can be forwarded electronically so others can view and analyze the data. XRY includes a free reader program for those forwarded reports.

XRY generates an encrypted file, making it impossible for unauthorized persons to view the information. XRY's file encryption is not included with other forensic tools. The individual file information can be hash verified, but the entire recovery file is not. This can assist in ensuring the acquired electronic evidence's integrity.

XRY also has a physical module for data acquisition from mobile phones and memory cards which can recover deleted information. XRY Physical is designed for an advanced forensic examiner. A common advanced forensic practice is to use non-forensic phone tools to acquire a memory dump and then manually examine the resulting files for evidence. XRY Physical is an alternative to these techniques and avoids compromising the evidence.

In my experience Microsystemation is more responsive than some other software vendors when it comes to engineering support for specific models of phones. When I was inundated with a model from a metropolitan provider, I found the existing tools only supported recovering the phonebook from the device. When asked, Microsystemation added support for the device within several weeks, creating a more comprehensive examination and evidence collection.

Microsystemation permits prospective buyers to test a fully functional version of the software and hardware interface. Contact the company to arrange an evaluation. Microsystemation invites interested parties to contact it, or a third party vendor, for pricing information.

FINALmobile Forensic by FinalData

FINALmobile Forensic is a relative newcomer to the forensic market. They advertise strong support for CDMA phones, setting it apart from European based software developers that have stronger support for GSM phones. In certain markets, CDMA phones are much more prevalent in the community than GSM models due to the marketing efforts of low-cost service providers. As a relatively new product, there is not a lot of anecdotal information from the forensic community regarding the software.

FINALmobile Forensic is able to acquire logical files from most major CMDA manufacturers Audiovoxx, LG, Motorola, Sanyo, and Samsung using the device manufacturers' standard cable and driver. FINALmobile Forensic can acquire physical memory dumps from some Motorola and LG models. Comparing FINALmobile to other software is difficult because the programmers designed the software to examine mobile phones. The software developers target specific chip sets of a mobile phone manufacturer.

Without comprehensive testing of every phone on the market it is challenging for them to create an accurate tally of devices the software will work with. However, the concept of targeting the phone's memory chip means the software will likely work with devices that use that particular chip. Instead of developing a software acquisition for every phone model, the software is able to read the underlying memory chip, regardless of cell service provider or model number variations. For example, the older Motorola Razr V3 series has numerous variations. Different forensic software applications were able to retrieve SMS text messages from the V3c but not the V3x. FINALmobile's product works universally on the chip, so no matter what variation of the phone you are examining, as long as the memory chip is the same, the software will support it. Reading the chip is accomplished through standard USB cables and not some of the more unique cables required by other operating systems.

When FINALmobile is able to acquire a physical memory dump of the device it is often able to either recover the four-digit handset security code or to read the memory by bypassing the security feature. Reading the raw, or physical memory, increases the likelihood of retrieving deleted data from a device. While the company does not specify exactly what deleted data is available from which specific device, the ability to recover deleted content, without advanced training, is a new innovation in the field. The company's focus on reading the chip separates FINALmobile from other cell phone forensic programs.

As with many other forensic software programs, FINALmobile is able to obtain many different types of data including text messages, multimedia messages, video files, image files, contact lists, user generated memos, and Internet connection logs. The software automatically creates reports, allows for data export for use with analytical tools, and generates hash codes for acquired data. FINALmobile is also able to acquire data from external storage media like Micro SD cards. FINALmobile has support for BlackBerry phones, as well as the iPhone, and those using the Windows Mobile operating system. FINALmobile can perform a physical memory dump of many devices.

FINALmobile is focused on acquisitions of CDMA phones. This is good news for investigators who encounter large numbers of phones using this technology. Several of the popular prepaid providers utilize CDMA technology, which may make FINALmobile a very useful tool.

FINALmobile is not the least expensive tool on the market. The initial cost is $2,995 for law enforcement and annual subscription rates are $599.

Cell Phone Analyzer by BK Forensics

I have a personal bias toward Cell Phone Analyzer by BK Forensics. The software was one of the first commercial tools to leverage existing hardware and software tools for repairing mobile phones and make it available to the law enforcement community. By doing so, they created a product that allowed for physical data dumps of phones without the inherent danger in using non-forensic tools.

Cell Phone Analyzer has a hardware intermediary to access a mobile device's programming or diagnostic mode. Previously, this was primarily used by repair technicians who were attempting to restore, recover, or replace the operating system of a phone or by hackers seeking to modify their phones. The intermediary hardware devices are commonly referred to as flasher boxes. "Flashing" a phone refers to re-writing the operating system of a particular device and flasher boxes are the generic term used to encompass many of these devices. Flasher boxes are also referred to by the name of one of the more popular devices, a Twister box. There are many different types of flasher boxes available. Most of them are brand specific such as the Smart-Clip, which works on Motorola phones. The devices could be generically targeted to CDMA or GSM phones.

When attached to a flasher box, the phone is receiving power but does not operate as a typical device when it is powered on. When the flasher box accesses the programming or diagnostic mode, it does so independently of the security lock and data can still be recovered. Similar to some other tools like BitPim, it is possible to use the hardware and software combination to obtain the handset security device and then examine the mobile phone with other logical acquisition tools. However, these logical tools do not usually recover deleted content from the device.

Cell Phone Analyzer reads the download generated by the flasher box to recover logical and physical memory files and create an easily interpreted report. The depth of a physical recovery can be astounding. In one case involving a very low-end Nokia device, I was able to not only recover all of the user's deleted call logs and text messages, but also the prior owner's deleted text messages. Similarly, with a BlackBerry device I was able to recover approximately 600 deleted emails that were still stored in the memory of the device.

Cell Phone Analyzer is stronger with certain brands of phones – specifically Nokia, Sony Ericsson, Samsung, and BlackBerry. Depending on the make and model of the device, Cell Phone Analyzer can retrieve the unique identifying information of the device, handset security lock codes, contact lists with names and phone numbers, missed, dialed, and received calls logs with dates and times, SMS text messages, image files, banner messages, email addresses and messages.

The strength of Cell Phone Analyzer is also one of its weaknesses. Specific training on using the flasher box intermediary is highly recommended before use. These tools are capable of completely rewriting the device's operating system or permanently altering and deleting data. I also speak from firsthand experience when I say it is also possible to turn a cell phone into a paper weight using a flasher box and other programming tools by pushing the wrong button. It's not necessarily fair to critique Cell Phone Analyzer for this, but the software won't work without the use of the flasher box to acquire the phone's memory.

The other aspects of the Cell Phone Analyzer I appreciate are the price, which costs $745.00 for law enforcement agencies, and that an agency can purchase the flasher boxes with cables from its company. Most flasher boxes are retailed over the Internet and some can come from sketchy countries. It is difficult to explain to commanding officers why you need their department credit card to make an Internet purchase in Russia.

As an added bonus, BK Forensics makes a fully functional version of the software available for download on its website so you can evaluate before purchase. The trial version inserts pesky little dots all over the report until you purchase it, but you can get a feel for the full capabilities of the software before committing your department's money.

Phone Image Carver by Getdata

Downloading a mobile phone's memory using a flasher box is commonly referred to as a "phone data dump," or simply a "dump." Phone data dumps using a flasher box or software tool result in a hexadecimal or binary image file of the phone's memory. As the name implies, Phone Image Carver extracts data from a mobile device's memory dumps. These memory dumps are the result of downloading a device's files using a flasher box. "Carving" is a term coined in the computer forensics world, referring to searching for files based on their content by looking for a file's headers or footers and then carving out the data in between. For example, in its raw state an operating system may designate a graphic image file (.gif) with the header 89a and a footer, or trailer, of 3Bh. This delineates the information in between as being part of a graphics file. By searching for data at this basic level it is possible to recover files, or fragments of files, irrespective of the operating system. This method of examination also allows for recovery of deleted or hidden data and also works when the operating system is corrupted or damaged.

As with any forensic software that relies upon using a flasher box, training is highly recommended before use. However, once completed, Phone Image Carver is a handy tool to have in your arsenal for forensic examination. Phone Image Carver works with the following types of data dumps: Hex, DD, Bin, RAW, and DMG. If you have no idea what those mean, Phone Image Carver is probably not for you. However, for those familiar with the concepts of computer forensics, Phone Image Carver may be a tool that appeals to you. Data dumping and evaluating the resulting file with Phone Image Carver parallels established computer forensic principals by imaging, carving, and preserving the original evidence in a familiar manner.

For those who know what they are doing, Phone Image Carver is one of the best deals out there at $69.00 per license.

CDMA Workshop

CDMA Workshop is a collection of software tools for flashing mobile phones. As flashing is not legal in all countries, CDMA Workshop developed a somewhat nefarious reputation as a black market or hacker tool. It doesn't help that the people who maintain and distribute the software are located in Russia, adding to a somewhat seedy aura. Make no mistake about it, the software was designed to replace, modify, or hack the operating system of most CDMA phones on the international market. Inevitably some of those uses are illegal and the developers are undoubtedly aware. However, CDMA Workshop has a couple of features that make it an attractive tool for forensic examiners.

As stated in the title, the software works on CDMA phones. Prepaid providers who focus their business model on the lower-end of the socioeconomic spectrum use CDMA technology, making this tool particularly useful. One of the features that brand CDMA Workshop a valuable tool is the ability to read and/or reset the handset security lock code and to facilitate a memory dump of CDMA mobile phones. CDMA Workshop is used by some of the larger law enforcement agencies solely for the handset security reading function.

CDMA Workshop is developed and distributed from Russia. The purchase process can be a little troubling and I would not recommend using your personal credit card. Perhaps the most secure method of payment involves sending the money via Western Union although this can cause the Finance Department of many law enforcement agencies to freak out.

The CDMA Workshop software is tied to the computer's motherboard when it is initially installed. There are no options for reloading the software onto a different machine without purchasing additional licenses. If you are going to go through the time and effort to obtain this software, make sure you download and install it onto the machine you intend to use it on.

Also, CDMA Workshop is a software tool that requires a hardware intermediary, namely a flasher box and cables. The developer does distribute the hardware, but it may be easier and more cost effective to use a U.S.-based distributor.

Frequently, a search will reveal results for free or unlocked versions of CDMA Workshop, MOBILEdit, and other forensic tools. It might be tempting to download one of these "free" tools for evaluation, or to try and save yourself and your department a few bucks. Make no mistake about it – these are hacked versions of the programs. We are in the business of enforcing laws, not breaking them. Knowingly downloading and using pirated software is illegal. Don't do it. It might be a little tough explaining on the stand how you downloaded pirated hacker software to assist your law enforcement investigation. How forensic is a pirated program?

If you still feel compelled to visit one of these websites and download an impromptu trial version of the software, you shouldn't be surprised to get several warnings from your anti-virus software, provided it's up to date.

There are other programming and diagnostic tools which have similar functions to CDMA Workshop circulating on the Internet. Cell phone manufacturers designed these tools to test, develop, and repair their products' operating systems. In most cases, these tools are proprietary and were stolen or misappropriated and then distributed. I would advise against using them for the same reasons as using "free" versions of CDMA workshop.

Platform Specific Tools

In addition to some of the free content management software programs described earlier, there are forensic applications designed to address a single specific operating system. These tools address the specific issues and challenges that arise with so-called smartphones. A smartphone is a communications device that combines the features of a mobile phone with the capabilities of a computer system. The development of specific forensic applications for smartphones merges the fields of computer and mobile forensics. The major smartphone operating systems in use in the United States today are Android, iOS (Apple's operating system), BlackBerry, and Windows Mobile.

When I teach law enforcement or corrections officers, I hear them scoff at the thought of criminal suspects either being able to afford these devices or being intelligent enough to use them. While it is likely true that an older generation may have difficulty in adapting to a new technology, the younger generation will embrace that same technology. Criminals will likely shy away from these devices due to the cost involved and the inability to use many of the advanced features. Smart crooks will not use these phones because of the staggering evidence and intelligence we can recover from them.

Smartphones

Smartphones are higher-end mobile devices offering enhanced computer power and Internet connectivity. Smartphones can combine the additional functions of personal data assistants, media players, video and still cameras, and GPS navigation, among other features. As these devices evolve they become more and more like miniature personal computers. The devices have an embedded operating system such as Google's Android, Microsoft Windows 7, Nokia Symbian (S60), Research in Motion's (RIM) BlackBerry, and Apple iOS.

Smartphones have certainly received the lion's share of media attention, but it is important to remember these devices only account for approximately 20 percent of the 620 million mobile phones shipped in the United States in 2010.[14] Most mobile phone users are not using these advanced devices, likely due to the cost of the device itself and the underlying service. Without a doubt smartphones will certainly increase market share in this country, but, for the time being, they do not represent the majority.

Currently, Google's Android operating system represents 33 percent of the smartphones in this country and appears to be gaining ground. RIM's BlackBerry has 29 percent and is losing market share while Apple's iOS remains steady at 25 percent. Additional platforms include Microsoft's Windows Phone 7 which has slipped to less than 8 percent of the market and Palm (you remember them don't you?) which retains less than 3 percent.

---

[14] http://www.visionmobile.com/blog/2010/10/smart-feature-phones-the-unbalanced-equation-100-million-club-series/

Apple/iOS

Due to their popularity a number of iOS forensic tools have been developed to acquire logical and physical acquisitions of these devices. Forensics experts Andrew Hoog and Katie Strzempka released a review of available forensic tools for the iPhone in November 2010. Their paper is the first side-by-side evaluation of 13 different tools and techniques for iPhone forensic examinations and is far better than anything I could write. Their report can be accessed at http://viaforensics.com/education/white-papers/iphone-forensics/.

The top rated technique in Hoog and Strzempka's evaluation was developed by Jonathan Zdziarski. Zdziarski pioneered techniques and tools, resulting in a full download of the data on a device using iOS with a tremendous amount of deleted data. His techniques can be used to bypass the handset security feature as well as recover data from phones with damaged operating systems. Even though his technique results in the most comprehensive recovery of data from an iOS device, he makes his tools and documentation available free to law enforcement officers who are involved in mobile forensics. I respect that Zdziarski could make a fortune by commercializing his products yet makes it freely available to the community because, in his opinion, it's the right thing to do. Zdziarski's method is not for the faint of heart and requires a comfort level with command line programming. His tools only work on Apple computers or those using a Linux operating system. Zdziarski will also perform forensic evaluations on iOS devices on a case-by-case basis. His technique has survived the scrutiny of the California court system.

Many of the well-developed commercial forensic tools such as Cellebrite's UFED, Microsystemations XRY, and Oxygen Forensic II offer support for iOS devices. This support usually has an acquisition of logical content from the device. However, the popularity of Apple products has resulted in developing a number of iOS specific tools such as Lantern by Katana Forensics at https://katanaforensics.com/ and Mobilyze by Black Bag Technology.

Perhaps one of the greatest challenges in examining iPhones, as well as all of the other brands of smartphones, is the abundance of applications that may contain evidence or intelligence not routinely collected by many of the forensic tools. Microsystemation's XRY has a relatively unique and developing capability to perform application forensics. Application forensics is examining the specific software application and attempting to extract information that might be relevant to a law enforcement investigation. Currently, XRY is able to analyze information from several popular iPhone applications such as Atomic Web Browser Lite, Dropbox, Facebook, TigerText, and Google Voice.

Lantern

Lantern by Katana Forensics bucks the trend of many forensic products and has developed a low-cost program that works extremely well. It is a comprehensive forensic solution focused primarily on iOS devices but has recently developed an emerging capability for Android devices. Lantern has strong features: comprehensive support for evidence collection from iOS devices and a low price point. Both of these features make it worthy of serious consideration for any law enforcement agency. In addition to logical data extraction, Lantern has enhanced capabilities to recover data from Facebook including contacts and messages, Facebook Messenger contacts and messages, and Facebook message geotagging information. Additional functions are recovery of Snap Chat videos and information from the applications Voxer, Text Now, Text Free, Text Plus, Snap Chat, Grindr, Meet24, ICQ, and Pinger.

The ability to recover information from the most popular free communications programs that operate outside of any service provider (and are subsequently much more difficult to investigate) make this an outstanding tool. Lantern's developing support for Android devices is an added bonus.

Not only does the software acquire data from iOS devices it also images the operating system using the Lantern Imager program. Imaging or creating an exact image of the file system has long eluded mobile phone forensics due to the variety of devices and operating systems. Now that smartphones have segmented into two main operating systems, iOS and Android, mobile phone forensic examiners have the same capability as their computer forensic counterparts. Lantern Imager is included with the main Lantern suite and could easily be a standalone tool worth the expense.

What I really appreciate from Katana forensic is its willingness to allow users to evaluate the software before purchasing it. Unlike some forensic product developers who refuse to let you try their products before purchase, Lantern is free to try.

The price point for Lantern makes it the obvious choice for many budget minded departments and administrators to purchase, especially after they have already spent thousands of dollars on other forensic tools. Katana is priced at $599.00 for law enforcement and $200.00 per year annual maintenance.

Now for the bad news: You knew it was coming after the buildup of enhanced collection capabilities, application forensics, imaging functionality, and low price. In order to work, you need an Apple computer. As I discussed in previous sections this can be a substantial challenge for law enforcement agencies that are addicted to the Windows operating system. Possible solutions for budget conscious departments is to see if there is not a Mac in property and evidence that can be converted to agency use or to approach your local Apple store and see if they will donate or discount a Mac to support the community.

Katana Forensics also has an additional bonus hidden on their webpage that you might not find on casual examination. If you have ever dealt with trying to convert Unix or Epoch time, especially Apple's version of Epoch time, which is different than everyone else's, you will find its Epoch Time Converter a very useful tool. It can be downloaded at http://katanaforensics.com/support/.

Lantern by Katana Forensics has been evaluated by NIST. The report can be viewed and downloaded from http://www.nij.gov/pubs-sum/241153.htm.

Android Operating System

Currently holding over a third of the U.S. market, the Android mobile phone operating system is one of the fastest growing and most popular platforms outpacing even Apple's product. The forensic examination of devices using the Android OS is frustrating without the proper tools or techniques. Perhaps the most comprehensive solution to Android forensics comes from Via Forensics. Via Forensics produces several tools focused exclusively on Android devices. Via Forensics makes one of its tools available free to law enforcement agencies and charges for another of its products. AFLogical is the free tool offered by Via Forensics. The software is loaded onto a forensic Micro SD card and inserted into the target device. The software is then run from the card and logical information is captured and recorded in .csv format. Via Forensics also offers viaExtract, a more comprehensive program with enhanced collection and reporting features. Data that can be collected are device information, Internet browser bookmarks, Internet browser searches, call logs, contacts and phone numbers, SMS and MMS messages, instant messaging accounts, contacts, messages, internal and external image, and video media.

The software costs $1,475, which comes with one year of support and updates. Additional licensing, support, and updates cost an extra $625.00 after the first

Chapter 9: Locked Devices

Locked or security protected handsets are problematic, but they are something law enforcement is going to have to deal with to obtain evidence and intelligence from suspects' phones. There are no easy technological methods for bypassing the handset security feature. Graphic processing unit (GPU) and distributed network password attacks have yet to be developed. A locked handset can be extremely difficult, and in some cases impossible, to circumvent. There are techniques that can try to bypass the security features of most devices.

Consent

The first password acquisition method is to attempt to obtain consent and cooperation from the device. For reasons that escape me, investigators don't even attempt to acquire the suspect's security code. The same techniques used in interviews and interrogations can be applied to suspects when getting the security code voluntarily. Start with this one because it's the easiest method out there. However, it does come with a warning; In 2010 the courts addressed this issue in the case *United States v. Rogozin*. The Courts found that obtaining a suspect's password or PIN to be testimonial and incriminating, so it requires Miranda admonition before questioning.

The last sentence should be footnoted with the following legal citation:

United States v. Rogozin, 2010 WL 4628520 (U.S. District Court for the Western District of New York 2010)

Deception

Someone somewhere is going to raise his eyebrows at the thought of law enforcement being less than completely honest with criminal suspects when it comes to searching their phones. One technique that has been met with varying degrees of success is allowing the suspect to access his phone to retreive phone numbers of family or friends to call from jail. An alternative technique has been to tell the suspect the jail phone system is inoperable but he can use his cell phone to make notification calls to loved ones. As soon as the suspect unlocks the phone it is immediately seized from them and, in the case of Android phones, the USB debugging option is accessed and selected. Once the USB debugging feature is activated it does not matter if the handset eventually relocks due to a set timer. Most major forensic tools can still access a password protected Android device once USB debugging has been enabled.

Guessing the passcode

Mobile phone users may not change the default security unlock codes on their devices. By default, most handset security codes are the last four digits of the device's phone number or simply "0000." Other password bypass methods can involve attempting to guess the security code. In a published article, a developer was able to capture over 200,000 passwords used to access an iPhone application. Of those 200,000 passwords the most common were 1234, 0000, 2580, 1111, 5555, 5683, 0852, 2222, 1212, and 1998. Most of the codes followed simple formats such as four identical digits, moving in a line up or down the pad, or repetition. The exception to the top ten passwords was 5683, the numeric representation of the word "L-O-V-E" on a keypad. Additional common passwords were the year of birth and the year of graduation.[15]

Guessing the security passcode can come with significant drawbacks. Devices might limit the number of incorrect password attempts and can, and will, delete the entire contents of the device if too many attempts are made. Even if the device does not delete any data, it may revert to a higher security code that will require a different security code, which is not even known by the owner. By default, smartphones limit the number of incorrect password attempts to 10, but some devices can be set to as few as three incorrect attempts before deletion begins. In an article regarding a murder investigation in Raleigh, NC a detective accidentally deleted data from the victim's phone.[16] There article noted:

"[The Detective] then showed photos he took while trying to recover data from [the victim's] phone. He showed how he went through more than 10 screens putting in a code to unlock the password-protected phone and how on his ninth attempt to unlock the phone he was warned that information would be erased if he tried again." Don't do the same thing. Limit the number of password guesses you make and remember as few as three incorrect password attempts can delete the data from the device. If you see a warning message, stop and turn the phone off.

Physical Screen Examination

Attempting to guess the security code can also destroy the very evidence that may lead to that code. Touch screen phones develop what I call "phone goo" or the coating of sweat and skin oils that accumulate on the glass screen. Some investigators have been able to recover security codes, particularly pattern unlock codes associated with Android OS phones, using standard criminal investigative techniques. The process is to examine the screen of the device using latent fingerprinting techniques. By applying fingerprint powder it may be possible to discern and replicate the pattern unlock or the position of the numbers pressed. Obviously, this technique will not work if someone has attempted to guess the pattern unlock or the keypad unlock code. Multiple failed attempts to guess the security code will destroy or obscure the original evidence.

The classic smudge attack does not work well with Apple devices or Android devices protected with a PIN code instead of the pattern unlock. This is due to the small surface area of the finger when it contacts the screen's glass surface. It is extremely difficult to differentiate the prints associated with the passcode and those associated with routine use of the device. One large east coast law enforcement agency has reported success using ultraviolet visible theft detection sprays on the screen. From a practitioner I have learned that if attempts to read the prints or dusting for fingerprints are unsuccessful they will clean the screen and then coat it with the UV antitheft spray. The phone is then returned to the suspect to look up phone numbers from his phone book or to make a call to friends and family. Then it is examined under black light to see which keys were pressed.

Personal Unlocking Keys

Cell phone manufacturers recognize their customers may forget passcodes and have programmed a master security code as a failsafe to unlock secured phones. This code is commonly referred to as the Personal Unlock Key/Code or PIN Unlock Key, also known as the PUK. The PUK is required after the security

---

[15] http://www.i-programmer.info/news/149-security/2668-passcodes.html

[16] http://www.newsobserver.com/2011/04/25/v-print/1153613/coopers-defense-witnesses-qu

code has been entered unsuccessfully several times and the phone has defaulted to a higher security code. The PUK is sometimes available from the service provider with the submission of proper legal process, usually a search warrant. The PUK is usually a six- or eight-digit code, entered after the user's security code has failed. PUK codes are not always infallible. I have obtained PUKs from cellular service providers that did not work. It is critical to remember that after 10 attempts to enter the PUK, the phone will become permanently locked and it is unlikely any evidence would be recovered using all but the most advanced forensic techniques.

Provider Unlock: Android Passcode Bypass Procedure

Apple and Google will both assist law enforcement with bypassing the handset security code on their respective devices. This is not as good as it sounds and each process comes with various procedures and problems.

The Password Reset Option

Google is by far the easiest to deal with, but that is not to say there are not some significant challenges that must be addressed. When customers purchase and activate a phone using the Android operating system they are prompted to create or log into a Gmail account. The email account's purpose is to be used as a last ditch password reset tool in the event the customer forgets the pattern unlock or PIN number code. After incorrectly entering the pattern unlock or PIN number 10 times the phone will prompt the user to enter his Google email account and password. With proper legal process, Google can reset the associated Gmail password and provide the new password to law enforcement so they can unlock the phone.

The Search Warrant

The following is an example of an affidavit and search warrant that was successful in obtaining the required assistance from Google. Check with your prosecutor to make sure it meets the legal standards for your jurisdiction before using it.

Affidavit

> *Pursuant to SW # ____________, which is attached and incorporated by reference (Attachment A), your affiant has been tasked with a forensic examination of a T-Mobile HTC/G-1 Android platform cellular telephone.*
>
> *The device is identified as T-Mobile HTC/G-1 Android platform cell phone, Model #_______, on the _______ network with access number _________, serial number __________, and FCC ID#_____________ (the "cellular device").*
>
> *Upon initial examination the cellular device your affiant discovered the device is locked with a pattern protected handset security code and requires a Google/Gmail user name and password to bypass the lock. At this time your affiant does not possess the equipment or training to successfully*

bypass this lock. To the best of your affiant's knowledge such technology does not currently exist at the Haywood Police Department or any other surrounding law enforcement agencies in Alameda County, California. Attempts to bypass the handset security lock using the existing tools and techniques available to your affiant have thus far failed.

Your affiant has spoken to representatives from the cellular service provider and Google. Your affiant was informed by representatives from both companies the only way to unlock the phone is to have the Gmail user name and password, which was required when the device was setup. Your affiant is aware from prior law enforcement training and experience, as well as, personal experience using a cellular device with the Android operating system, when a user purchases and activates an Android operating system phone they are prompted to create or enter a Google Gmail account. The purpose of this account is to enable Google to remotely reset the handset security lock feature in the event the user forgets their original code.

Therefore, your affiant believes Google can provide the Gmail account name associated with the device, as well as, reset the password and provide a new, temporary password to you affiant or his designee. As established previously, I believe there is probable cause to believe evidence of the crime under investigation exists in the cellular device described in Exhibit 1A. In order to recover evidence of the crime under investigation your affiant requests Google be directed to provide information and technical assistance, by way of example and not limitation, including the Gmail user name associated with the device and resetting the password to the associated account.

Search Warrant

It is hereby ordered that Google Inc. provide the username for the cellular device identified as T-Mobile HTC/G-1 Android platform cell phone, Model #________, on the ________ network with access number __________, serial number ___________, and FCC ID#_______________ (the "cellular device") by providing the associated user name and by resetting the cellular devices password.

It is hereby further ordered that Google shall reset the passcode for this cellular device and provide the passcode to law enforcement.

Provider Unlock: iPhone/iPad/iPod Touch Passcode Bypass Procedure

Apple has a number of challenges that you should be aware of and must be addressed prior to enlisting them for assistance. Foremost, Apple says it is technologically impossible for it to reset the handset security code on any device using iOS version 4.0. However, it can read the raw data around the handset security code and provide investigators with the data contained on the device. Apple has a lot of requirements to send it a device and this is where the fun begins. For reasons that will become clear later, I would strongly suggest reading this entire section before embarking on the path of having Apple assist you.

Properly Identifying the Device

First, you are going to need to properly identify the device by model number, network (the cellular service provider), phone number or International Mobile Equipment Identified (IMEI) and Federal Communications Commission (FCC) ID number. To get the information from most phones you would access the settings menu. This is a bit of a problem if the handset is locked, which is probably the reason you're asking Apple for its assistance in the first place. You could always remove the battery cover, but you might have noticed removing the battery cover on an Apple is not something it wants you to do. Apple does not want consumers tinkering with their devices and replacing their own batteries – they want the phone sent to an Apple repair facility. Law enforcement investigators think the numbers located on the bottom of the back cover of an Apple device are the serial numbers and/or the international mobile equipment identifier (IMEI). They are not, unless it is an iPhone 5. For all other models the numbers located at the bottom are the FCC identifiers for the model of phone and every iPhone model has the same numbers. This is one of the pieces of information you need but not the most critical.

Determine the IMEI

So if the number on the bottom back of the phone is not the IMEI, where is it? Apple came up with an ingenious method for placing the IMEI and the serial number by placing them on the bottom of the SIM card holder for the device. If you examine the top of an Apple device, equipped with a SIM card, you will see a small hole located next to the audio jack. This hole is used to remove the SIM card and the SIM card holder. You could use a specially designed tool for this, which is available on the Internet, or you could insert a small paper clip. Use caution because the SIM card holder can eject with surprising velocity. Once you have removed the SIM card holder you will see the SIM card held within. If you turn the holder over you will find the IMEI and the serial number imprinted in very tiny font. These are the numbers Apple needs to assist you.

There's one small catch with this method. If the service provider is Verizon you have a problem. Verizon is a CDMA network and as such its devices do not use SIM cards. Verizon is transitioning to GSM so some of its newer phones may have SIM card slots and SIM cards, but most models do not. If there is no SIM card slot you are going to have to use some alternative methods. One method is to plug the iPhone into a computer that does NOT have iTunes installed. Once the phone is connected, Windows will detect it as a camera and the phone will show up as a connected device. **Right click** on the icon for the device and select **Properties**.

The serial number and the IMEI may be found on any packaging material and sales documents, which accompanied the device when it was purchased. If all else fails you may be forced to provide Apple with the device's phone number and ask it to locate the information in its records.

The last method is to seize all computers the device might have been synched with. This seems like a draconian method just to obtain the IMEI number. However, there is information contained in the backup file on the synched computer that you might be able to access. The methods for this are detailed in subsequent sections, but as long as you are writing the search warrant for the phone, you might want to include the suspect's residence or business for any computers the phone was paired with.

Determine the Model

There are a number of variations out there. The iPhone 5 is pretty easy to identify, but it can get confusing with the various models of the iPhone 3G, 3GS, 4, and 4S. An Internet search will help you confirm the phone's physical characteristics, but as a general rule these are some of the major differences between the models:

iPhone 4S:

Square in shape. Glass back. Front & back camera. Model number on back cover is A1387.

iPhone 4:

AT&T: Looks just like the iPhone 4S except the serial number on the back is A1332. It also has a SIM tray on the right side.

iPhone 4:

Verizon/Sprint: Looks just like the iPhone 4S except the serial number on the back is A1349 and it does not have a SIM card tray on the right side.

iPhone 3GS:

Plastic back. No front camera. Back text is the same shiny silver as the Apple logo.

iPhone 3G:

Looks just like the 3GS except the back text is a dull silver instead of shiny

Determine the Phone Number

First ask the suspect what the phone number is. If he won't tell you or if you are worried law enforcement interest in the device might cause him to remotely delete the contents, try calling 911 using the emergency calling feature from the lock screen. The call taker at the receiving public safety answering point should be able to tell you the number displayed on their system after you identify yourself.

The Search Warrant

Now that you have properly identified the device you can get your search warrant drafted and signed. But make no mistake about it – you do need a search warrant. Apple does not care if your suspect has a Fourth Amendment waiver permitting a warrantless search of the suspect's possessions as a condition of supervised probation or parole. You need a warrant. Not only do you need a warrant, but you need to include some specialized language in the warrant to get Apple's cooperation. Also, Apple would like to review the warrant before it gets signed by a magistrate. This does not mean you need to send it the affidavit of probable cause supporting the warrant. Apple does not need to see that nor does it have the right to. What it would like to review is the section regarding the places to be searched, items to be seized, and additional instructions. As of the time of publication this is the most current language straight from Apple's legal department:

> *"It is hereby ordered that Apple Inc. assist law enforcement agents conduct the search of one Apple iOS device, Model #____________, on the _______ network with access number (phone number) _________, serial number __________, and FCC ID#_____________; It is hereby further ordered that Apple shall provide reasonable technical assistance to enable law enforcement agents to obtain access to unencrypted data ("Data") on the Device.*
>
> *It is further ordered that, to the extent that data on the iOS device is encrypted, Apple may provide a copy of the encrypted data to law enforcement but Apple is not required to attempt to decrypt, or otherwise enable law enforcement's attempts to access any encrypted data.*
>
> *Apple's reasonable technical assistance may include, but is not limited to, bypassing the iOS Device's user's passcode so that the agents may search the device, extracting data from the Device and copying the data onto an external hard drive or other storage medium that law enforcement agents may search, or otherwise circumventing the Device's security systems to allow law enforcement access to Data and to provide law enforcement with a copy of encrypted data stored on the iOS Device.*
>
> *Although Apple shall make reasonable efforts to maintain the integrity of data on the Device, Apple shall not be required to maintain copies of any user data as a result of the assistance ordered herein; all evidence preservation shall remain the responsibility of law enforcement agents."*

As long as you are writing all of this additional language in the search warrant you might want to consider adding not only the cellular service provider for its subscriber, call detail, and financial information, but also Apple for any associated iTunes subscriber information. And as long as you are adding that you should probably demand the contents of any iCloud backup accounts associated with the device and subscriber.

Contact Information

The point of contact at Apple is:

Joann Chang
Apple Litigation Group
1 Infinite Loop, MS 169-2NYJ
Cupertino, CA 95014
(Phone): 408-783-0933
(Fax): 408-974-9316
joann_chang@apple.com<mailto:joann_chang@apple.com>

Waiting for Apple

Once the search warrant is signed you can fax or email it to the Apple legal department. Make sure the name of the magistrate on the search warrant is legible. Do not send the device to Apple until it advises you to do so. After an internal approval process, it will contact you to schedule an appointment for the passcode bypass. They recommend that the device be delivered in person, because of a potential loss through shipping. This creates an obvious problem for investigators who work for law enforcement agencies outside of Northern California. Check with your command structure and your prosecutor's office to determine the proper steps to establish the appropriate chain of custody.

You should also be aware that, as of the time of publication, Apple's backlog was reportedly 60 days. Apparently, there is exactly one technician at Apple who performs this procedure and this is the estimated time before any newly submitted device will be examined.

Storage Media

Provided the device is running iOS4 or later, Apple requires you provide adequate storage media to enable it to give you a copy of the data. The storage media should be large enough to store twice the rated capacity of the device you are submitting. Considering the iPhone 5 can have up to 64 gigabytes of storage, you will need to provide them with a storage media of up to 128 gigabytes. A couple of DVDs are not going to suffice so you will likely be looking at an external hard drive to store that amount of information. Once you do obtain the evidence copy of the device's files, make a forensic copy of the file for the working version. It would be an epic mistake to start browsing through the files on the storage media returned by Apple without first making a duplicate and booking the original into evidence.

The Aftermath

After jumping through hoops to identify the device, get the warrant drafted, reviewed, and signed, obtaining the storage media, waiting over two months for your appointment to send the device in, you will receive the examination's results back from Apple. Hopefully, you take my advice and make a working copy of the files and book the original storage media into evidence.

Now that you have a working copy you open it up like a kid on Christmas morning and find another problem. You see, when Apple reads around the handset security lock and copies the file system the resulting data is stored in its native formats; namely file types known as plists and other database files. This data is not human readable without the use of additional software to open and read the plist and database files. These files require specialized software to read and the results can be confusing.

If you've ever owned or used an iPhone and were expecting the results from Apple to appear the same as what you are used to seeing on the screen you would be horribly mistaken. For example, most people are familiar with Apple's text message screen. They don't know if the data they are viewing is actually stored in different databases. In other words, in the results sent back from Apple you might find the text message results are actually stored in multiple databases; one for the contact number, one for the message, and one for the date/time. Each set of data may be in its own file and will need to reconcile the results from the other files to make one coherent text message. Not only that, you will find yourself dealing with issues surrounding the date/time stamps that are in Unix or Epoch time. While the rest of the computing world calculates Unix time as the number of seconds from January 1, 1970, Apple calculates its time as the number of seconds from January 1, 2001. So most Unix time calculators will not work because Apple chose a different date.

At this point you might be asking yourself "Why bother?" It is true this is a long and frustrating path to get to the evidence we need to investigate crimes, turn suspects into convicts, and send them to prison for the better part of their adult lives. While there currently is no easier way to get the information, there are alternatives. Specifically, attacking the backup file created when the iPhone was synched with an authorized computer. Read the subsequent chapter on iPhone backup files and see why it might make more sense to go after the computer instead of the phone.

Smartphone Passcode Bypass

Imagine processes that circumvent the handset security of most mobile phones. In addition, these processes are able to recover the device's complete physical contents, and make many current forensic tools irrelevant. These new techniques also work on phones which may have sustained physical damage such as being smashed before the suspect can be taken into custody. These two new techniques are known as chip-off and JTAG, but I cover them the reader should note that I'm going to address these technology-based methods for circumventing the passcodes in general terms.

Because I have no way of controlling the distribution of this book I want to make sure the specifics of these methods are kept within the law enforcement community. More information is available online at a number of different law-enforcement-only websites, blogs, and user groups. I should also point out both of these techniques were spearheaded by Victoria Police Department (Canada) Detective Bob Elder, who was assisted by other unnamed and/or unknown law enforcement officers, forensic technicians, and cellular industry personnel. My gratitude to Detective Elder and everyone else involved in the process for helping the law enforcement community overcome what was becoming an insurmountable problem – locked smartphones.

Joint Test Action Group (JTAG)

Joint Test Action Group (JTAG), resulted in a standard established by the Institute of Electrical and Electronics Engineers (IEEE) that electronics manufacturers voluntarily implement. The standard is designed for engineers to check the capabilities and debug the device's hardware and software by accessing Test Access Points (TAPS). JTAG TAPs are found in consumer electronics such as game consoles, DVD and Blue Ray players, and GPS navigation devices. With TAPs equipment manufacturers, test engineers, and repair technicians can access the components and embedded software or a device's operating system without activating the user interface. Specific to mobile phones, the TAPs are used by the original equipment manual for testing and de-bugging. Cellular repair technicians use the TAPs when there is a problem with the devices operating system or they want to repair or "flash" a device.

For law enforcement the JTAG process opens many doors which were previously closed. For example, most forensic techniques require the device to be powered on in order to work. If there is damage to the screen or the data port, that $10,000 forensic tool is just an expensive dust collector.

But a challenge for forensic examiners is the pattern-locked Android operating system phone. Without the other techniques discussed previously, it has been extremely difficult to bypass these locks. To perform a forensic examination on a phone using the Android operating system the USB Debugging option must be selected on the device. To my knowledge, all Android phones ship with the USB Debugging turned off as this is an advanced feature that was not designed for the casual consumer to use. If USB Debugging is not turned on there are very few techniques that will work to overcome the passcode lock.

Those that do work may change the underlying operating system of the phone, altering the underlying evidence – something we are trying to avoid during a forensic examination.

With the JTAG technique an investigator is able to access the TAPs to obtain a read of the full physical memory of the device, including the encrypted PIN lock and/or gesture key files. Through other law enforcement investigative techniques it is possible to decrypt these passwords enabling access to the device. This technique works when the phone is passcode locked, physically or water damaged, or has USB debugging turned off.

Chip-Off

For several years some federal law enforcement agencies in the United States and other foreign governments have been using the "chip-off" technique to recover information from locked or damaged phones in high profile criminal cases and terrorism investigations. However, it wasn't until recently this technique was made available to state and local law enforcement agencies.

The process involves removing, via de-soldering, the NAND flash memory chip from the mobile device. NAND flash memory is the same type of memory found in Micro SD card that is present in most smartphones. However, these same chips can also serve as the primary memory for many types of devices including some types of mobile phones, memory cards. USB flash drives, and other specialized electronics.

The chip-off procedure currently focuses on one version of NAND memory, known as Thin Small Outline Packages (TSOP). TSOP NAND is used in flash memory devices and some cell phone models. However, many phones use what is known as Ball Grid Array (BGA), NAND chips. The research regarding removing and examining BGA NAND chips is still ongoing and currently the training and technology focuses primarily on data from the TSOP NAND, and some NAND chips.

This research field is exciting for a number of reasons. First, reading the NAND chip directly may bypass any security features incorporated into the handset itself, such as those employed by smartphones. However, even if the data on the chip is encrypted it may be possible to use decryption techniques such as GPU and distributed network password decryption to read the data. "Chip-off" examinations can also reveal a complete physical read of the NAND chip. This is particularly exciting due to the potential to recover a significant amount of deleted content. As noted in the section regarding data recovery from Micro SD cards, deleted data can exist for a while on a chip due to the structure, formatting, and the flash memory's wear leveling processing. It may be possible to recover an incredible amount of deleted data from the NAND chip. Lastly, flash memory chips are notoriously robust and able to survive immersion in fresh and salt water as well as survive extreme heat and other conditions, making other chips completely unreadable. The "chip-off" procedure may enable data recovery when the enclosing handset has been physically destroyed.

The Downside of JTAG and Chip-off

If it was easy, everyone would be doing it. But there are some limitations to what both JTAG and the chip-off procedures can accomplish. Neither technique will work on every phone. I'm not going to list what phones the processes work on and which ones they don't because the list changes and I don't want the criminals to know where our blind spots are. The JTAG and chip-off processes also come with some fiscal considerations. If you are currently involved with cell phone forensics investigations you know it is already an expensive endeavor with popular software and hardware tools starting in the $5–10,000 range. Multiple tools are required to effectively process the majority of phones on the market and each tool has an annual license fee. Both JTAG and chip-off add to that expense, putting it out of the reach of small and medium sized law enforcement agencies.

First, the training course itself can be cost prohibitive in the $3,700 range for a one week class, not including travel and lodging. The processes also require purchasing specialized equipment adding anywhere from an additional $2,000–4,000 in expenses just to get started. As new phones are introduced to the market and investigative techniques are developed to overcome the passcodes, additional expenses are incurred to keep up with the latest technologies.

However, the most significant hurdle to overcome when attempting to implement these procedures is the time involved in the examination process. These are not quick techniques and an investigator who performs phone forensics as an ancillary duty or “other duties as assigned” will find difficulty in dedicating the time necessary for the examination. A smartphone with multiple gigabytes of onboard memory can literally take several days just to obtain an image of the memory. Carving out data and recovering the passcode can add significantly to this time. You will not be able to set it and forget it like a rotisserie chicken as the techniques can be temperamental and require constant supervision and troubleshooting. While the JTAG and chip-off procedures are able to defeat the passcode on many of today’s cell phones the process, costs, and time involved will likely limit the widespread implementation of the techniques.

Chapter 10: iPhone Backup Files

Evidence, including data contained in iTunes backup files, should be recovered forensically whenever possible. However, reality needs to factor in. Many jurisdictions have an in-house forensics capability or access to it. Unfortunately, most in-house examiners and task forces are not lacking for work. You might wait several months before a seized computer can be examined. The challenge has led to a development of forensic triage tools that allow an officer to recover data. From there, he can make a preliminary decision to seize and submit a computer for forensic examination.

The techniques discussed below are not forensically sound. An investigator should reconcile the need to collect this information in a timely manner versus the need to collect the data in a forensically sound manner.

iPhone Synch Process

An iPhone synchronizes with an authorized computer that has iTunes on it every time it is plugged in, unless the user interrupts the process. Connecting to an Internet-enabled computer is also required to download larger applications and files, although this connection may be wireless or through a data cable. The synching process creates a backup file on the computer designed for users to reinstall contacts, text messages, calendar entries, photos, videos, and application data in the event their phone is damaged, malfunctioning, lost, or stolen. Investigators can examine these files for valuable information when the phone itself is password protected.

By default, iTunes stores the backup file created during the synch process in set locations based on the underlying operating system. For this book, we are presuming the majority of readers are using a version of Windows operating system. The location of the iPhone backup file generated by iTunes is as follows:

For computers using Windows XP:

\Documents and Settings\USERNAME\Application Data\Apple Computer\MobileSync\Backup\

For computer using Windows Vista or Windows 7:

\Users\USERNAME\AppData\Roaming\Apple Computer\MobileSync\Backup\

Replace **Username** with your personal account/computer name.

iCloud

iCloud is a storage service Apple started offering in June 2011. With the service customers can store data from their devices on remote computers servers, allowing them to access and download the stored content on multiple devices. Users can also sync contacts, email, calendars, notes, to-do lists, and other application data in one central location. iCloud serves as a storage center and backup for purchased music, applications, electronic books, movies, and television shows. Each of Apple's 150 million iCloud users is eligible to store up to five gigabytes of data using their service. If your suspect uses iCloud exclusively and had deselected the option of making a local backup copy of his files on a home or work computer, the techniques in this chapter will not work. To obtain the information you will need to do so from Apple via legal process such as a search warrant.

Seizing the Backup File

The software discussed below is designed to be installed on a home user's computer to examine the backup files. This will not work for law enforcement investigators, as installing software to a suspect's computer violates every tenant of forensics. There are two methods to avoid installing software onto the suspect's computer and they involve portable applications. You might be tempted to copy the files onto a USB thumb drive and examine them back at the station. While this technique will work in a pinch it may be difficult to establish critical elements in the chain of custody, consequently reducing the likelihood it can be admitted into evidence.

The first method to use the tools described below is to make them portable. This process is best suited in situations where the presence or absence of relevant criminal information in the iTunes backup file is unknown, such as sex offender compliance monitoring and probation or parole searches. Each of the programs can be installed onto a computer and are not natively portable. To install them onto a portable USB drive you will need to use third party software. With the popular program Cameyo, it is possible to make almost any application portable. Install and run the Cameyo software and then install the iTunes backup extraction program of your choice. Once completed with the installation of the extraction program, re-run Cameyo. The software will be packaged in a portable application format you can transfer onto a portable USB drive.

An alternative method is to use a free software tool from Access Data called FTK Imager. The company and the product are well known in the computer forensics world but have not been widely used for mobile phone forensic applications. It is possible to recover the iTunes backup file using FTK Imager to create a verifiable process for recovering the information for later analysis. To get started download FTK Imager from Access Data's website at:

http://www.accessdata.com/support/product-downloads#.UdnsdKwkyrg

Copy and Install the FTK Imager Program to a portable USB drive. Once the program has been installed in on the USB drive it can be used on different computers without being installed on the targeted system. Plug the USB containing FTK Imager into the target computer. Select **File** and then **Add Evidence Item** from the dropdown. This should bring up the **Select Source** dialogue box. Accept the default **Physical Drive** option and select **Next**.

The software will have automatically identified all of the hard drives installed in the computer and any additional media attached to it such as USB drives. Select the hard disk drive and **Finish**. At the completion of the process you should see the hard drive loaded in the left pane.

Manually navigate to the location where the backup file is stored by clicking the **+** symbol next to the Physical Drive. Depending on the operating system of the computer the iTunes backup file will be located at the following locations:

For computers using Windows XP:

\Documents and Settings\USERNAME\Application Data\Apple Computer\MobileSync\Backup\

For computer using Windows Vista or Windows 7:

\Users\USERNAME\AppData\Roaming\Apple Computer\MobileSync\Backup\

Replace **USERNAME** with your personal account/computer name.

The file will be named with a string of alphanumeric characters. In this case the file is named *c96a294ed6a06e43064517cef2fe772e3299cf02.*

Right click on the file name and select **Export Files**. This will bring up a dialogue box asking where you want to export the files to. I would strongly recommend exporting the iTunes backup file to a secondary USB drive. At the completion of the process you should see a notification box letting you know the process was successful.

Copying and pasting in the first place would not necessarily transfer some important data such as modified, accessed, and created times. There is also no verification the files transferred are the identical to the files they were copied from. To authenticate the integrity of the files you need to use an additional function in the menu to calculate the hash values.

Right click on the file again and select **Export File Hash List**. This will create a comma separate value (.csv) file with the MD5 and SHA1 hash values of the components of the iTunes backup file.

Experienced examiners might wonder why I chose to use this process instead of using the standard forensic practice of duplicating the file, mounting it, and then parsing the files from the backup file from there. I am going to discuss using an inexpensive commercial program that does not work using the standard imaging procedures.

Force Synching

Imagine a scenario where the suspect has an iPhone that is passcode protected. Among the techniques you try, consider using a judicially authorized search warrant to force synch the phone to a known device. While many people lock their iPhone handsets, most fail to encrypt the backup file generated by iTunes when the phone is synched. If you know the suspect synchs his phone with a home or work computer you can attach the iPhone to the computer, synch it, and then either seize the computer for forensic examination or use one of the improvised techniques outlined below. This technique works because of the trusted connection that is established when an Apple product is hooked up to a device where the user has already entered a password.

The first time a user connects his Apple device to a new computer with iTunes installed the software requires him to enter his passcode. In all subsequent attachments the user is no longer required to re-enter their passcode because they have already entered it once, establishing the trusted connection. Naturally, force synching is not without its drawbacks. Foremost investigators need to know where their computers are. This may be a challenge for those who live in areas with a transient criminal population. Also, when the phone is synched, the prior backup file is overwritten so any older data that was on the backup file is now gone. However, it will reveal what is on the phone at the time the synch is performed.

Encryption

It is possible the user encrypted the backup file iTunes created. Encryption involves transforming data using an algorithm to make the files unreadable to anyone except those possessing the key or password. There are a couple of ways to circumvent the password protection and encryption and neither of them are pretty. The first is to compel the suspect to give you the password. This alerts him to your interest, though, and anything he wasn't supposed to have on his phone will be gone as soon as he can get to a computer. You may also attempt to guess the password to the backup file if you know the passwords your suspect frequently uses. Luckily, there are software tools that will break or bypass the security lock on Apple.

To decrypt a password protected iPhone backup file you will need to purchase the Phone Password Breaker from Elcomsoft Software for $79.00. Elcomsoft is the maker of password recovery and decryption tools used by law enforcement agencies around the world. The purchase of this software can be troublesome for some government agencies because the company is located in Russia. However, I have found them to be reputable with no spam, viruses, or other malicious content from the site or the company.

The Phone Password Breaker uses a computer's processing power attempt to break the password by rapidly guessing possible combinations or words, numbers, and special characters. The success of this process depends heavily on the speed of your computer's processor(s) and the complexity of the password. Fortunately, many suspects use simple, all lower case passwords based off of words instead of random combinations of upper case and lower case letters, numbers, and special characters. Our next challenge is time. For example, there are just under 57 billion possible six-character alphanumeric passwords. If your computer were able to process 2,000 password attempts a minute it would take about 20,000 days or 55 years to decrypt. Fortunately, technology can make this go much faster. A computer with a dual core processor can attempt about 500 password attempts per second. A computer with a quad core processor can attempt about 1,100 passwords per second. You can find out what type of processor you have by right clicking on the **My Computer** icon on your computer and selecting **Properties**.

Some hardware you may already have on your computer can significantly speed up password decryption. Computers use Graphics Processing Units (GPUs) to speed up videos and graphics video games commonly use. The proper GPU can increase the number of password attempts to just under 12,000 per second. It is possible to achieve more, but it gets increasingly expensive to use. If you're really serious about this and want to pick up a GPU to decrease the decryption time, you can find good deals on Amazon.com in the $80.00 to $120.00 price range. Just make sure your purchase is compatible with the software by checking the list at http://developer.nvidia.com/cuda/cuda-gpus.

The Federal Bureau of Investigation has a cryptoanalysis unit that can offer assistance to state and local law enforcement. Their assistance is based on the crime type and their available resources. Requests for FBI assistance with decryption should go through your local field office.

Reincubate Labs iPhone Backup Explorer

One of the first tools to use to look at a backup file is Reincubate Lab's iPhone Backup Explorer. While it does not capture everything, it is one of the best tools on the market for recovering information and the price is affordable.

However, the software's trial version is very limited in what it will recover. You will have to purchase a license for the software to get the full functionality of the program. You will find free versions on some websites, but I don't recommend downloading them for a couple of reasons. First, many of the sites that host the fully functional "free" version of this software also host other pirated or hacked versions as well as pirated movies and child pornography. I don't patronize those places and you shouldn't either. Software from those sites are also intentionally inflicted with viruses and other malicious programs that will damage your computer or steal your personal information.

Prior to purchasing the software, install and run software designed to make application portable such as Cameyo or ThinApp. After purchasing, install and run the program and transfer the resulting package onto your portable USB drive. You can now take the software with you to multiple locations.

Once the USB drive is inserted into a suspect's computer and runs it will automatically detect the version of Windows installed on the computer. It will also detect multiple backups from different devices on the same computer so it is possible to run the software on a computer where multiple people synch their Apple devices to. This works well if you have more than one suspect or if your suspect has multiple Apple devices like an iPhone and an iPad.

Simple Mode

Simple mode is the easiest method for viewing an iphone backup file's contents. From the simple user interface an investigator can check the photos, videos, and image files, the contacts/electronic phone book, voicemail messages, calendar entries, notes, text messages, call history, and possibly location data. To view the data in any of these files, click on the blue underlined number after the entry. If there is no blue underlined entry then there is no data for that field. Clicking on an underlined number will open a dialog box that states **Browse for Folder**. This is asking you where you want to save the information from the backup file. It is probably best to store the resulting data on a separate partition with the USB you installed the software on or a separate USB thumb drive specifically for capturing the data.

Before you start clicking madly away at the blue underlined numbers, there is some important information about the software's preferences that you must understand. By default the iPhone Backup Extractor will transfer information in certain predefined formats. For example, if you select **Contacts** by clicking the blue underlined number the program will automatically export every contract as a vCard. A vCard is an electronic business card format that can be attached to email messages or exchanged via text messages. vCards commonly contain information such as name, address, phone numbers, email addresses, web site addresses, and sometimes, photographs or graphics files. While this is a convenient way to store and share information from the Contacts backup file, it actually obscures some information of law enforcement interest.

Contacts

A better way to extract and view the contents of the Contacts file using the iPhone Backup Extractor, is use the option found in the **Extract** menu to place the contacts into **CSV format**. Comma Separated Value (CSV) is one fomat used to view information in Excel. A vCard can only be opened using Microsoft's Outlook email. An advantage to using the export to CSV function instead of using the default export to vCard function is the loss of some precious information. When vCards are exported they are automatically sorted alphabetically. When the contacts are exported using the export to CSV function they are sorted in a hybrid format that can show which contacts were added recently.

For example, when looking at the results of the contacts exported in the CSV formats you may see the contacts are not sorted alphabetically. This may show which contacts were stored chronologically and added recently. You will not be able to see that in the vCard format because of the way they are extracted. Groups of contacts with no phone number, but only an email address, are those contacts that were automatically added to the file when the user responded to an email. Again, these are chronologically and not alphabetically. So the most recent contact with only an email address stored was one of the most recent contacts the user responded to an email from. This does not mean it was the last email the contact sent, only that when he responded to the email it was not already stored in the phone. Because he responded to the email, the iPhone automatically added them as a contact.

Calendar

Calendar entries should be viewed using the same technique as the Contacts file. Instead of clicking on the blue underlined numbers next to Calendar, you should use the **Extract** button in the menu bar to view them as a **CSV file**. Otherwise, the calendar entries will be extracted as an iCal file. iCal is a calendar function for Apple manufactured devices. It is best to export the calendar function as a CSV file.

The exported field will include the summary of the event the user entered, the location, any additional description, the start and end date and time, the time zone, and whether the event was scheduled to be all day long.

Summary	Location	Descriptic	Start date	End date	Time Zone	All Day	Frequency
New Event			5/24/2012 13:00	5/24/2012 14:00	US/Pacific	FALSE	OneTime
Bridal shower			6/2/2012 19:00	6/2/2012 20:00	US/Pacific	FALSE	OneTime
Jack and Robyns			6/16/2012 19:00	6/16/2012 20:00	US/Pacific	FALSE	OneTime
MD appointment			6/4/2012 13:00	6/4/2012 14:00	US/Pacific	FALSE	OneTime
Dentitst Appointment			6/25/2012 12:00	6/25/2012 13:00	America/Los_Angeles	FALSE	OneTime

SMS/Text Messages

It's not clear exactly how many text messages an iPhone can hold in the backup folder using iTunes Backup Extractor. Some say there is the potential to store a limitless number of text messages but law enforcement forensic examiners have encountered a maximum of 5,000 messages. Whatever the number, the text messages that were on the phone at the time it was synched will be stored in the backup file.

The iPhone Backup Extractor will automatically download all text messages that were on the device when it was last synched with the computer. Unlike the Contacts and Calendar fields you do not need to use the Extract tab to get the messages into CSV format. Clicking the blue underlined number will download a spreadsheet with all of the text message content, the sender, and the date/time the message was received.

However, the spreadsheet will only list the phone number the messages were sent from and it does not automatically fill in the name from the Contacts file. The spreadsheet also lists each text message separately and it does not show the conversation in a single step like you might be used to seeing on your phone. To match the name with the phone number you may have to switch between the Contacts spreadsheet and the SMS spreadsheet. Any easy way to do this is to use the **Replace** function found in the **Find and Select** feature in the Home screen of Excel.

1. Find the phone number you want to replace with the contact name. Click in the cell or box and highlight the phone number. To copy the phone number you can either click the right mouse button and select **Copy** or press the **Control (CTRL) button** and the **C key**.
2. Highlight the columns that contain the phone numbers you want to replace with the name of sender or receiver. This will be the first two columns in Excel labeled **From** and **To**. Select both

columns at the same time by holding the **Control (CTRL) button** while using the **left mouse button** then click on the **From** and **To** columns.

3. Select the **Find and Select** button and drop down to the **Replace** function.
4. Click inside the box that says **Find What:** and either **right click** and select **Paste** or press the **Control (CTRL) button** and the **V key** to paste the phone number.
5. Open up the **Contacts** spreadsheet and find the name associated with the phone number. You should know that the phone number stored in the Contacts file is not always the same as the phone number in the SMS file. The SMS file contains the full 11-digit phone number, including the number 1 before the area code, and it is in a different format than the phone numbers stored in the contacts file. The phone numbers in the SMS file appear as 15105551212 while the phone numbers in the Contacts are formatted as (510) 555-1212. To find the phone number and the contact name you will open the **Find and Select** function again. Instead of **Replace**, you will select the **Find** tab. The phone number you entered should still be there, but because it may be stored in a different format you should search for the last seven digits of the phone number, such as 5551212. Highlight the first four digits, the number 1 and the area code, and delete them by pressing the **Backspace** key or the **Delete** key. You will also need to put a dash in between the prefix, the first three numbers, and the last four numbers of the phone number. Now you can search for the name associated with the phone number.
6. Select **Find All** in the **Find and Replace** box. Excel will automatically highlight the phone number with a black box around the number.

7. Go back to the SMS worksheet and replace the phone number with the Contact name using the **Find and Replace** function. The phone number should still be there, but it will be in the seven-digit format you changed it to. Bear with me, we are almost done! Put the full 11-digit phone number in the **Find What** box without using the dash (for example 15105551212). Place the Contact's name in the **Replace With** field.
8. Press the **Replace All** button and all of the phone numbers in the **From** and **To** columns will be replaced with the Contact's name.

Deleted Text Messages

It is not easy to retrieve deleted text messages from an iPhone or any other cell phone. Most major service providers do not store text message content once it is delivered and the only other method is usually by forensic examination of the phone. The iPhone Backup Extractor cannot retrieve text messages if they were deleted prior to a backup of the phone being made by iTunes.

Recordings

Recordings are sound files created using the Apple application installed on the device. For a variety of reasons, this application is not frequently used and is replaced by fancier applications from the iTunes store. Don't discount these files as they may contain snippets of your suspect's conversations.

If there are no recordings listed in the Simple Mode of the iPhone Backup Extractor, you may have check the Expert Mode to see if there are any other applications which may be used to record voice memos or conversations. Click on the **Advanced Mode** button on the iPhone Backup Extractor.

Next to the word **Application** should be a small + symbol. Clicking that expands the tree and allows you to view all of the applications installed on your suspect's phone. This can be an eye opener by itself. To see if there are any other applications that for making voice recordings you will have to examine each application individually. Some of them are pretty self-explanatory such as Angry Birds. But others may require you to search Google or iTunes to see exactly what the application is or does.

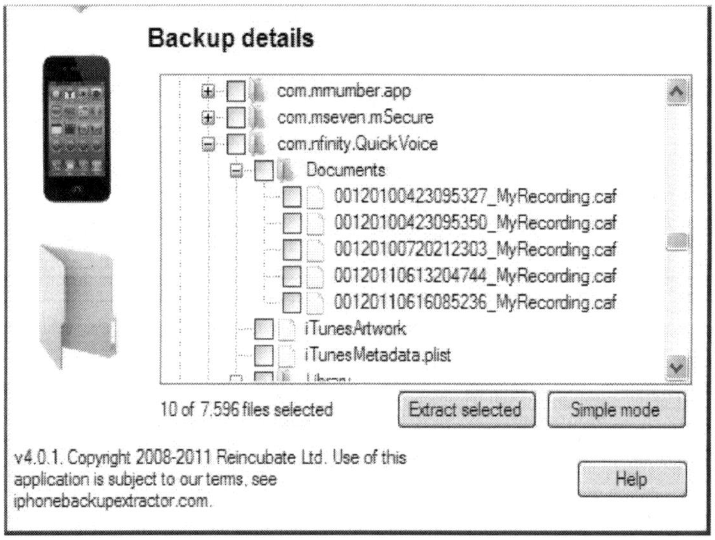

In this example you can see we have discovered an application called Quick Voice. By clicking on the + symbol next to the application name we open up a tree that includes a file entitled **Documents**. Inside the Documents file are several files labeled myrecording.caf. The extension .caf stands for Core Audio Format, which is a file type Apple developed to overcome some limitations in older audio file types.

To recover these files, place a check mark in each box. Clicking in the box above the files will automatically select all of the files below it. Next press the **Extract Selected** button and navigate to where you want to extract the files to.

In most cases you will need to either download the software or use an online service to convert it to a format that can be played from your computer. Apple's QuickTime Player allows you to play many types of audio and video files associated with their products, but to save them into a universal file you are required to purchase the upgraded version. Audacity is a free, fully functional software for playing and converting audio files. It was designed for more than just simple software conversions and it can appear very intimidating at first glance. You can download the software from http://audacity.sourceforge.net/.

Follow the prompts to install the software. When you are finished, open the software and select **Open** from the **File** tab. Press the green triangle button to play the recording. If you decide you need to save the recording as a file type that can be shared with and listened to by others, such as law enforcement, you can convert the CAF file into something more user friendly. Select **Save As** from the File tab and you can save the file as an MP3 or WAV file, two of the most common media formats.

Video

Video files taken with an Apple device are usually saved as 3GP and 3G2 formats. These are playable by most video players including QuickTime but, as with audio files, you may need to convert them into something simpler. There are a couple of ways to do this including installing software such as Format Factory or VLC Media Player, allowing you to play and convert video files between any number of common formats. To expose you to a wide array of tools, I'd like to introduce you to Zamzar.

Zamzar is a free online media conversion service that works on a large number of document, audio, and video file formats. Follow the prompts in **Step 1** to point to where you've recovered the video file. **Step 2** is to decide which format you want the video file converted to – MP4 and WMV are the most common and almost universally playable file types. **Step 3** is worrisome – it wants your email address. Firsthand experience says that there is no spam as a result of using this service. That doesn't mean it's totally safe, but the risk seems pretty low. I've been using this service for a long time and have yet to receive any unsolicited email or spam from them. If you are really worried about it, sign up for an alternate email, using Gmail or Yahoo. **Step 4** is clicking on the **Terms of Service** button and agreeing. Again there are no surprises here and none of my converted files has ever ended up on YouTube. After agreeing to the terms by clicking the **Convert** button a link will be emailed to you. This can sometimes take a few minutes, so don't worry. From the email you will be able to download and save the converted file. If giving out your email address or uploading a video file with your suspect on it still causes you concern, try using one of the free software programs such as Format Factory or VLC Media Player. Format Factory is a straight conversion utility that changes the file type from one video file to another. VLC allows you to play the video and then **Save As** a different file type.

Voicemail

Not many people know the iPhone is one of the only phones on the market that stores incoming voicemails on the device itself. Most voicemail messages are stored on the computers of the cell phone company and require the user to call in to the network to retrieve his messages. This means that all voicemail messages stored on the phone at the time it is synched with an authorized computer can be retrieved and listened to. To give you the widest variety of tools in your investigative toolkit, I'm going to recommend you try using an online conversion service found at http://media.io/.

Navigate to the file you want to convert, upload it, and decide on the format you want to download it as. In this case I use MP3 because it's universal. This service also works for 3GP media files, which you might find while in other areas of your suspect's backup file.

Notes

There are note taking applications available to users of Apple iPhones, iPads, and iTouch, but one is included standard with every device. Remember, your suspect may have purchased or downloaded a free note taking application in addition to the preloaded version, so you may have to use the Expert Mode to find it. By default the iPhone Backup Explorer will download the files from the Notes application in the same computer language used for web page – HyperText Markup Language (HTML), which is readable by any Internet browser.

Call History

I'm not exactly sure how many call records the iPhone has, but it appears there is a difference in what is available using the iPhone Backup Extractor and other law enforcement tools. The Backup Extractor is only able to retrieve 100 of the most recently dialed or received calls while other forensic tools are able to acquire much more.

	A	B	C	D	E
1	From	To	Start date	Duration	
2	9.25E+09	Me	########	0h 00m 34s	
3	9.25E+09	Me	########	0h 02m 158s	
4	9.25E+09	Me	########	0h 02m 148s	
5	5.1E+09	Me	########	0h 01m 67s	
6	9.26E+09	Me	########	0h 00m 00s	
7	9.25E+09	Me	########	0h 00m 00s	
8	Me	9.26E+09	########	0h 10m 633s	
9	8.01E+09	Me	########	0h 00m 00s	
10	Me	5.11E+09	########	0h 01m 66s	
11	9.25E+09	Me	########	0h 03m 236s	
12	4.15E+09	Me	########	0h 00m 00s	
13	Me	9.25E+09	########	0h 00m 35s	
14	Me	4.15E+09	########	0h 10m 607s	
15	9.25E+09	Me	########	0h 01m 95s	

The files are automatically downloaded into CSV format when you click the blue underlined number. However, you may see some data that does not make sense or even appear. When the files are downloaded into the Excel spreadsheet and exceed the length of the cell or box, the program reduces the number in size. The simple fix is click on the far right side on the column that contains the data. You should see a solid black line with an arrow on either side of it. By **double clicking** you will increase the size of all of the cells to accommodate the length of the information contained in the field.

	A	B	C	D	E
9	8E+09	Me	6/22/2012 13:53	0h 00m 00s	
10	Me	5.1E+09	6/22/2012 17:11	0h 01m 66s	
11	9.3E+09	Me	6/22/2012 17:43	0h 03m 236s	
12	4.2E+09	Me	6/22/2012 21:34	0h 00m 00s	
13	Me	9.3E+09	6/23/2012 11:44	0h 00m 35s	
14	Me	4.2E+09	6/23/2012 14:34	0h 10m 607s	
15	9.3E+09	Me	6/23/2012 14:50	0h 01m 95s	
16	Me	1.5E+10	6/24/2012 14:09	0h 07m 467s	

Location Data

You may remember when the story broke in 2011 about the secret cache of location data Apple's products were storing. The media attention was furious and probably overblown, but it caused Apple to make changes. When Apple introduced its operating system upgrade version 4.3.3 they limited the quantity of data stored to about a week and stopped storing location information altogether in future backups. If you never had a chance to look at the data, it wasn't as scary or intrusive as initially portrayed. The location information was often inaccurate and used purely for application and services installed on the device. You should view the location data obtained by the iPhone Backup Extractor cautiously as it can be very, very inaccurate.

Location Services

If you are concerned about location data being stored on your personal phone you can turn the services off by going into **Settings** and then **Location Services**. The Location Services are turned on by default, but you can turn it off if you don't want to use this feature. You can also individually control which application have access to Location Services information. However, if you turn Location Services off, you'll need to turn it on again the next time an application attempts to use the feature, such as when looking up a location in the maps application.

Photos

Photos can be some of the most telling items an investigator looks at. Suspects seem to take pictures of everything, including images of themselves doing things they shouldn't or being with people they aren't supposed to be with. The iPhone Backup Extractor offers an easy way to recover the digital images stored on the phone. Click on the blue underlined number and direct the software to the file you want to recover the images to.

This is one of the few processes that can take a long time with the iPhone Backup Extractor, so be patient. You will see small green squares covering the icons on the phone in the main screen to show you how far and fast it is progressing. Using a test database it took about two minutes to download 757 images.

While the images themselves can be important, there is important information contained in the image for investigators. By right clicking on any image and scrolling down to **Properties** you can see the date and time the image was created, modified, and last accessed. In this example, you can see the dates and times the file was created is listed as the date the iPhone Backup Extractor was run and the image files recovered. The modified data and time is closer to the actual date and time but does not necessarily show the actual date and time the image was made. If your suspect used any applications to modify the image, such as adding effects or cropping the image, the modified date and time will be different than the actual date and time the image was taken.

We have to delve deeper into the file for the true data and time an image was taken as well as additional information such as the Global Positioning System (GPS) coordinates where the image was shot. Specifically we are going to be looking for EXIF data. EXIF data is sometimes referred to as data about data. More precisely it is data about the container information it is stored in. Image files contain data that the user cannot normally view and can include the camera or phone's make and model, the date and time the picture was taken, and, in some cases, the GPS coordinates where the picture was taken. Refer to the previous chapter for information on EXIF viewers and their use.

Before we begin you are going to need to know which image or images you want to examine. You may need to write down the file name and or the image name so you can come back to it later. The file the iPhone Backup Extractor saved the images to should will be named **Media**.

Inside the Media file will be another Media file containing yet another folder named DCIM. DCIM stands for Digital Camera Image and, depending on how many pictures your suspect took, will likely contain a number of other files labeled 100Apple, 101Apple, 102Apple, etc. These files are arranged chronologically so the file folder with the highest number will contain the most recent images. An easy way to scan all of the images in a folder is to select **View** at the top of the screen and then **Film Strip**. This will allow you to easily view all of the images in a folder and quickly identify any that need further investigating.

We wil need to install another free software program to view the EXIF data. There are a number of EXIF reader programs available for free on the Internet, but the one I like using most is Opanda iExif, which can be downloaded from http://www.opanda.com/en/iexif/. After installing the program, double click the new icon on your desktop to open it.

Select **Open** from the top of the Opanda program and navigate to the folder where you extracted the pictures to. When you have found the picture you are interested in, click on it and Open it in Opanda. The data displayed on the right side of the screen includes some of the information we are looking for.

Pictures: Taken with the Suspect's Camera Phone or Sent by a Friend?

One of the most common excuses suspects use for having pictures on their phone that they are not supposed to have is that the message was sent to them from a friend via text message (known as MMS or Multi-Media Message Service) or email. While it is possible for a picture to be sent to the suspect, which they then save to their phone, it is stored in a different location than the DCIM folder. Remember, images found in the DCIM folder were taken with the camera for that device.

To find images that were sent via text message we need to look in a different place. In the Media folder where we found the Media subfolder and the DCIM files is another folder labeled Library.

By clicking on the **Library** folder we will drill down through folders named SMS and then a folder named **Parts**. Inside the **Parts** folder may be a number of subfolders. Each subfolder contains one or more media files, including pictures that were sent or received as an attachment to a text message. Photos received via text message are downloaded to a different file. Depending on the operating system the Apple device received photos can either be found by entering the Expert Mode and clicking **Media**, then **PhotoData**, and then **100Apple** or by drilling down through Expert Mode to **Media** to **PhotoData**, to **MetaData** to **DCIM** to **100Apple**. Older operating systems lump photos received or sent from incoming and outgoing messages into the **Parts** file listed above. Here's the bad news. The EXIF data is partially removed from the picture when it is sent from one phone to another.

Expert Mode

Expert mode scares a lot of investigators off from looking at the data. But there can be some important clues hiding in those files. As we saw in the Recordings section, data can be stored, and sometimes hidden, within applications downloaded onto the device. The files obtained using the iPhone Backup Extractor are not the full applications themselves. That would take up a lot of room and possibly hours to download. Instead, recovered files contain the information stored inside the application or program. For example, the data the Angry Bird application stores contains the levels completed, scores, and achievements – not the actual files needed to play the game. Examining the information contained within the applications can be a real eye opener, but you need to know what to research and investigate.

The first step is the enter **Expert Mode** and then select the **+** symbol located next to **Applications**. This will populate a tree of every installed application.

Each application is listed by its domain first (such as .com or .org) and then the developer's name followed by the name of the application. Pay attention to the application names and either check them in Google or from the iTunes store to see exactly what it does.

If you find a suspicious application you can click on the **+** symbol next to it to expand the sub tree and see what information is contained in the file. You can see there are a number of .JPG files that are usually pictures or graphics files.

One of the nice things about using the Expert Mode is that you are able to access the underlying information used by the application without activating the application itself. In this case, if you were to look at the actual phone you might not find the suspicious application because it has a feature that lets the user conceal the application as something innocuous and non-threatening.

Even if you did locate the application, it is protected by a password, which our suspect might not be willing to give up. Placing a check box in the file above the .JPG picture files will select all files contained in the sub tree. Some sophisticated security applications will actually encrypt the data and it may be impossible to recover, but in this case you can obtain the images. They can be copied to your USB drive folder by pressing the **Extract Selected** button. Similar to retrieving photos, this process can take a while. The iPhone Backup Extractor will display green boxes over the icon on the screen of the phone in the main window to show you the progress it is making.

There is some EXIF data contained in the images that can be accessed using the techniques discussed in the Photos section.

PLISTs, DATs, and DBs

The Expert Mode in the iPhone Backup Extractor allows you to search for and recover a significant amount of information stored in your suspect's phone. These files are stored in three main types of database files that require specialized tools to read. The three main file types you will encounter while in Expert Mode are:

Property List Files (PLIST): these are used to store serialized data including a user settings. To read these files you will need to first download a PLIST editor from http://www.icopybot.com/plist-editor.htm.

An example of information you can view from a PLIST database is the information contained in **Library** and then **Maps**. Using the PLIST editor you will be able to view recent queries into Apple's mapping application. The results will show you the destination location the user queried. Export the Map file from within **Expert Mode** by selecting the **History.plist** file. There should be a green check mark next to the box. Export the file to your Desktop folder. Open the PLIST editor and select **Open** from the File menu. Navigate to where you exported the History.plist file, select it, and press **Open**. The destination addresses will be listed in plain text.

You can also use the PLIST editor to view files such as Internet browsing history, Internet bookmarks, and Internet Cache files. From the **Expert Mode** select **Library** and then **Safari** (Apple's Internet browser). This will automatically select the PLIST files below. View them in the PLIST editor and you will be able to see your suspect's Internet browsing history and bookmarks.

DAT: A DOS operating system database file. DAT databases can be difficult to view. The simplest method is to open the file using Windows Notepad (you can usually find it by pressing the **Start** button, **Accessories,** and then **Notepad**). Then copy the information from the Notepad into Windows Excel or a similar spreadsheet program.

I don't know why but if you try and open the file using Excel directly it does not seem to work as well. An example of information viewable using this technique can be found by navigating to the **Library** in Expert Mode, selecting **Keyboard**, and then selecting the file that reads **dynamic-text.dat**. This file contains information from the custom dictionary of your suspect's phone and has snippets of conversations he may have had via text message or email. A word of caution: there is no context for this information and there are no dates and times associated with the information. However, some pieces of conversation can be especially useful even without context or dates and times.

DB: A SQLite database file. To view these database files you will need a software program available at http://sourceforge.net/projects/sqlitebrowser/files/latest/download. Most of the files viewable using the SQLlite browser are recovered by the iPhone Backup Extractor, but it's nice to have a backup in case you find an interesting file using the Expert Mode.

Apple Time

A note about Apple and times: You may note there are not very many dates and times listed in some of the database files. Instead you will probably see what appear to be random, nonsensical numbers such as 363321407.4. This represents an epoch chosen as the origin of a particular era. The "epoch" serves as a reference point from which time is measured. In computing, most systems use an epoch date of 1/1/1970, commonly called UNIX time and the number 363321407.4 represents the number of seconds since epoch of UNIX time. Unfortunately, Apple had to be different and it programmed its systems using an epoch of 1/1/2001 to commemorate the release of Mac OS X 10.0. This makes most of the online time calculators ineffective because they are based off the UNIX epoch of 1/1/1970 and not 1/1/2001. The simplest way to calculate the date based off Apple's epoch is to go to the website Wolframalpha.com. Wolfram Alpha is sort of like Google, but on steroids. Lots and lots of steroids. It is designed to handle scientific calculations and complex problems using basic commands. Fortunately, you can enter the date 1/1/2001 plus 363321407 seconds and it will return the date of 7/7/2012.

Wide Angle Software's Ibackup Extractor

Wide Angle Software's iBackup Extractor is a great tool for examining the files created when an Apple product is synched to a computer. One of the things I like the most about the software is that the trial version allows me to see much more data than the iPhone Backup Explorer. However, on the test file I used I found the iBackup Explorer failed to acquire or missed some important data. To be fair, I used an old iPhone 3G for the test, but several key pieces of information that were recovered by the iPhone Backup Explorer were missed by the extractor. However, the reverse was also true. I was able to obtain some information from the iBackup Extractor that the other missed.

The software is easy to download and install. Once installed, open the software by double clicking. You should notice there are a couple of unusual items of data at the top. In this case the software detected the backup file as an iPhone 1.2. This presumably refers to the generation of iPhone as there was never an iPhone 2 – they went straight to the iPhone 3G, which is the test subject for this software. Also, you can see the Date of Backup lists a date of January 1, 0001. I don't know why the program wasn't able to capture the correct date of the backup.

File Explorer

There is a file explorer option similar to the other software but it did not appear to be functional during my test. This may be because the software was in Demo mode at the time and was not the fully licensed version.

Call Logs

I ran into the same problem using the demo version in that it did not locate or recover any of the call logs stored in the backup file.

Calendar

The software recovered all of the stored calendar entries and I liked the monthly display of the data. Press **Extract Calendar** and you can click on the particular entry to see the details.

Contacts

Contacts were a bit of a disappointment. The software recovered all of the contacts in the backup file. Unfortunately, the only option for exporting the data was either using a vCard or importing the data into Microsoft Outlook. Keep in mind this software was designed to let people recover their own contacts and was not designed specifically for law enforcement investigators. An important distinction of the iBackup Extractor is the contacts are in alphabetical order and are not recovered chronologically like with the other software.

Photos

Similar to the other software, the iBackup Extractor was able to locate and recover the images stored in the camera folder of the backup file. It did not recover any images stored in other applications on the phone.

Internet

The iBackup Extractor recovered bookmarks and search history from the backup file, a feat the other software couldn't finish. For clarification, bookmarks are exactly what they sound like, placeholders that allow a user to quickly return to a specific webpage without having to type the address or navigate through a search engine. Press the bookmark and you are automatically directed to the webpage. These can be handy for investigators as the user sets them up although it is possible for them to be accidentally set.

While bookmarks are useful, Internet browsing history is especially useful. The iBackup Extractor was able to locate and recover the complete chronological browsing history in the backup file. This information can be extremely helpful when investigating what the suspect was searching for online from his phone.

Text Messages

SMS (Short Message Service or text messages) and MMS (Multimedia Messaging Service) recovery was far superior using the iBackup Extractor. It is also capable of recovering iMessages, Apple's free text messaging service for its devices. This feature is not available on the other software. The messages were much easier to read and sort as opposed to using the other software.

Voicemail Messages

The iBackup Extractor can play the messages from within the software. You don't need to extract the messages and then use a third party software tool to listen to them.

iBackup BoT

Another option for backup reader software is called iBackup Bot. This software has a very minimalist appearance and does not seem as accessible as some of the other programs. However, the software's functions make it extremely useful.

Unlike many of the other programs I have discussed iBackup Bot features a fully functional demonstration version of the software. It has incorporated a nag screen that pops up every time you click on a file and the software tries to direct you to the purchase page for its website. Despite those annoyances, this software is one of the most versatile.

Getting Started

After installation the program automatically begins searching for the backup file. The launch and the initial screen is a little confusing – it doesn't tell you much nor does it prompt you to do anything. By exploring, I was able to find that even though the software has detected the backup file, it has not loaded anything for review. Presumably this is because there may be more than one backup file in iTunes such as when multiple devices are synched to one computer. From this screen you must **double-click** on the backup folder.

The results of double-clicking on the backup folder were a little confusing at first because the only thing initially available was what appears to be some random data in the box labeled **Path**. However, clicking on the + symbol located next to the backup folder name populated a list of all of the applications installed on the device when it was backed up.

Accessing the Data

iBackup Bot actually has two methods for looking at the messages, contacts, notes, call history, and other files. The first way is to click on the **Plugin** tab at the very top of the window

Text Messages

The first available data from the **Plugins** menu is **Messages**. Press the word **Messages** and iBackup Bot automatically displays all of the text messages from the backup file. Outgoing messages are highlighted in green.

Clicking on the + symbol next to a name or phone number will expand the list to show all of the text messages. This is convenient, but the screen can get crowded and confusing so I like to export the messages. To begin, click on the icon that looks like a blank sheet of paper with a small white arrow on it.

Unlike some of the other software programs, iBackup Bot gives you a couple of alternatives for exporting and saving the data. I prefer to use Excel files because there can literally be thousands of text messages and Excel is an easy option for finding and sorting text messages.

Contacts

The next set of data available from iBackup Bot is the contacts menu. Similar to the text messages, select **Contacts** from the Plugins menu. If you see a + symbol next to a name it means there is more than one phone number or additional information for the particular contact. What I really like about this feature in iBackup Bot is the software displays the creation date. This is particularly helpful for identifying new contacts as many of the other software programs export them in alphabetical order. Data can be exported the same way as the text messages.

Notes

iBackup Bot discovered all of the notes stored in the stock program found on Apple phones. Double-clicking on **note** brings up its contents in a pane on the right side. To export the data follow the process outlined earlier.

Call Logs

The call log information from iBackup Bot is very nice compared to other programs that recovered the same information. The contact name and number are displayed along with the date and time, the type of call, and duration.

Recent Email

This was a curious feature of iBackup Bot. The software displays the date and time of the email as well as the recipient's address but does not include any of the message information. This information could provide some clues for an investigator, but it doesn't really matter without the message content.

Media Browser-Pictures, Multimedia Messages, Voice Memos, and Voicemails

Apparently by design iBackup Bot places all of the media related files in one folder called Media Catalog. Within the file are several sub files including pictures taken with the phone's camera, multimedia messages, voice memos, voicemails, and a file titled "others."

There was compelling information contained in the camera role folder that was not recovered by any of the other software programs. You will see a list of the image files contained in the folder Examining information contained in the pane labeled **Path**. You should be able to see there are two types of file extension – .JPG and .PNG. The .JPG or jpeg files are pictures taken with the phone's camera. Portable network graphic (or .PNG) is usually an icon or image. However, in this case the images are actually those taken with the screen capture feature of the iPhone. You can take a picture of the screen of an iPhone by pushing the power and the home button at the same time.

I didn't realize all of the other software tools had failed to find these images until iBackup Bot did. There is an abundance of information contained in these and if you had only been using other software tools you would have missed all of the data. Apparently, the other programs are only looking for the .JPG or picture files and missed the .PNG files.

The .JPG and .PNG images are exportable and you can use the techniques outlined earlier to attempt to determine the date and time the picture was taken as well as the phone's GPS location at the time.

Multimedia Messages

As I discussed earlier, many times suspects will deny taking a picture and will tell you a friend sent it to them. As a reminder these pictures are stored in different files. A picture taken with the phone's camera will be in the **Camera Roll** folder (or DCIM folder for some of the other programs) and pictures or videos sent to them will be in the multimedia messages folder. While iBackup Bot found all of the pictures, it does not differentiate between pictures that were sent from the phone or to the phone.

Voice Memos

iBackup Bot found the voice memos made using Apple's stock program. However, when I attempted to play the messages from inside the program I received an error message that told me I needed to install iTunes and the Quick Time media player; both were already installed on the computer. I was able to play the voicemail messages by exporting the files and playing them outside the program using both Quick Time and several other media players.

Exporting Media Files

Exporting the data was extremely simple. From the main Media Browser screen select the same export box. iBackup Bot will give you the option to select which files you want to export and where you want to export them to. This process takes significant time and memory depending on how many pictures and other files exists.

Applications

iBackup Bot really differentiates itself software programs in the ability to examine data within the applications stored in the backup file. If you started from the beginning you know I used iPhone Backup Extractor to view data within applications, including naughty photos hidden in a password protected application called PhotoSafe. However, iBackup Bot's interface is much cleaner and allowed me to examine data in multiple other applications, which I missed on the first go around.

For example, the application QuickVoice is an alternative to the voice recording software included with every iPhone. By clicking on the application in the left side of the screen the software brought up the information in the right side of the screen, including the audio files made using the application. These recordings were not discovered by iBackup Bot or any other software I tried because the files are stored inside the application and were not found in the recordings file where Apple's application normally saves them.

One of the features I really liked was that iBackup Bot showed me the date the application was purchased in the right side of the screen and the account used to purchase it. It's helpful information if the suspect denies knowledge of the applications existence on his phone. Clicking on the **.CAF file** (Apple's file format) you will be presented with a screen that looks similar to this:

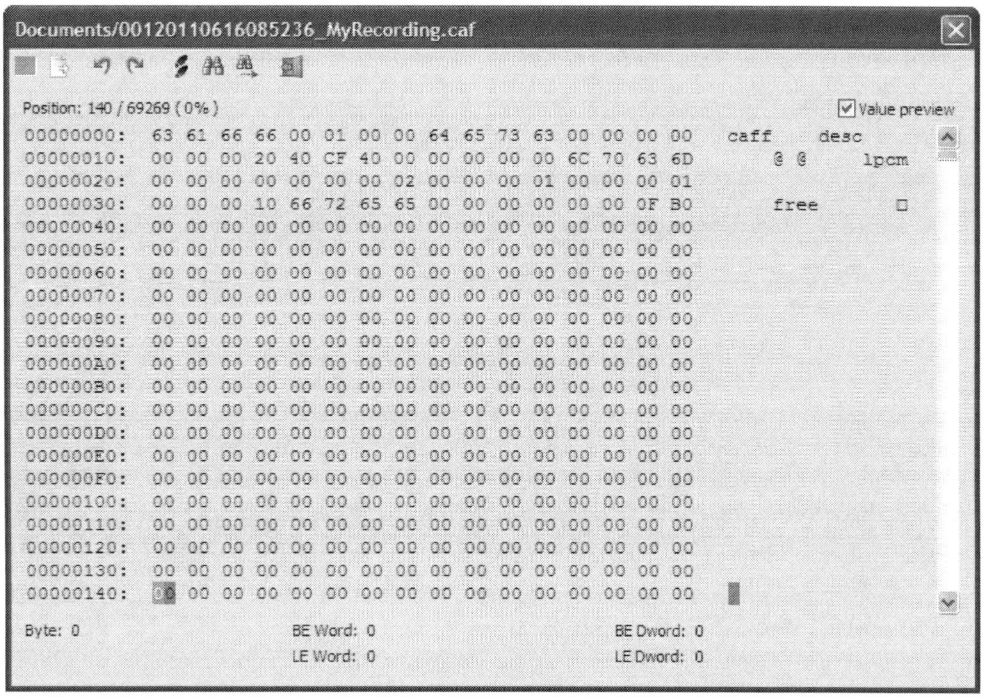

This is what is known as the hex view of a file. You won't be able to play the recording from inside the message, but you will be able to export it using the same technique discussed earlier and then play it using your favorite media player.

Now the detective work really begins. Other applications contained useful information such as the application 2Do. This is a task management application and by browsing through the files on the right side of the screen I could view the details of the information contained within the application. What I really liked was that iBackup Bot appears to have a native database reader for the files used by Apple and I did not have to download or use any additional software to read the files as I did with various programs.

Another example was found in the search of other applications that are ostensibly anonymous. The backup file examined below contained data for an application titled Anonymous D. A quick look at iTunes tells us this app is designed for anonymous web searching and web browsing without leaving pesky traces in the Internet search history. Someone should tell the developer his application is no match for a properly motivated and trained investigator because exploring the files in the application showed us a whole lot of data including sites visited "anonymously."

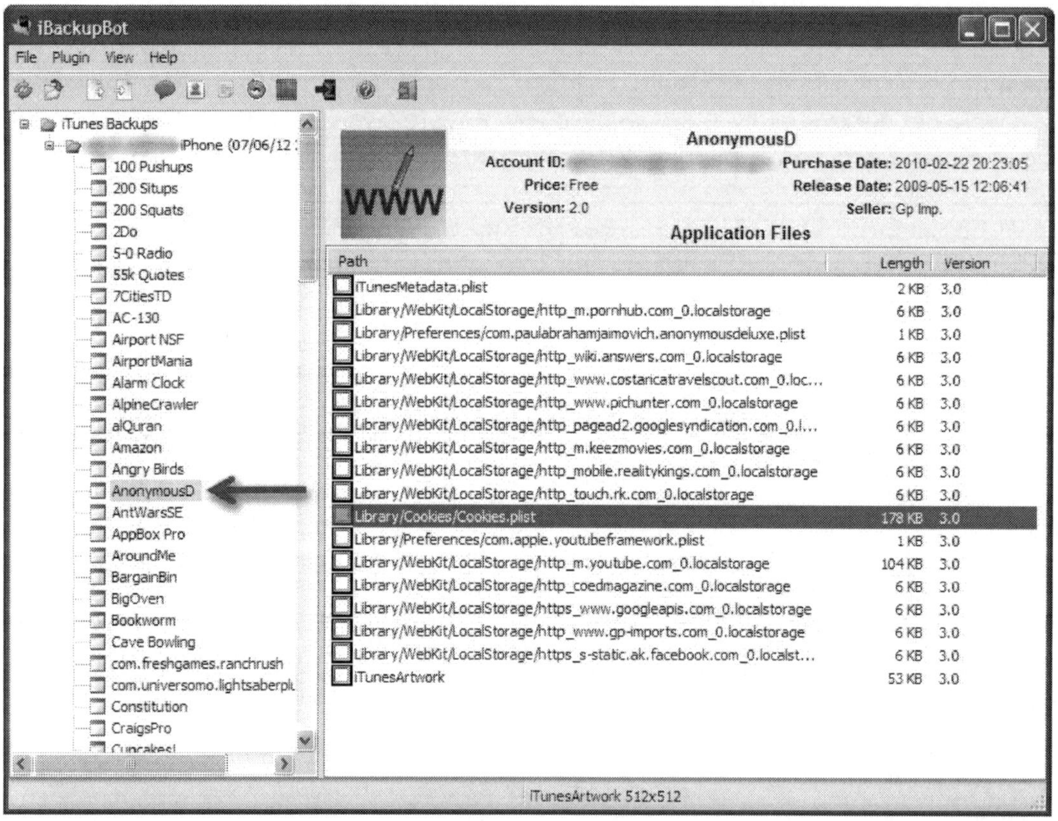

Again iBackup Bot saved the day by having a native database reader for the PLIST files the app stored the data in. From the software I was able to scan down the list of visited websites and find the ones the suspect visited. I was also able to discover interesting information in other common applications such as CraigsPro, an application used as an interface to craigslist.com. I could see postings that were accessed and the text content of the posting. Same thing with eBay's application. By clicking on the application on the left side I could view data stored in the application including thumbnail images of items viewed and searches made.

Of course I found the same troubling PictureSafe app that contained all of the pornographic images I found using the other software. Interestingly, the applications full name was listed in the other software but abbreviated PS when I used iBackup Bot. As a reminder, you may want to check and identify every application because the name may seem innocuous and its purpose may not be apparent.

Chapter 11: Sample Search Warrants

Normal Investigative Procedures

Based upon your affiant's training and experience, as well as the experience of other (list the Special Agents of the and/or state/local officers of), and based upon all of the facts set forth herein, it is your affiant's belief that the interception of electronic communications is the only available technique with a reasonable likelihood of securing the evidence necessary to prove beyond a reasonable doubt that (list the violator/s), and others as yet unknown are engaged in the above-described offenses.

Numerous investigative procedures usually employed in the investigation of this type of criminal case have been tried and have failed, reasonably appear to be unlikely to succeed if they are tried, or are too dangerous to employ.

Include a discussion of specific problems with using alternative investigative techniques in this investigation. Then cover standard problem areas, as synopsized below, modifying the statements to agree with the actual circumstances of your case.

Physical Surveillance

Physical surveillance has been attempted on many occasions in this investigation. Although it has proven valuable in identifying some of the targets' activities and associates, physical surveillance, if not used in conjunction with other techniques including electronic surveillance, is of limited value. Even if highly successful, physical surveillance does not always succeed in gathering evidence of the criminal activity under investigation. It is an investigative technique used to confirm meetings between alleged conspirators, and usually only leads investigators to speculate as to the purpose of the meeting(s). It is also a technique used to corroborate information obtained from confidential informants. Further, physical surveillance of the alleged conspirators will not establish conclusively the elements of the subjects' violations and has not and most likely will not establish conclusively the identities of various conspirators. Prolonged or regular physical surveillance of the targets would most likely be noticed, causing them to become more cautious in their illegal activities, to flee to avoid further investigation and prosecution, to cause a threat to the safety of the informant(s) and undercover agent(s), or otherwise to compromise the investigation.

With regard to this investigation, physical surveillance is unlikely to establish conclusively the roles of the named conspirators, to identify additional conspirators, to identify the conspirators' sources of supply, or otherwise to provide admissible evidence in regard to this investigation because:

- *Conspirators are using counter-surveillance, such as erratic driving behavior to detect surveillance or have evinced that they suspect law enforcement surveillance of their activities;*
- *The nature of the neighborhood forecloses physical surveillance [e.g., a close-knit community; cul-de-sac, dead end, or large apartment building; and/or the neighbors all know each other and call the police when surveillance is spotted];*
- *Further surveillance would only serve to alert the conspirators of the law enforcement interest in their activities and compromise the investigation.*

Use of Grand Jury Subpoenas

Based upon your affiant's experience and conversations with Assistant United States Attorneys for the_______________ District of_______________ who have experience prosecuting violations of criminal law, your affiant believes that subpoenaing persons who are believed to be involved in this conspiracy, or their associates before a Federal Grand Jury would most likely not be completely successful in achieving the stated goals of this investigation. The targets of this investigation, their co-conspirators, and other participants, should they be called to testify before the Grand Jury, would most likely be uncooperative and invoke their Fifth Amendment privilege not to testify. It would then be unwise to seek any kind of immunity for these persons because granting such immunity might foreclose prosecution of the most culpable members of this conspiracy, and could not ensure that such immunized witnesses would provide truthful testimony before the Grand Jury. Additionally, the service of Grand Jury subpoenas upon the targets or their co-conspirators would only alert the targets to the existence of this investigation, thereby causing them to become more cautious in their activities, to flee to avoid further investigation or prosecution, to threaten the lives of the informant(s) and the undercover agent(s), or otherwise to compromise this investigation.

Add specific information about any persons who have been subpoenaed before the Grand Jury, especially when the Fifth Amendment was invoked or when the witness later advised the targets.

Confidential Informants and Cooperating Sources

Reliable confidential informants/cooperating sources have been developed and used, and will continue to be developed and used, in regard to this investigation, but these sources:

- *Exist on the fringe of this organization and, therefore, have no direct contact with mid- or high-level members of the organization, or such contact is virtually impossible because the sources have no need to communicate with such individuals;*
- *Refuse to testify before the Grand Jury or at trial because of fear for personal or family safety or their testimony would be uncorroborated or otherwise subject to impeachment [due to prior record, criminal involvement, etc.];*
- *Are no longer associated with the targets of this investigation and their information is included for historical purposes only.*

None of the confidential informants described in this affidavit are able to furnish information that would identify fully all members of this ongoing criminal conspiracy or define the roles of those conspirators sufficiently for prosecution or that would identify sufficiently [the source(s) of supply or all details of delivery, quantities, financial arrangements, and the like].
Your affiant believes that information the confidential sources provided, even if all sources agreed to testify, would not, without the evidence available through the requested electronic surveillance, result in a successful prosecution of all of the participants.

Undercover Police Officers and Agents

> *Undercover police officers and/or agents have been unable to infiltrate the inner workings of this conspiracy due to the close and secretive nature of this organization. Your affiant believes that there are no undercover officers/agents who can infiltrate the conspiracy at a level high enough to identify all members of the conspiracy or otherwise satisfy all the goals of this investigation.*

Indicate if infiltration is not feasible because the confidential informant(s) is not in a position to make introductions of undercover officers to mid- or high-level members of the organization. Details on using undercover officers should have been provided in the body of this affidavit, with this section indicating the limitations.

Interviews of Subjects or Associates

> *Based upon your affiant's experience, your affiant believes that interviews of subjects or their known associates would produce insufficient information concerning the identities of all of the persons involved in the conspiracy, the source of the drugs, financing, the location of records, drugs, or other pertinent information regarding the subject crimes. Your affiant also believes that any responses to the interviews would contain a significant number of half-truths and untruths, diverting the investigation with false leads or otherwise frustrating the investigation. Additionally, such interviews would likely result in non-targeted interviewees alerting the members of the conspiracy, thereby compromising the investigation and resulting in the possible destruction or concealment of evidence and the possibility of harm to cooperating source(s), the identity of whom may become known or whose existence may otherwise be compromised.*

This portion of the affidavit is sometimes merged with the discussion on using the Federal Grand Jury. Any actual interviews conducted and any resulting problems should also be discussed here.

Search Warrants

> *The execution of search warrants in this matter has been considered. However, use of such warrants would, in all likelihood, not yield a considerable quantity of narcotics or relevant documents, nor would the searches conducted pursuant to such warrants be likely to reveal the total scope of the criminal operation and the identities of the co-conspirators. [It is unlikely that all, or even many, of the principals of this organization would be at any one location when a search warrant was executed.] Your affiant believes that search warrants executed at this time would be more likely to compromise the investigation by alerting the principals of the investigation, thereby, allowing unidentified co-conspirators to insulate themselves further from successful detection as well as to otherwise frustrate the purposes of this investigation.*

If search warrants were executed, then discuss the results and why this information is not enough to satisfy the investigation's goals.

Pen Registers/Telephone Tolls/Trap and Trace

Telephone toll/pen register/trap and trace information has been used in this investigation, as described above.

Provide a synopsis of the results reviewing these phone records; describe why this information is insufficient to identify fully other coconspirators or fulfill the needs of the investigation.

Other Limitations

Based upon the foregoing, it is your affiant's belief that the interception of electronic communications is an essential investigative means in obtaining evidence of the totality of the offenses in which the subject(s) and others as yet unknown are involved.

Provide details as to violence and other situations present in this investigation that limit the effectiveness of normal investigative techniques.

Prior Applications

Based upon a check of the records of the Federal Bureau of Investigation (and any other pertinent agency), no prior applications for an order authorizing the interception of wire, oral, or electronic communications have been made involving the persons, premises, or facilities named herein.

Include additional information concerning prior or ongoing electronic surveillance, such as person named, court that issued the order, date and relevance, if any, to the current investigation.

Conclusion

Your affiant believes that the facts alleged herein establish that the targets of this investigation are engaged in an ongoing criminal enterprise and that the evidence sought will be intercepted on a continuing basis following the first receipt of the particular communications that are the object of this request. Therefore, it is requested that the interception not be required to terminate when the communications described herein are first intercepted, but be allowed to continue until communications are intercepted, which fully reveal the scope of the enterprise, including the identities of all participants, their places and methods of operation, and the various criminal activities in which they are engaged which are in furtherance of the enterprise, not to exceed thirty (30) days measured from the earlier of the day on which investigative or law enforcement officers first begin to conduct an interception under this Court's Order or ten (10) days after the Order is entered.

____

(NAME)
Special Agent
(Agency)

Pen Registers/Traps and Traces

Except as provided in 18 U.S.C. § 3121, no person may install or use a pen register or a trap and trace device without first obtaining a court order under 18 U.S.C. § 3123 or under the Foreign Intelligence Surveillance Act (50 U.S.C. 1801, et seq.). The application may be made by an attorney for the government or a state law enforcement or investigative officer, and must certify that the information likely to be obtained is relevant to an ongoing criminal investigation. Unlike Title III pleadings, a pen register application need not establish probable cause and does not require prior Department approval.

The order is valid for 60 days, but may be extended for additional 60-day periods. It must specify the identity, if known, of the person who is leased, the name listed with the telephone line, or other facility to which the pen register or trap and trace device is to be applied; the identity, if known, of the person who is the subject of the criminal investigation; the attributes of the communications to which the order applies, including the number or other identifier and, if known, the location of the telephone line or other facility to which the pen register or trap and trace device is to be attached or applied, and, in the case of an order authorizing installation and use of a trap and trace device under subsection 3123(a) (2) (State court order), the geographic limits of the order; the offense(s) to which the information to be obtained from the pen register or trap and trace will relate; and direct, upon the request of the applicant, the furnishing of information, facilities, and technical assistance necessary to accomplish the installation of the pen register or trap and trace device.

The order should also direct that the application and order be sealed until otherwise ordered by the court, and that no disclosure of the existence of the pen register or trap and trace or the existence of the investigation be made to the subscriber or other persons until directed by the court. See generally Fregoso, supra ("The judicial role in approving use of trap and trace devices is ministerial in nature"); In re Application of United States for Order Authorizing Installation and Use of Pen Register and Trap and Trace Device, 846 F. Supp. 1555 (M.D. Fla. 1994) (the court must issue a pen register order on mere statutory certification by the government). A pen register/trap and trace order is executable anywhere within the United States and, upon service, the order applies to any person or entity providing wire or electronic communication service in the United States whose assistance may facilitate the execution of the order.

Whenever such an order is served on any person or entity not specifically named in the order, upon request of such person or entity, the attorney for the Government or law enforcement or investigative officer that is serving the order shall provide written or electronic certification that the order applies to the person or entity being served—18 U.S.C. § 3123(a).

Section 3121(c) requires that a government agency authorized to install and use a pen register or trap and trace device use technology reasonably available to it that restricts the recording or decoding of electronic or other impulses to the dialing, routing, addressing, and signaling information utilized in the processing and transmitting of wire or electronic communications so as not to include the contents of any wire or electronic communications.

Section 3127(3) defines a "pen register" as "a device or process which records or decodes dialing, routing, addressing, or signaling information transmitted by an instrument or facility from which a wire or electronic communication is transmitted, provided, however, that such information shall not include the contents of any communication but such term does not include any device or process used by a provider or customer of a wire or electronic communication service for billing, or recording as an incident to billing, for communications services provided by such provider or any device or process used by a provider or customer of a wire communication service for cost accounting or other like purposes in the ordinary course of its business."

Section 3127(4) defines a "trap and trace device" as "a device or process which captures the incoming electronic or other impulses which identify the originating number or other dialing, routing, addressing, and signaling information reasonably likely to identify the source of a wire or electronic communication, provided, however, that such information shall not include the contents of any communication."

Pen register and trap and trace devices may obtain any non-content information—all "dialing, routing, addressing, and signaling information"—utilized in the processing and transmitting of wire and electronic communications. Such information includes IP addresses and port numbers, as well as the **To** and **From** information contained in an email header. Pen/trap orders cannot, however, authorize the interception of the content of a communication, such as words in the "subject line" or the body of an Email. Agents and prosecutors with questions about whether a particular type of information constitutes content should contact the Office of Enforcement Operations in the telephone context (202-514-6809) or the Computer Crime and Intellectual Property Section in the computer context (202-514-1026).

Digital analyzers/cell site simulators/triggerfish and similar devices may be capable of intercepting the contents of communications and, therefore, such devices must be configured to disable the interception function, unless interceptions have been authorized by a Title III order.

Because section 3127 of Title 18 defines pen registers and trap and trace devices in terms of recording, decoding or capturing dialing, routing, addressing, or signaling information, a pen register/trap and trace order must be obtained by the government before it can use its own device to capture the ESN or MIN of a cellular telephone, even though there will be no involvement from the service provider.

Templates

Subscriber, billing, credit, toll records, text messages (including content) and call detail information (cell site and address locations and GPS location information if enabled) for the monthly billing periods that covers September 1, 2011 through November 9, 2011 through 30 days after this court order is issued for the telephone number listed below. Please deliver the requested data in an "Excel" or delimited "CSV" (non-pdf) format preferably via email to [insert email]

It is also ordered as follows:

GPS Locations: *Metro PCS is directed to provide real time cell tower information on request as well as to query the location of the phone if it is enabled with a Global Positioning System (GPS) interface.*
Release of Information: *MetroPCS, Inc. shall furnish affiant with all information gathered pursuant to this order in a reasonable time (five business days) while this order is in effect.*
Non-disclosure: *Having determined that there is probable cause to believe that the disclosure of the existence or execution of this order would impede a criminal investigation, it is ordered that MetroPCS, Inc., its officers, employees and agents not disclose to any person any information regarding the existence or execution of this order, or any information obtained pursuant to this order, unless ordered to do so by this Court, as such disclosure would impede the investigation.*
Jurisdiction: *This court has jurisdiction to issue this order pursuant to 18 USCS ss 3121 et seq, California Penal Code ss 1523 et seq, California Code of Civil Procedure ss 128, 187 and United States v. New York Telephone company (1977) 434 US 159.*

Telephone Records (Expanded)-Search Warrant

Subscriber information: This should give you the name, address, phone numbers, and other personal identifying information relating to the subscriber.
Account comments: Anytime the provider has contact with the customer or modifies the customer's account, a notation will be made by a service representative on the account.
Credit information: Many providers run a credit report on customer prior to activating the account
Billing records: Do not ask for toll information; that is a landline term for long distance. Specify the time period desired. Billing records are intended for the customer and typically appear different from other call detail records as they do not show information such as cell towers.
Outbound and inbound call detail: This is the real time, current activity that is not yet on the customer's bill and includes information they might not normally see such as cell tower information.
Call origination / termination location: For some carriers this information may only be available for a limited time (45 days) and gives location information on cell sites used, length of call, date, time, and numbers dialed.
Physical address of cell sites and RF coverage map: Needed to determine where cell site is located when you receive inbound & outbound or call origination & termination location. The RF coverage map models the theoretical radio frequency coverage of the towers in the system. You will want to limit this request to a specified geographical area. Most providers do not routinely send this with their response so you may need to re-contact them to obtain it.
Any other cellular telephone numbers that dial the same numbers as (xxx) xxx-xxxx: If you want to know who calls the same number the target calls (for example a pager or landline number). Depending on the provider this information may only be available for a limited time.
Subscriber information on any cellular numbers that (xxx) xxx-xxxx dials
Subscriber information on the carrier's network that is dialing the target
All of the above records whether possessed by cellular service provider [target of warrant] or any other cellular service provider

If you anticipate the suspect may be roaming or if the number is roaming in the providers market, you may be able to obtain information from other cellular carriers if you include this language in your description of records.

All stored communications or files, including voice mail, email, digital images, buddy lists, and any other files associated with user accounts identified as: account(s) xxxxxx, mobile numbers (xxx) xxx-xxxx, or email account bigdog123@gmail.com.

Cellular service providers now offer similar services to an Internet service provider (ISP) and maintain the same type of records such as text messaging, email, and file storage for the transfer of data including digital pictures. To access stored voicemail messages the provider may have to reset the user's password, which might alert them to the investigation. All connection logs and records of user activity for each such account including:

1. Connection/Disconnection dates and times.
2. Method of connection (e.g., telnet, ftp, http)
3. Data transfer volume.
4. User name associated with the connections.
5. Telephone caller identification records.
6. Any other connection information, such as the Internet Protocol (IP) address of the device

The above is a standard request made to ISP to track connection information. Remember with the type of cellular service offered today the user can send a message from the phone or from the associated account via a computer or other access device.

Any other records or accounts, including archived records related or associated to the above referenced names, user names, or accounts and any data field name definitions that describe these records.

This is the catch all to use when you want everything. This request also includes "archived" information. Many companies now "archive" records, allowing for the preservation of subscriber records for a significant time. Archived records are usually stored in a spread sheet format encompassing a variety of data fields. You must request the data field name definitions to understand the spreadsheet.

Telephone Records-Affidavit (General)

Through experience and training, your affiant knows cellular service providers maintain records related to subscriber information, account registration, credit information, billing and airtime records, outbound and inbound call detail, connection time and dates, Internet routing information (Internet Protocol number), and message content, that may assist in the identification or last known locations of person/s accessing and utilizing the account.

Through experience and training, your affiant knows that the cellular service provider maintains records that include cell site information and GPS location. Cell site information shows which cell site a particular cellular telephone was within at the time of the cellular phone's usage. Some model cellular phones are GPS enabled which allows the provider and user to determine the exact geographic position of the phone. Further, the cellular service provider maintains cell cite maps that show the geographical location of all cell sites within its service area. Using the cell site geographical location or GPS information, officers would be able to determine the physical location of the individuals using the listed cell phone numbers.

Telephone Records-Affidavit (Robbery)

Based upon my training and experience I know it is common for subjects to mask or hide their identity through multiple cellular and telephone accounts in order to conceal the identity of co-conspirators directly or indirectly involved in criminal activity. It is also common for witnesses, victims' and/or suspects of violent crimes such as a robbery to receive or make calls to other witnesses, victims and/or suspects as well as other persons that may have knowledge of or had acted in the robbery. The information may also answer questions as to the suspect(s) and witnesses whereabouts prior to this robbery as well as other investigative inquires.

Telephone Records-Affidavit (Narcotics/Gangs/Weapons/Trafficking)

Based on my training and experience I know that narcotics/gangs/weapons trafficking is a conspiratorial scheme involving many persons. Communications between co-conspirators is most frequently accomplished by use of telecommunications devices, including pagers, traditional landline telephones and cellular telephones.

Based on my training and experience, which includes not only my own participation in narcotics/gangs/weapons trafficking investigations, but also the information I have learned from other experienced investigators in this field, I know that a review of the telephone business records maintained by the cellular telephone service provider for telecommunications devices used by narcotics/gangs/weapons traffickers has resulted in the identification of dealers, distributors, smugglers, purchasers, and money launderers using the telephone, of previously unidentified co-conspirators, locations utilized by co-conspirators in the furtherance of illegal activities and other evidence of the crimes being investigated.

Voicemail Records/Password Reset-Search Warrant

IT IS HEREBY ORDERED Metro PCS shall change the voice mail password for ###-###-#### and to further provide law enforcement officers with digital copies of all messages stored in this account at the time of execution of the warrant.

Based on my training and experience, I know that voice mail messages for a cellular telephone are maintained on the service provider's computer server and are retrievable via the suspect's cellular telephone handset. I know that co-conspirators using cellular telephones commonly receive and/or store incoming voice mail messages, including messages from other gang members and gang leaders. I believe a review of the content of conversations stored in the voice mail messages stored via the suspect's cellular telephone in this investigation will provide investigative leads and evidence relevant to this ongoing criminal investigation.

Based on my training and experience, I know that a cellular telephone subscriber/user can access his/her voice mail messages from this server by prompting a previously established password on the keypad of the cellular telephone. Based on my training and experience, I know that upon receipt of a court order commanding such action, the service provider can change the existing password (usually created by the subscriber to maintain security and protection of his voice mail messages) to a new password that, when provided to law enforcement, will allow law enforcement to access the target cellular telephone's voice mail messages. I believe a review of the content of conversations stored in the voice mails for__________ will provide investigative leads and evidence relevant to this ongoing criminal investigation.

Telephone Records: Calls-to-Destination Search Warrant

A "calls-to-destination" search requires all cellular telephone service providers who have been served with the search warrant to check their records for any phone calls made to a particular phone number at a certain time. For example, a school receives a bomb threat from a caller ID blocked phone number. The investigators would a search warrant to *every* communications provider in the area, cellular and landline, asking them to check if any of their customers called the school at the time the call was received. Having a narrow time frame is extremely important as a location, such as a school, may receive multiple phone calls in a very short time frame.

Another scenario where this type of search may be useful is during narcotics investigations. If an informant receives a phone call from an upper level dealer, but the caller ID is blocked, a calls-to-destination search will reveal the phone number that called. The suggested search warrant language is:

> *IT IS HEREBY ORDERED, ________ (insert name of company) conduct a calls-to-destination search of its call detail records for the period ______ to ________ (limited time frame) on ______ (date) and identify any and all _______(insert name of company) telephone numbers used to place calls to the following telephone number _______ (insert phone number that received the call).*
> *IT IS HEREBY ORDERED, upon locating a corresponding record _______ (insert name of company) shall provide the then current subscriber information and its corresponding call detail records for incoming and outgoing calls for 30 days. _______ (insert name of company) shall also provide any other phone numbers on the same account and any other accounts billed to any subscriber at the same address as for any of the numbers located during the above search.*

Calling Card-Search Warrant

For calling cards where you have the actual card:

> *IT IS HEREBY ORDERED _________ (insert name of calling card company) shall provide telephone call detail records (date, time, duration, originating number, and destination number,) for all calls placed using the prepaid calling card _______ (describe card) administered by _________ (insert name of calling card company) associated with the "scratch off" personal identification number ___________ (insert PIN).*

If you know the suspect called an access number for a particular calling card company but don't know the PIN or the destination number:

> *IT IS HEREBY ORDERED_________ (insert name of Calling Card Company) shall provide telephone call detail records (date, time, duration, originating number, and destination number,) for all calls placed using its administered prepaid calling cards based on the following criteria:*
> *All calls placed via any of the ________ (insert name of calling card company) prepaid access phone numbers: 800-555-1234 from the following origination numbers _________ (insert target phone number).*
> *If any of the prepaid access numbers listed as being _________ (insert name of Calling Card Company) have in fact been resold to other, the company shall provide the name, address, and telephone contact information for the company they were sold to.*

Pen Register-Search Warrant

As defined in Section 3127 of the Federal Electronic Communications Privacy Act, the term "pen register" means a device that records or decodes electronic or other impulses that identify the numbers dialed or transmitted on the telephone line to which such device is attached. The term "trap-and-trace device" is one that captures the incoming electronic or other impulses which identify the originating number of an instrument or device from which a wire or electronic communication was transmitted. Thus, the installation of both a pen register and a trap-and-trace device allows law enforcement officers to determine telephone numbers being called from and to a particular telephone number.

The Federal Electronic Communications Privacy Act regulates the use of pen registers and trap- and-trace devices (18 U.S.C. Sections 3121-3126). Section 3122 expressly authorizes a state investigative or law enforcement officer to apply for an order, or an extension of an order, authorizing the installation and use of a pen register or a trap-and-trace device, in writing under oath, to a court of competent jurisdiction of the state. Section 3123 requires the applicant to justify that the information likely to be obtained is relevant to an ongoing criminal investigation (as opposed to probable cause required for a warrant). The order shall not exceed 60 days. Extensions may be granted, but only upon application, and may not exceed 60 days. The order shall direct that the order be sealed until otherwise directed by the court, and it shall direct the person owning the line or assisting in the installation not to disclose until further order of the court. An order for a pen register or trap-and-trace device may be issued by a state court of "general criminal jurisdiction"; (Cal Const. Art. VI, Section 10), so a Superior Court judge can issue an order.

No California case has discussed the propriety of obtaining such an order. However, a California Attorney General Opinion, No. 03-406, has concluded that the federal statutes governing the installation of pen registers and trap-and-trace devices do not provide authority for issuance of a state court order permitting a state law enforcement officer to install or use pen registers or trap-and-trace devices. The rationale of the Opinion is that in California telephone call information is protected by a right of privacy so that such records may only be obtained by a law enforcement officer upon a showing of probable cause. In view of this Opinion, California state law enforcement officers should only use the federal statute if the court order includes a provision that the affidavit establishes probable cause.

A pen register and/or a trap-and-trace device may be installed if a search warrant has been obtained. (*People v. Larkin* (1987) 194 Cal.App.3d 650, 654. In contrast to the procedure described above, the warrant is valid for only 10 days pursuant to Penal Code section 1534, and the warrant must establish probable cause.

Pursuant to 18 U.S.C. § 3123, Applicant has certified that the information likely to be obtained by such use is relevant to an ongoing criminal investigation being conducted by the River City Police Department (hereinafter the "Investigative Agency") in connection with possible violations of 11378/11379 H&S by Kathleen Lama.

Pursuant to 18 U.S.C. §§ 2703(c)(1)(B) and 2703(d), Applicant has offered specific and articulable facts showing that there are reasonable grounds to believe that records or other information identifying subscribers or customers (not including the contents of communications) for telephone numbers identified through the pen register and trap and trace devices on the Target Device, changes in service regarding the Target Device, cell site information' regarding the Target Device, and records or other information pertaining to subscriber(s) or customers) (but not including the contents of communications) for the Target Device are relevant and material to an ongoing criminal investigation of the specified offenses.

> *THEREFORE, IT IS HEREBY ORDERED, pursuant to 18 U.S.C. § 3123, that agents of the Investigative Agency may install, or cause to be installed, and use a pen register to record or decode dialing, routing, addressing, or signaling information transmitted' from the Target Device, to record the date and time of such dialings or transmissions, and to record the length of time the telephone receiver in question is "off the hook" or connected for incoming or outgoing calls and attempts, for a period of sixty (60) days from the date this Order is filed by the Court,' provided, however, that such information shall not include the contents of any communication;*

IT IS FURTHER ORDERED, pursuant to 18 U.S.C. § 3123, that agents of the Investigative Agency may install, or cause to be installed, and use a trap and trace device on the Target Device to capture and record the incoming electronic or other impulses which identify the originating numbers or other dialing, routing, addressing, or signaling information' reasonably likely to identify the source of a wire or electronic communication, and to record the date, time, and duration of calls created by such incoming impulses, for a period of sixty (60) days from the date this Order is filed by the Court, provided, however, that such information shall not include the contents of any communication;

IT IS FURTHER ORDERED that the Investigative Agency is authorized to obtain from the telephone Service Providers: (1) any cell site information that might be available when the Target Device is turned "on" but a call is not in progress; (2) information that would allow it to triangulate' multiple antenna tower locations and thereby attempt to determine the precise location of the user of the Target Device; or (3) Global Positioning System (GPS) information regarding the location of the Target Device. Provider shall, upon direction of the investigating officer or his designee, initiate a signal to determine the location of the target telephone on the service provider's network or use other reference points which may reasonably be available to assist in the location of the target telephone. This may be performed at any time of day or night, owing to the potential need to locate the target phone outside of normal business hours.

IT IS FURTHER ORDERED pursuant to 18 U.S.C. §§ 2703(c)(1)(B), 2703(c)(2), 2703(d), 3122, and 3123 that Adelphia Communications, Adelphia Long Distance, Allegiance Telecom of California, Inc., Astound, AT&T California, AT&T Local Service, AT&T Long Distance, AT&T Midwest, AT&T Nevada, AT&T Southwest, Bell South Telecommunications, Broadwing Communications, Cellco Partnership doing business as Verizon Wireless, Cellular One, Central Wireless Partnership doing business as Sprint PCS, Cingular Wireless, Comcast, Cox Communications, Dobson Cellular, Dobson Communications, Edge Wireless LLC, Electric Lightwave Inc., Ernbarq, Ernest Communications, Evans Telephone Company, Frontier: A Citizens Communications Company, Genesis Communications International, Google, ICG Communications, ICG Telecom Group, Locus Communications, Metrocall, Metro PCS, Mpower, Nationwide Paging, Navigator Telecommunications LLC, Network Services LLC, Nextel Communications, NIX Communications, Pac West Telecomm Incorporated, Qwest Communications, RCN Communications, Roseville Telephone Company, Slcype, Sprint-Nextel, TelePacific Communications, Teligent, Time-Warner Telecom, T-Mobile USA Inc., TracFone Wireless, USA Mobility, US Cellular, US TelePacific Corp. doing business as TelePacific Communications, Verizon California, Verizon District of Columbia, Verizon Maryland, Verizon New Jersey, Verizon New York, Verizon Northwest, Verizon Texas, Virgin Mobile, Vonage, Weblink Wireless, West Coast PCS LLC doing business as Sure West Wireless, Western Wireless Corporation, XO Communications, and any and all other telephone service providers (including any Internet service provider or other electronic communications provider providing voice-over IP telephony [VoIP (collectively, "the Telephone Service Providers"), and any and all other persons or entities providing wire or electronic communications service in the United States whose assistance may facilitate the execution of this Order, shall disclose or provide, upon oral or written demand by agents of the Investigative Agency:

1. *For the Target Device, records or other information pertaining to subscriber(s) or customer(s), including cell site information and toll or call detail records (including in two-way radio feature mode) for the sixty (60) days prior to the date this Order is filed by the Court (but not including the contents of communications);*

2. *For the Target Device, after receipt and storage, records or other information pertaining to subscriber(s) or customer(s), including (1) the means and source of payment for the service and (2) cell site information, provided to the United States for (a) the origination of a call from the Target Device or the answer of a call to the Target Device and (b) the termination of the call (but not including the contents of the communications);*
3. *Pursuant to 18 U.S.C. § 2703(c), after receipt and storage, the following subscriber records and other information for all published, non-published, or unlisted dialing, routing, addressing, or signaling information captured by the pen register and trap and trace device on the Target Device:*

   i. *name;*
   ii. *address;*
   iii. *local and long distance telephone connection records, or records of session times and durations with Target Device;*
   iv. *length of service (including start date) and types of service utilized; and*
   v. *telephone or instrument number or other subscriber number or identity, including any temporarily assigned network address; and*

4. *Any and all changes (including additions, deletions, and transfers) in service regarding the Target Device, including telephone numbers; other unique identifiers such as Electronic Serial Numbers (ESNs), Subscriber Identity Modules (SIMs), International Mobile Subscriber Identifiers (IMSIs), International Mobile station Equipment Identities (IMEIs), and/or Urban Fleet Mobile Identifiers (UFMIs); and subscriber information (published, non-published, listed, or unlisted) associated with these service changes.*

IT IS FURTHER ORDERED that this authorization for the installation and use of a pen register and trap and trace device applies not only to the Target Device, but also to any changed telephone numbers subsequently assigned to an instrument bearing the. same ESN, SIM, IMSI, IMEI, and/or UFMI as the Target Device, or any changed ESN, SIM, IMSI, IMEI; and/or UFMI subsequently assigned to the same telephone number as the Target Device, and any additional changed telephone number, ESN, SIM, IMSI, IMEI, and/or UFMI, whether the changes occur consecutively or simultaneously, listed to the same subscriber or account number as the Target Device within the 60-day period authorized by this Order and further, that, pursuant to 18 U.S.C. § 3123(b)(1)(C), the tracing operations authorized by this Order be without geographical limits.

IT IS FURTHER ORDERED that the Target Device's Telephone Service Providers shall furnish the results of the pen register and trap and trace devices to agents of the Investigative Agency as soon as practicable, twenty four (24) hours a day for the duration of the Order.

IT IS FURTHER ORDERED that the Target Device's Telephone Service Providers be reasonably compensated by the Investigative Agency for reasonable expenses directly incurred in providing information, facilities, and assistance.

IT IS FURTHER ORDERED, pursuant to 18 U.S.C. §§ 2705(b) and 3123(d), that this Order and the Application be SEALED until otherwise ordered by the Court, that the identity of any targets of the investigation and the possible violations thereof may be redacted from any copy of the Order served on any service provider or other person, and that the Target Device's Telephone Service Providers and any other Telephone Service Provider which provides service to a telephone number that either places telephone calls to, or receives telephone calls from, the Target Device, shall not disclose in any manner, directly or indirectly, by any action or inaction, the existence of this Order, in full or redacted form, of the pen register or trap and trace devices, or of this investigation, to the listed subscribers for the Target Device, or to any other person unless otherwise ordered by this Court.

IT IS FURTHER ORDERED that night service of this order is authorized, owing to the need to utilize one or more provisions of this Order outside of normal business hours.

IT IS FURTHER ORDERED, this Order shall remain in effect from 5/8/2012 through 7/7/2012.

Pen Register for PCTDD-Affidavit

Post cut through dialed digits (PCTDD) are any key pressed after the completion of the call. For example, a suspect who calls an 800 phone number to access his calling card will be required to enter the PIN and the phone number to be dialed. Without a pen register it is extremely difficult to obtain the PIN and dialed number. Some jurisdictions consider this content, similar to a text message, and require a Title III wiretap to obtain it. Others allow it to be obtained through the use of a pen register. Check with your legal advisor regarding the legality of collecting PCTDD.

Based on my training an experience I know it is nearly impossible to obtain outgoing phone calls when the call is routed through a prepaid phone card without the use of a pen register. The only other alternative known to your affiant is to identify the personal identification number (PIN) printed on the card and seek an additional Court Order for phone records made on that particular prepaid calling card. This requires accessing the actual physical card and would be extremely difficult without alerting the suspect to law enforcement's interest.

Application for Electronic Communications Interception

UNITED STATES DISTRICT COURT
__________DISTRICT OF__________

IN THE MATTER OF THE APPLICATION)
OF THE UNITED STATES OF AMERICA FOR)
AN ORDER AUTHORIZING THE INTERCEPTION)
OF ELECTRONIC COMMUNICATIONS)

APPLICATION FOR INTERCEPTION OF ELECTRONIC COMMUNICATIONS

______________, Assistant United States Attorney__________________, District of______________/Special Attorney, United States Department of Justice, being duly sworn, states:

1. I am an investigative or law enforcement officer of the United States within the meaning of Section 2510(7) of Title 18, United States Code, that is, an attorney authorized by law to prosecute or participate in the prosecution of United States federal felony offenses. I am also an attorney for the Government as defined in Rule l(b) (1) of the Federal Rules of Criminal Procedure, and, therefore, pursuant to Section 2516 (3)of Title 18, United States Code, I am authorized to make an application to a Federal judge of competent jurisdiction for an order authorizing the interception of electronic communications.
2. This application is for an order pursuant to Section 2518 of Title 18, United States Code, authorizing the interception of electronic communications for a thirty (30) day period of (name the interceptees) and others as yet unknown to (and from) the (telephone/digital-display paging device (s)/facsimile machine/computer/Internet account number__________) (bearing or using the telephone number(s)_________ ,subscribed to by_________) concerning federal

felony offenses, that is, offenses involving violations of (list the section(s) of the United States Code and briefly describe the applicable offense(s)) .

3. I have discussed all of the circumstances of the above offenses with Special Agent___________ of the_____________, who has directed and conducted this investigation, and have examined the Affidavit of Special Agent_____________ of this date (attached to this application as Exhibit__, and which is incorporated by reference). Whereof your applicant states upon information and belief that:

   a. there is probable cause to believe that (name the violators) have committed, are committing and will continue to commit violations of (list the offenses);
   b. there is probable cause to believe that particular electronic communications of (name the interceptee/s) concerning the above-described offenses will be obtained through the interception for which authorization is here in applied. In particular, there is probable cause to believe that the communications to be intercepted will concern the telephone numbers of associates of (name the violators) and the dates, times and places for commission of the aforementioned federal felony offenses when (name the interceptee/s) communicate with their co-conspirators, aiders and abettors, and other participants in the conspiracy, thereby identifying the co-conspirators and aiders and abettors of (name the violators) and others as yet unknown, their places of operation. In addition, these communications are expected to constitute admissible evidence of the above-described offenses;
   c. normal investigative procedures have been tried and have failed, reasonably appear to be unlikely to succeed if tried, or are too dangerous to employ, as are described in further detail in the attached affidavit of Special Agent and
   d. there is probable cause to believe that (list the facilities from which, or the place where, the electronic communications are to be intercepted) are being, and will continue to be used in connection with the commission of the above-described offenses.

The attached affidavit contains a full and complete statement of facts concerning all previous applications that have been made to any judge of competent jurisdiction for authorization to intercept, or for approval of interception of wire, oral or electronic communications involving any of the same individuals, facilities, or places specified in this application.

On the basis of the allegations contained in this application and on the basis of the attached affidavit of Special Agent____________________.

IT IS HEREBY REQUESTED that this Court issue an order, pursuant to the power conferred on it by Section 2518 of Title 18, United States Code, authorizing the (name the investigative agency/agencies) to intercept electronic communications to (and from) the above-described (telephone/digital display paging device, facsimile machine, computer, Internet account), and providing that such interceptions not terminate automatically after the first interception that reveals the manner in which the alleged co-conspirators and others as yet unknown conduct their illegal activities, but continue until all communications are intercepted which reveal fully the manner in which the above-named persons and others as yet unknown are committing the offenses described herein, and which reveal fully the identities of their confederates, their places of operation, and the nature of the conspiracy involved therein, or for a period of thirty(30) days measured from the day on which investigative or law enforcement officers first begin to conduct an interception under this Court's order or ten (10) days after this order is entered, whichever is earlier.

IT IS REQUESTED FURTHER that in the event that the target facility is transferred outside the territorial jurisdiction of this Court, interceptions may take place in any other jurisdiction within the United States.

IT IS REQUESTED FURTHER that this Court issue an order pursuant to Section 2518(4) of Title 18, United States Code, directing that (list the communications service provider(s)), a communication service provider as defined in Section 2510(15) of Title 18, United States Code, shall furnish, and continue to furnish, the applicant and investigative agency with all information, facilities and technical assistance necessary to accomplish the interceptions unobtrusively and with a minimum of interference with the services that such providers are according the persons whose communications are to be intercepted, and to ensure an effective and secure installation of electronic devices capable of interception of electronic communications to (and from) the above-described (telephone/digital display paging device/facsimile machine/computer/Internet account), with the service provider to be compensated by the applicant for reasonable expenses incurred in providing such facilities or assistance.

IT IS REQUESTED FURTHER that, to avoid prejudice to this criminal investigation, the Court order the said providers of electronic communication service and their agents and employees not to disclose or cause a disclosure of this Court's order or the request for information, facilities and assistance by the identify the investigative agency/agencies or the existence of the investigation to any person other than those of their agents and employees who require said information to accomplish the services hereby requested. In particular, said providers and their agents and employees should be ordered not to make such disclosure to a lessee, telephone subscriber, or any interceptee or participant in the intercepted communications.

IT IS REQUESTED FURTHER that this Court direct that this order be executed as soon as practicable after it is signed and that all monitoring of communications shall be recorded and examined by monitoring agents or attorneys to determine the relevance of the intercepted electronic communications to the pending investigation and that the disclosure of the contents or nature of the electronic communications intercepted be limited to those communications relevant to the pending investigation, in accordance with the minimization requirements of Chapter 119 of Title 18, United States Code. The interception of communications authorized by this Court's order must terminate upon attainment of the authorized objectives or, in any event, at the end of thirty (30) days measured from the earlier of the day on which investigative or law enforcement officers first begin to conduct an interception under this Court's order or ten (10) days after the order is entered, whichever is earlier.

IT IS REQUESTED FURTHER that the Court order that either Assistant United States Attorney/Special Attorney________________, or any other Assistant United States Attorney/Special Attorney familiar with the facts of this case, provide to the Court a report on or about the (tenth), (twentieth) and (thirtieth) days following the date of this order showing what progress has been made toward achievement of the authorized objectives and the need for continued interception. If any of the aforementioned reports should become due on a weekend or holiday,

IT IS REQUESTED FURTHER that such report become due on the next business day thereafter.

IT IS REQUESTED FURTHER that the Court order that its orders, this application and the accompanying affidavit and proposed order(s), and all interim reports filed with the Court with regard to this matter be sealed until further order of this Court, except that copies of the order(s), in full or redacted form, may be served on the (identify the investigative agency/agencies) and the service provider(s) as necessary to effectuate the Court's order as set forth in the proposed order(s) accompanying this application.

DATED this day _____________ of_________ 20__

____

Assistant United States Attorney

SUBSCRIBED and SWORN to before me
This_____________ day of_________20__

____

UNITED STATES DISTRICT COURT JUDGE
(District)

Affidavit for Electronic Communications Interception

UNITED STATES DISTRICT COURT
____________DISTRICT OF_____________

IN THE MATTER OF THE APPLICATION)
OF THE UNITED STATES OF AMERICA)
FOR AN ORDER AUTHORIZING THE) MISC. NO.
INTERCEPTION OF ELECTRONIC)
COMMUNICATIONS)

AFFIDAVIT IN SUPPORT OF APPLICATION

___________________, being duly sworn, deposes and states as follows:

1. I am a Special Agent with the United States Department of Justice. I have been so employed by the___________ for the past ( ) years. I have participated in investigations involving (organized crime/drug trafficking, etc.) activities for the past ( ) years. (Describe present assignment.)
2. I am an investigative or law enforcement officer of the United States within the meaning of Section 2510(7) of Title 18, United States Code, in that I am empowered by law to conduct investigations and to make arrests for federal felony offenses.
3. This affidavit is submitted in support of an application for an order authorizing the interception of electronic communications occurring (specify the facility or facilities to which the application and affidavit are directed).
4. I have participated in the investigation of the above offenses. As a result of my personal participation in this investigation, through interviews with and analysis of reports submitted by other (Special Agents of the ___________and/or other state/local law enforcement personnel), and by the analysis of (surveillance logs/pen register information, etc.), I am familiar with all aspects of this investigation. On the basis of this familiarity, and on the basis of other information which I have reviewed and determined to be reliable, I allege that:
    a. There is probable cause to believe that (list the violators) have committed, are committing, and will continue to commit (list the offense(s)—can be any federal felony offense).
    b. There is probable cause to believe that particular electronic communications of (list the interceptees) concerning the above offenses will be obtained through the interception of such communications to (and from) the (telephone/digital pager/facsimile machine/computer/Internet account) (assigned/using/bearing account/telephone number/s____________________), subscribed to by_____________(and if applicable, the facility's physical location)). In particular, there is probable cause to believe that the communications to be intercepted will concern the (telephone numbers of associates of (list the violator(s)) and the dates, times, places, and plans for commission of the aforementioned federal felony offenses when (list the interceptees) communicate with their co-conspirators, aiders and abettors, and other participants in the conspiracy, thereby identifying the co-conspirators and aiders and abettors of(the violators), and others as yet unknown, their places of operation, (etc.). In addition, these communications are expected to constitute admissible evidence of the above-described offenses.

c. Normal investigative procedures have been tried and have failed, reasonably appear to be unlikely to succeed if tried, or are too dangerous to employ, as is described herein in further detail.

d. There is probable cause to believe that (list the facilities over which the electronic communications are to be intercepted) are being, and will continue to be, used in connection with the commission of the above offenses.

Persons Expected to be Intercepted

Include a short description of each expected interceptee; if appropriate, explain why certain participants in the offenses are not expected to be interceptees.

Facts and Circumstances

Provide an in-depth discussion of the facts that support the probable cause statements above. If informant information provides a basis for probable cause for any of the required information, provide adequate qualifying language for each informant.

In drug cases, if appropriate, include a "facts and circumstances" paragraph regarding use of pagers, e.g., "I know from my training, experience, and discussions with other experienced agents that narcotics traffickers frequently use paging devices to further their illicit business. Pagers permit co-conspirators to contact each other with virtually no possibility that their communications will be intercepted. For example, the type of paging device used in this matter allows a conspirator to signal a confederate, identify himself through a numerical code, and convey the number of a secure or non-suspect telephone, usually a pay telephone, at which he can be contacted. The conspirator receiving this information can then go to a secure or non-suspect telephone, return the call, and engage in a criminal discussion with his confederate, which, under normal circumstances, will be incapable of interception by law enforcement authorities."

Order for Interception of Electronic Communications

UNITED STATES DISTRICT COURT
___________DISTRICT OF__________

IN THE MATTER OF THE APPLICATION)
OF THE UNITED STATES OF AMERICA FOR)
AN ORDER AUTHORIZING THE INTERCEPTION)
OF ELECTRONIC COMMUNICATIONS)

ORDER AUTHORIZING THE INTERCEPTION
OF ELECTRONIC COMMUNICATIONS

Application under oath having been made before me by________________, Assistant United States Attorney __________________, District of_______________, /Special Attorney, United States Department of Justice, an "investigative or law enforcement officer" of the United States within the meaning of Section 2510(7) of Title 18, United States Code, and an attorney for the Government as defined in Rule l(b) (1) of the Federal Rules of Criminal Procedure, for an Order authorizing the interception of electronic communications pursuant to Section 2518 of Title 18,United States Code, and full consideration having been given to the matter set forth therein, the Court finds:

a. There is probable cause to believe that (list the violators) have committed, are committing, and will continue to commit violations of (list the offenses—can be any federal felony offense);

b. There is probable cause to believe that particular electronic communications of (list the interceptees) concerning the above-described offenses will be obtained through the interception for which authorization is herein applied. In particular, there is probable cause to believe that the communications to be intercepted will concern the telephone numbers of associates of (the violator/s) and the dates, times, places and plans for commission of the aforementioned federal felony offenses when (list the interceptee/s) communicate with their co-conspirators, aiders and abettors and other participants in the conspiracy, thereby identifying the co-conspirators and others as yet unknown, their places of operation, (etc.). In addition, these communications are expected to constitute admissible evidence of the above-described offenses;

c. It has been established adequately that normal investigative procedures have been tried and have failed, reasonably appear to be unlikely to succeed if tried, or are too dangerous to employ;

d. There is probable cause to believe that (list the facilities over which the electronic communications are to be intercepted) have been, are being and will continue to be used in connection with the commission of the above-described offenses.

WHEREFORE, IT IS HEREBY ORDERED that Special Agents of the (name the investigative agency/agencies) are authorized to intercept electronic communications over the above-described facilities.

PROVIDED that such interception(s) shall not terminate automatically after the first interception that reveals the manner in which the alleged co-conspirators and others as yet unknown conduct their illegal activities, but may continue until all communications are intercepted which fully reveal the manner in which the above-named persons and others as yet unknown are committing the offenses described herein, and which reveal fully the identities of their confederates, their places of operation, and the nature of the conspiracy involved therein, or for a period of thirty (30) days measured from the day on which investigative or law enforcement officers first begin to conduct an interception under this Order or ten (10) days after this Order is entered, whichever is earlier.

IT IS ORDERED FURTHER that, pursuant to 18 U.S.C. § 2518(3), in the event that the target facility is transferred outside the territorial jurisdiction of this court, interceptions may take place in any other jurisdiction within the United States.

IT IS ORDERED FURTHER that, based upon the request of the Applicant pursuant to Section 2518(4) of Title 18, United States Code, (name the communication service provider/s), communication service provider(s) as defined in Section 2510(15) of Title 18,United States Code, shall furnish, and continue to furnish, the Applicant and the investigative agency/agencies with all information, facilities, and technical assistance necessary to accomplish the interceptions unobtrusively and with a minimum of interference with the services that such provider(s) is according the persons whose communications are to be intercepted, with the service provider(s) to be compensated by the Applicant for reasonable expenses incurred in providing such facilities or assistance.

IT IS ORDERED FURTHER that, to avoid prejudice to the Government's criminal investigation, the above provider(s) of electronic communication service and its agents and employees are ordered not to disclose or cause a disclosure of this Order or the request for information, facilities, and assistance by the (name the investigative agency/agencies) or the existence of the investigation to any person other than those of its agents and employees who require said information to accomplish the services hereby ordered. In particular, said provider(s) and its agents and employees shall not make such disclosure to a lessee, telephone or paging device subscriber or any interceptee or participant in the intercepted communications.

IT IS ORDERED FURTHER that this Order shall be executed as soon as practicable and that all monitoring of the electronic communications shall be recorded and examined by the monitoring agents or attorneys to determine the relevance of the intercepted electronic communications to the pending investigation and that the disclosure of the contents or nature of the electronic communications intercepted be limited to those communications relevant to the pending investigation, in accordance with the minimization requirements of Chapter 119 of Title 18, United States Code. The interception of communications must terminate upon the attainment of the authorized objectives, not to exceed thirty (30) days measured from the earlier of the day on which investigative or law enforcement officers first begin to conduct an interception under this Order or ten (10) days after the Order is entered.

IT IS ORDERED FURTHER that Assistant United States Attorney/Special Attorney__________ or any other Assistant United States Attorney/Special Attorney familiar with the facts of this case shall provide this Court with a report on or about the (tenth), (twentieth) and (thirtieth) days following the date of this Order showing what progress has been made toward achievement of the authorized objectives and the need for continued interception. If any of the above-ordered reports should become due on a weekend or holiday, IT IS ORDERED FURTHER that such report shall become due on the next business day thereafter.

IT IS ORDERED FURTHER that this Order, the application, affidavit, and proposed Order(s), and all interim reports filed with this Court with regard to this matter shall be sealed until further order of this Court, except that copies of the Order(s), in full or redacted form, may be served on the (investigative agency/agencies) and the service provider(s) as necessary to effectuate this Order.101

____

UNITED STATES DISTRICT COURT JUDGE
(District)

Application for Wire and/or Oral Interceptions

UNITED STATES DISTRICT COURT
______________ DISTRICT OF _____________

IN THE MATTER OF THE APPLICATION)
OF THE UNITED STATES OF AMERICA FOR)
AN ORDER AUTHORIZING THE INTERCEPTION)
OF (WIRE) (ORAL) COMMUNICATIONS)

APPLICATION FOR INTERCEPTION OF (WIRE) (ORAL) COMMUNICATIONS

______________Assistant United States Attorney, _________District of___________, being duly sworn, states:

1. I am an investigative or law enforcement officer of the United States within the meaning of Section 2510(7) of Title 18,United States Code, that is, an attorney authorized by law to prosecute or participate in the prosecution of offenses enumerated in Section 2516 of Title 18, United States Code.
2. This application is for an order pursuant to Section 2518 of Title 18, United States Code, authorizing the interception of (wire) (oral) communications until the attainment of the

authorized objectives or, in any event, at the end of thirty (30) days from the earlier of the day on which the investigative or law enforcement officers first begin to conduct an interception under the Court's order or ten (10) days after the order is entered, of (list those persons who will be intercepted over the telephone or within the premises, "interceptees") and others as yet unknown (if wire: "to and from the telephone(s) bearing the number(s)______________, subscribed to by________________ and located at/billed to________________") (if oral: "occurring inside the premises located at__________________________" or "occurring in and around a (describe the make, color and year of the vehicle) bearing the license plate number___________ and the vehicle identification number___________") concerning enumerated in Section 2516 of Title 18, United States is, offenses involving violations of (list section(s) offenses Code, that of the U.S. Code and describe briefly the applicable offense(s)) that are being committed by (list the interceptees and those persons who are also part of the conspiracy but may not necessarily be intercepted over the target facility or within the target premises/vehicle, collectively they are referred to as "violators,") and others as yet unknown.

3. Pursuant to Section 2516 of Title 18, United States Code, the Attorney General of the United States has specially designated the Assistant Attorney General, any Acting Assistant Attorney General, any Deputy Assistant Attorney General or any acting Deputy Assistant Attorney General of the Criminal Division (or, in the case of a roving interception, the Assistant Attorney General or Acting Assistant Attorney General in the Criminal Division) to exercise the power conferred on the Attorney General by Section 2516 of Title 18, United States Code, to authorize this Application. Under the power designated to him by special designation of the Attorney General pursuant to Order Number (currently 2407-2001) of (currently March 8, 2001), an appropriate official of the Criminal Division, (insert official's name and title), has authorized this Application. Attached to this Application are copies of the Attorney General's order of special designation and the Memorandum of Authorization approving this Application.

4. I have discussed all of the circumstances of the above offenses with Special Agent____________ of the (name the investigative agency), who has directed and conducted this investigation and have examined the Affidavit of Special Agent______________, which is attached to this Application and is incorporated herein by reference. Based upon that Affidavit, your applicant states upon information and belief that:

   a. there is probable cause to believe that (list the violators) and others as yet unknown have committed, are committing, and will continue to commit violations of (list the offenses—must be enumerated in Section 2516 of Title18, United States Code);

   b. there is probable cause to believe that particular(wire) (oral) communications of (name the interceptee(s))concerning the above-described offenses will be obtained through the interception of (wire) (oral) communications. In particular, these (wire) (oral) communications will concern the (characterize the types of criminal communications expected to be intercepted). In addition, the communications are expected to constitute admissible evidence of the commission of the above-stated offenses;

   c. normal investigative procedures have been tried and failed, reasonably appear to be unlikely to succeed if tried, or are too dangerous to employ, as is described in further detail in the attached Affidavit.

   d. there is probable cause to believe that (identify fully the telephone(s) from which, or the premises where, the wire or oral communications are to be intercepted)is/are being used and will continue to be used in connection with the commission of the above-described offenses.

   If a roving interception, add the following language:

"The attached affidavit contains information demonstrating, within the meaning of Title 18, United States Code, Section 2518(11) (a) and/or (b), that (if oral: "specification of the place(s) where communications of (name the interceptees) are to be intercepted is not practical"
(if wire: "that (name the person/s)use of various and changing facilities could have the effect of thwarting interception from a specified facility")

5. The attached Affidavit contains a full and complete statement of facts concerning all previous applications which are known to have been made to any judge of competent jurisdiction for approval of the interception of the oral, wire or electronic communications of any of the same individuals, facilities, or premises specified in this Application. (If there has been no previous electronic surveillance, state: "The applicant is aware of no previous applications made to any judge for authorization to intercept the oral, wire or electronic communications of any of the persons or involving the (facilities) (premises) specified in this application.")

WHEREFORE, your applicant believes that there is probable cause to believe that (name the violators) and others as yet unknown are engaged in the commission of offenses involving (cite to the offenses), that (name the interceptees) and others yet unknown are using (the telephone bearing the number___________, subscribed to by_______________ and located at/billed to_____________) and/or (the premises or vehicle described as______________) in connection with the commission of the above-described offenses; and that (wire) (oral) communications of (name the interceptees) and others yet unknown will be intercepted (over the above-described telephone facility) and/or (within the above-described premises or the above-described vehicle).

Based on the allegations set forth in this application and on the affidavit of Special Agent_____________ attached, applicant the requests this court to issue an order, pursuant to the power conferred upon it by Section 2518 of Title 18, United States Code, authorizing the (investigative agency) to intercept (wire communications to and from the above-described facility(ies)) and/or (oral communications from the above-described premises) until such communications are intercepted that reveal the manner in which the named violators and others unknown participate in the specified offenses and reveal the identities of (his/their) co-conspirators, place(s) of operation, and nature of the conspiracy, or for a period of (not to exceed thirty (30) days) measured from the day on which the investigative or law enforcement officers first begin to conduct the interception or ten days from the date of this order, whichever occurs first.

If interception of oral communications is requested, add:

IT IS REQUESTED FURTHER that this Court issue an order pursuant to Section 2518 of Title 18, United States Code, authorizing Special Agents of the (name investigative agency) to make all necessary surreptitious and/or forcible entries to effectuate the purposes of this Court's Order, including entries to install, maintain, and remove electronic listening devices from (describe the premises/vehicle). The applicant shall notify the Court of any surreptitious entry.

If interception of wire communications is requested, add:

IT IS REQUESTED FURTHER that the authorization given be intended to apply not only to the target telephone number(s) listed above, but to any changed telephone number subsequently assigned to the same cable, pair, and binding posts utilized by the target telephone(s) within the thirty (30) day period. (If the telephone is a cellular telephone, the language should state: "The authorization given be intended to apply not only to the target telephone number(s) listed above, but to any changed telephone number or any other telephone number subsequently assigned to or used by the instrument bearing the same electronic serial number as the target cellular phone within the thirty (30) day period.") It is also requested that the authorization be intended to apply to background conversations intercepted in the vicinity of the target telephone(s) while the telephone(s) is off the hook or otherwise in use.)

If multi-jurisdictional authorization for a portable/mobile facility is requested, add:

IT IS REQUESTED FURTHER that in the event that the target facility/vehicle is transferred outside the territorial jurisdiction of this Court, interceptions may take place in any other jurisdiction within the United States.

If a roving interception is requested, add:

IT IS REQUESTED FURTHER that this Court issue an order authorizing the roving interception of the (wire) (oral) communications of (name the target(s)) (if wire: "from various and changing telephone facilities), (if oral: "from various locations in (name the jurisdiction) that are not practical to specify) as provided in Title 18, United States Code, Section2518 (11)(a) or (b) and as specifically authorized by the (Acting)Assistant Attorney General of the Criminal Division, for a thirty (30) day period).

If wire communication, add:

IT IS REQUESTED FURTHER that this Court issue an order pursuant to Section 2518(4) of Title 18, United States Code, directing the (name the communications service provider(s)), an electronic communications service provider as defined in Section2510(15) of Title 18, United States Code, to furnish and continue to furnish the (investigative agency) with all information, facilities and technical assistance necessary to accomplish the interceptions unobtrusively and with a minimum of interference with the services that such providers are according the persons whose communications are to be intercepted, and to ensure an effective and secure installation of electronic devices capable of intercepting wire communications over the above-described telephone. The service provider shall be compensated by the Applicant for reasonable expenses incurred in providing such facilities or assistance.)

IT IS REQUESTED FURTHER, to avoid prejudice to this criminal investigation, that the Court order the providers of electronic communication service and their agents and employees not to disclose or cause a disclosure of this Court's Order or the request for information, facilities, and assistance by the (investigative agency) or the existence of the investigation to any person other than those of their agents and employees who require this information to accomplish the services requested. In particular, said providers and their agents and employees should be ordered not to make such disclosure to a lessee, telephone subscriber, or any interceptee or participant in the intercepted communications.

IT IS REQUESTED FURTHER that this Court direct that its Order be executed as soon as practicable after it is signed and that all monitoring of (wire) (oral) communications shall be conducted in such a way as to minimize the interception and disclosure of the communications intercepted to those communications relevant to the pending investigation, in accordance with the minimization requirements of Chapter 119 of Title 18, United States Code. The interception of (wire) (oral) communications authorized by this Court1s Order must terminate upon attainment of the authorized objectives or, in any event, at the end of thirty (30) days measured from the day on which investigative or law enforcement officers first begin to conduct an interception or ten (10) days after the Order is entered.

Monitoring of conversations must immediately terminate when it is determined that the conversation is unrelated to communications subject to interception under Chapter 119 of Title 18, United

States Code. Interception must be suspended immediately when it is determined through voice identification, physical surveillance, or otherwise, that none of the named interceptees or any of their confederates, when identified, are participants in the conversation unless it is determined during the portion of the conversation already overheard that the conversation is criminal in nature.

IT IS REQUESTED FURTHER that the Court order that either (Applicant/AUSA) or any other AUSA familiar with the facts of the case provide the Court with a report on or about the (tenth), (twentieth) and (thirtieth) days following the date of this Order showing what progress has been made toward achievement of the authorized objectives and the need for continued interception. If any of the aforementioned reports should become due on a weekend or holiday, it is requested further that such report become due on the next business day thereafter.

IT IS REQUESTED FURTHER that the Court order that its Orders, this application and the accompanying affidavit and proposed Order(s), and all interim reports filed with the Court with regard to this matter be sealed until further order of this Court, except that copies of the Order(s), in full or redacted form, may be served on the (name the investigative agency/agencies) and the service provider(s) as necessary to effectuate the Court1s Order as set forth in the proposed order(s) accompanying this application.

DATED this ___________ day of_____________, 20__.

____

(Name and title of the applicant)

(NAME)
Assistant United States Attorney

SUBSCRIBED and SWORN to before me
This__________ day________ of_______, 20__

UNITED STATES DISTRICT COURT JUDGE
(District)

Affidavit for Oral and/or Wire Interception

UNITED STATES DISTRICT COURT
____________DISTRICT OF____________

IN THE MATTER OF THE APPLICATION)
OF THE UNITED STATES OF AMERICA FOR)
AN ORDER AUTHORIZING THE INTERCEPTION)
OF (WIRE) (ORAL) COMMUNICATIONS)

Affidavit In Support Of Application

Introduction

______________, being duly sworn, deposes and states as follows:

1. I am a Special Agent with the United States Department of Justice. I have been so employed by the (name the agency) for the past ( ) years. I have participated in investigations involving (organized crime/drug trafficking/money laundering, etc.) activities for the past (_) years. (Describe present assignment.)
2. I am an investigative or law enforcement officer of the United States within the meaning of Section 2510(7) of Title 18, United States Code, and am empowered by law to conduct investigations and to make arrests for offenses enumerated in Section 2516 of Title 18, United States Code.
3. This affidavit is submitted in support of an application for an order authorizing the interception of (wire) (oral) communications occurring (describe the facility or premises to which the application and affidavit are directed).
4. I have participated in the investigation of the above offenses. As a result of my personal participation in this investigation, through interviews with and analysis of reports submitted by other (Special Agents of the___________ and/or other state/local law enforcement personnel), I am familiar with all aspects of this investigation. On the basis of this familiarity, and on the basis of other information which I have reviewed and determined to be reliable, I allege the facts to show that:
    a. there is probable cause to believe that (name the violators) are committing, and will continue to commit violations of (list the offenses—must be ones enumerated in Section 2516of Title 18, United States Code);
    b. there is probable cause to believe that particular (wire) (oral) communications of (name the interceptees) concerning the above offenses will be obtained through the interception of such communications (if wire: "to and from the telephone(s)bearing the number(s)___________, subscribed to by_____________ and located at/billed to______________"; if oral: "occurring within premises located at_____________" or "occurring in and around a (indicate the make, model and year of the vehicle) bearing the license plate___________ and vehicle identification number_________"); if a roving wire interception: "over various and changing facilities within (identify the jurisdiction) used by (name the particular interceptee(s)—do not include the language "and others yet unknown"); if a roving oral interception: "within presently unknown premises used by (name the particular interceptee(s)—do not include the language "and others yet unknown") that it is impractical to specify.")

In particular, these communications are expected to concern the specifics of the above offenses, including (I) the nature, extent and methods of the (describe the illegal activity) business of (name the violators) and others; (ii) the nature, extent and methods of operation of the business of (name the violators) and others; (iii) the identities and roles of accomplices, aiders and abettors, co-conspirators and participants in their illegal activities; (iv) the distribution and transfer of the contraband and money involved in those activities; (v) the existence and location of records; (vi) the location and source of resources used to finance their illegal activities; (vii) the location and disposition of the proceeds from those activities; and (viii) the locations and items used in furtherance of those activities. In addition, these (wire) (oral) communications are expected to constitute admissible evidence of the commission of the above-described offenses.

The statements contained in this affidavit are based in part on information provided by Special Agents of the (name the investigative agency/agencies), on conversations held with detectives and officers from the (identify the local/state police department), on information provided by confidential sources, and on my experience and background as a Special Agent of the___________. Since this affidavit is being submitted for the limited purpose of securing authorization for the interception of (wire) (oral) communications, I have not included each and every fact known to me concerning this investigation. I have set forth only the facts that I believe are necessary to establish the necessary foundation for an order authorizing the interception of (oral) (wire) communications.

Persons Expected to be Intercepted

Include a short description of each known violator; if appropriate, explain why certain participants in the offenses are not expected to be interceptees. If applicable, note which persons are currently facing pending state or federal criminal charges.

Facts and Circumstances

Provide an in-depth discussion of the facts in support of the probable cause statements set forth above. If informant information provides a basis for any of the required information, provide adequate qualifying language for each informant. Remember that you must show probable cause 1) that the alleged offenses are being committed; 2) that the named subjects and others unknown are committing them; and 3) that the targeted telephone(s) and/or premises is/are being used to commit these offenses. It is Department of Justice policy that pen register or telephone toll information for the target telephone(s), or physical surveillance of the targeted premises, standing alone, is generally insufficient to establish probable cause. Probable cause to establish criminal use of the facilities, or premises requires independent evidence of use in addition to pen register or surveillance information, e.g. informant or undercover information. On rare occasions, criminal use of the target facilities or premises may be established by an extremely high volume of calls to other known or suspected co-conspirators that coincides with incidents of illegal activity, or by a regular pattern of telephone or premises use involving known or suspected co-conspirators going back for a period of years.

When requesting a roving wire interception, you must establish that the specifically targeted subject uses various and changing facilities in such a way that has the effect of thwarting law enforcement's ability to intercept the subject's communications from a specified facility. See 18 U.S.C. §2518(11) (b) (ii). The effect on the government's ability to intercept a subject's calls can be demonstrated by the subject's actions over a period of time (e.g., physical surveillance and phone record analysis establishing that the subject travels from payphone to payphone to call other co-conspirators, or the analysis of phone records demonstrating that the subject uses different cellular phones in succession for brief periods of time (usually three weeks or less) to contact other coconspirators, in furtherance of criminal activity). Roving wiretaps will be authorized for public and cellular telephones only; and only when it is clear that the telephones cannot be identified in advance, and that the subject's actions are having the effect of preventing the government from conducting interceptions over his phones.

In roving oral interceptions (see 18 U.S.C. §2518 (11) (a) (ii)), you must establish probable cause that it is not practical to specify the place where the oral communications of the targeted individual(s) are to be intercepted. Once again, a roving oral interception will generally be authorized only for public facilities, vehicles, hotel rooms, or similar locations, and a pattern of activity demonstrating the impracticability of naming specific premises must be established.

Need for Interception

Based upon your affiant's training and experience, (as well as the experience of the other (Special Agents of the and/or state/local officers), and based upon all of the facts set forth herein, it is your affiant's belief that the interception of (wire) (oral) communications is the only available technique that has a reasonable likelihood of securing the evidence necessary to prove beyond a reasonable doubt that (name the violators), and others as yet unknown are engaged in the above-described offenses.

Your affiant states that the following investigative procedures, which are usually employed in the investigation of this type of criminal case, have been tried and have failed, reasonably appear to be unlikely to succeed if they are tried, or are too dangerous to employ.

Alternative Investigative Techniques

The following is an example of language that discusses the use of physical surveillance in general, you should also discuss the effectiveness of this, and the following other investigative techniques, as they are applicable to your particular case.

Physical surveillance has been attempted on numerous occasions during this investigation. Although it has proven valuable in identifying some activities and associates of (list the violators), physical surveillance, if not used in conjunction with other techniques, including electronic surveillance, is of limited value. Physical surveillance, even if highly successful, has not succeeded in gathering sufficient evidence of the criminal activity under investigation. Physical surveillance of the alleged conspirators will not (has not) established conclusively the elements of the violations and has not and most likely will not establish conclusively the identities of various conspirators. In addition, (continued) surveillance is not expected to enlarge upon information now available; rather, such prolonged or regular surveillance of the movements of the suspects would most likely be noticed, causing them to become more cautious in their illegal activities, to flee to avoid further investigation and prosecution, to cause a real threat to the safety of the informant(s) and undercover agent(s), or to otherwise compromise the investigation.

Physical surveillance is also unlikely to establish conclusively the roles of the named conspirators, to identify additional conspirators, or otherwise to provide admissible evidence in regard to this investigation because (discuss any of the following which are applicable to the case):

- *The subjects are using counter-surveillance techniques, such as erratic driving behavior, or have evinced that they suspect that law enforcement surveillance is being conducted against them; and/or*
- *It is not possible to determine the full nature and scope of the aforementioned offenses by the use of physical surveillance; and/or*
- *The nature of the neighborhood forecloses physical surveillance; (e.g., close-knit community, physical location (cul-de-sac, dead-end, large apartment building), observant neighbors); and/or*
- *Further surveillance would only serve to alert the suspects of the law enforcement interest in their activities and compromise the investigation.*

Use of Grand Jury Subpoenas

> *Based upon your affiant's experience and conversations with Assistant United States Attorney______________, who has experience prosecuting violations of criminal law, your affiant believes that subpoenaing persons believed to be involved in this conspiracy and their associates before a Federal Grand Jury would not be completely successful in achieving the stated goals of this investigation. If any principals of this conspiracy, their co-conspirators and other participants were called to testify before the Grand Jury, they would most likely be uncooperative and invoke their Fifth Amendment privilege not to testify. It would be unwise to seek any kind of immunity for these persons, because the granting of such immunity might foreclose prosecution of the most culpable members of this conspiracy and could not ensure that such immunized witnesses would provide truthful, testimony. Additionally, the service of Grand Jury subpoenas upon the principals of the conspiracy or their co-conspirators would only (further) alert them to the existence of this investigation, causing them to become more cautious in their activities, to flee to avoid further investigation or prosecution, to threaten the lives of the informant(s) and the undercover agent(s), or to otherwise compromise the investigation.*

Add specific information regarding any persons who have been subpoenaed before the Grand Jury, especially when the Fifth Amendment was invoked or when the witness later advised the targets.

Confidential Informants and Cooperating Sources

> *Reliable confidential informants/cooperating sources have been developed and used, and will continue to be developed and used, in regard to this investigation. However, these sources (discuss only those that are applicable):*

- *exist on the fringe of this organization and have no direct contact with mid- or high-level members of the organization, or such contact is virtually impossible because the sources have no need to communicate with such individuals; and/or*
- *refuse to testify before the Grand Jury or at trial because of fear of personal or family safety, or their testimony would be uncorroborated or otherwise would be subject to impeachment (due to prior record, criminal involvement, etc.);and/or*
- *are no longer associated with the subjects of this investigation (and their information is included for historical purposes only); and/or*
- *are unable to furnish information which would identify fully all members of this ongoing criminal conspiracy or which would define the roles of those conspirators sufficiently for prosecution.*

In addition, discuss whether the information confidential sources provided, even if all sources agreed to testify, would not, without the requested electronic surveillance, result in a successful prosecution of all of the participants.

Undercover Agents

> *Undercover agents have been unable to infiltrate the inner workings of this conspiracy due to the close and secretive nature of this organization. Your affiant believes that there are no undercover*

agents who can infiltrate the conspiracy at a level high enough to identify all members of the conspiracy or otherwise satisfy all the goals of this investigation.

Indicate if infiltration is not feasible because the confidential informant(s) is not in a position to make introductions of undercover agents to mid- or high-level members of the organization. Details on using undercover agents should have been provided in the body of the affidavit, with this section indicating the limitations involved.

Interviews of Subjects or Associates

Based upon your affiant's experience, I believe that interviews of the subjects or their known associates would produce insufficient information as to the identities of all of the persons involved in the conspiracy, the source of (the drugs, financing, etc.), the location of (records, drugs, etc.), and other pertinent information regarding the named crimes. Your affiant also believes that any responses to the interviews would contain a significant number of untruths, diverting the investigation with false leads or otherwise frustrating the investigation. Additionally, such interviews would also have the effect of alerting the members of the conspiracy, thereby compromising the investigation and resulting in the possible destruction or concealment of documents and other evidence, and the possibility of harm to cooperating sources whose identities may become known or whose existence may otherwise be compromised.

This portion of the affidavit is sometimes merged with the discussion regarding the use of the Federal Grand Jury. Any actual interviews conducted, and any resulting problems, should also be discussed here.

Search Warrants

The execution of search warrants in this matter has been considered. However, use of such warrants would, in all likelihood, not yield a considerable quantity of (narcotics, money, or other identified contraband) or (relevant documents) nor would the searches be likely to reveal the total scope of the illegal operation and the identities of the co-conspirators. (It is unlikely that all, or even many, of the principals of this organization would be at anyone location when a search warrant was executed.) The affiant believes that search warrants executed at this time would be more likely to compromise the investigation by alerting the principals to the investigation and allowing other unidentified members of the conspiracy to insulate themselves further from successful detection.

Pen Registers/Telephone Toll Records/Traps and Traces

Pen register (and/or trap and trace) information has been used in this investigation, including pen register(s) (and/or traps and traces) on the target telephone(s), as described above. The pen register (and/or trap and trace) information has verified frequent telephone communication between the target telephone(s) and other telephones. Pen registers (and/or traps and traces), however, do not record the identity of the parties to the conversation, cannot identify the nature or substance of the conversation, or differentiate between legitimate calls and calls for criminal purposes. A pen register (and/or trap and trace) cannot identify the source or sources of the controlled substances, nor can it, in itself, establish proof of the conspiracy. Telephone toll information, which identifies the existence and length of telephone calls placed from the target telephone to telephones located outside of the local service zone, has the same limitations as pen registers (and/or traps and traces), does not show local calls, and is generally available only on a monthly basis.

Other Limitations

Provide details concerning violence, such as murdered or hurt witnesses, threats, etc., and other situations present in your investigation that limit the effectiveness of normal investigative techniques.

Based upon the foregoing, it is your affiant's belief that the interception of (wire) (oral) communications is an essential investigative means in obtaining evidence of the offenses in which the subject(s) and others as yet unknown are involved.

Prior Applications

Based upon a check of the records of the (Federal Bureau of Investigation, the Drug Enforcement Administration, and any other appropriate agency), no prior federal applications for an order authorizing or approving the interception of wire, oral, or electronic communications have been made involving the persons, premises or facilities named herein.

If the facts warrant, add information concerning prior or ongoing electronic surveillance, including the dates of the interception, the jurisdiction where the order was signed and the relevance, if any, to the instant application. While there is no obligation to conduct a search of state law enforcement electronic surveillance indices, information about prior state taps must be included if the government has knowledge of them through other means.

Minimization

All interceptions will be minimized in accordance with the minimization requirements of Chapter 119 of Title 18, United States Code, and all interceptions conducted pursuant to this Court's Order will terminate upon attainment of the authorized objectives or, in any event, at the end of thirty (30) days measured from the earlier of the day on which investigative or law enforcement officers first begin to conduct an interception under the Court's Order or ten (10) days after the Order is entered. Monitoring of conversations will terminate immediately when it is determined that the conversation is unrelated to communications subject to interception under Chapter 119 of Title 18, United States Code. Interception will be suspended immediately when it is determined through voice identification, physical surveillance, or otherwise, that none of the named interceptees or any of their confederates, when identified, are participants in the conversation, unless it is determined during the portion of the conversation already overheard that the conversation is criminal in nature.

If pertinent, add additional language concerning the use of foreign languages and other minimization considerations particular to the case, such as targeting the use of public facilities or premises or non-interception of privileged communications of interceptees who have pending criminal charges.

____
(NAME)
Special Agent
(Agency)

Sworn to before me this
_________day of______ , 20__

____
UNITED STATES DISTRICT COURT JUDGE

(District)

Order for Interception of Wire and/or Oral Communications

UNITED STATES DISTRICT COURT
_________DISTRICT_________

IN THE MATTER OF THE APPLICATION)
OF THE UNITED STATES OF AMERICA FOR)
AN ORDER AUTHORIZING THE INTERCEPTION)
OF (WIRE) (ORAL) COMMUNICATIONS)

ORDER AUTHORIZING THE INTERCEPTION OF (WIRE)
(ORAL) COMMUNICATIONS

Application under oath having been made before me by__________________, Assistant United States Attorney,_______________ District of _____________,an investigative or law enforcement officer of the United States within the meaning of Section 2510(7) of Title 18, United States Code, for an Order authorizing the interception of (wire) (oral) communications pursuant to Section 2518 of Title 18, United States Code, and full consideration having been given to the matter set forth therein, the Court finds:

The following lettered paragraphs should be virtually identical to the probable cause paragraphs contained in the application and affidavit.

a. there is probable cause to believe that (list the violators) have committed, and are committing, and will continue to commit violations of (list the offenses—must be ones enumerated in Section 2516 of Title 18, United States Code);

b. there is probable cause to believe that particular(wire) (oral) communications of (name the interceptees) concerning the above-described offenses will be obtained through the interception for which authorization has herewith been applied. In particular, there is probable cause to believe that the interception of (wire communications to and from the telephone bearing the number___________, subscribed to by____________ and located at/billed to______________ (oral communications occurring in the premises located at and/or______________ in and around the vehicle described as _______________), will concern the specifics of the above offenses, including the manner and means of the commission of the offenses(s);

(If roving interception is applied for: "the application has also demonstrated adequately within the meaning of Title 18, United States Code, Section 2518(11) (a) and/or (b),that (if oral: "specification of the place(s) where the oral communications of (name the interceptee(s)) are to be intercepted is not practical") (if wire: "(name the interceptee(s) use of various changing facilities could have the effect of thwarting interception from a specified facility.");

c. it has been established that normal investigative procedures have been tried and have failed to reasonably appear to be unlikely to succeed if tried t or are too dangerous to employ; and

d. there is probable cause to believe that (identify the facilities from which or the place where, the wire or oral communications are to be intercepted) have been and will continue to be used in connection with commission of the above-described offenses.

WHEREFORE, IT IS HEREBY ORDERED that Special Agents of the(name the investigative agency/agencies; also indicate if state and local officers are participating in the investigation, particularly if they will be monitors) are authorized, pursuant to an application authorized by a duly designated official of the Criminal Division, (insert official's name and title),United States Department of Justice, pursuant to the power delegated to that official by special designation of the Attorney General and vested in the Attorney General by Section 2516 of Title 18, United States Code, to intercept (wire) (oral) communications (if wire: "to and from the above-described telephone(s)")

If oral: "in the above-described premises (or vehicle)."

PROVIDED that such interception(s) shall not terminate automatically after the first interception that reveals the manner in which the alleged co-conspirators and others as yet unknown conduct their illegal activities, but may continue until all communications are intercepted which reveal fully the manner in which the above-named persons and others as yet unknown are committing the offenses described herein, and which reveal fully the identities of their confederates, their places of operation, and the nature of the conspiracy involved therein, or for a period of thirty (30) days measured from the day on which investigative or law enforcement officers first begin to conduct an interception under this order or ten (10) days after this order is entered, whichever is earlier.

If a mobile or cellular telephone or facility, add:

IT IS ORDERED FURTHER that in the event that the target facility/vehicle is transferred outside the territorial jurisdiction of this court, interceptions may take place in any other jurisdiction within the United States.

If oral communications, add:

IT IS ORDERED FURTHER that Special Agents of the (name the agency/agencies) may make all necessary surreptitious and/or forcible entries to effectuate the purposes of this order, including but not limited to entries to install I maintain and remove electronic listening devices within (describe the premises or vehicle). Applicant shall notify the Court of each surreptitious entry.

If interception of wire communications is requested, add:

IT IS ORDERED FURTHER that the authorization apply not only to the target telephone number(s) listed above,but to any changed telephone number subsequently assigned to the same cable, pair, and binding posts utilized by the target telephone(s) within the thirty (30) day period. (In the case of a cellular telephone: "but to any changed telephone number or any other telephone number subsequently assigned to or used by the instrument bearing the same electronic serial number as the target cellular phone within the thirty (30) day period.") It is also ordered that the authorization apply to background conversations intercepted in the vicinity of the target telephone(s) while the telephone(s) is off the hook or otherwise in use.

If a roving interception is being ordered, add:

IT IS ORDERED FURTHER that the authorization to intercept (wire) (oral) communications shall include the interception of the (wire) (oral) communications of (name the interceptee(s)) ((if wire: "from various and changing telephone facilities," pursuant to 18 U.S.C. § 2518 (11) (b)),(if oral: "from presently unknown premises used by (name the interceptee(s)) that it is not practical to specify, pursuant to 18 U.S.C. § 2518(11)(a)).

If wire communications, add:

IT IS ORDERED FURTHER that I based upon the request of the Applicant pursuant to Section 2518(4) of Title 18, United States Code, the (name the communication service provider(s)), an electronic communication service provider(s) as defined in Section 2510(15) of Title 18, United States Code, shall furnish the (investigative agency) with all information, facilities and technical assistance necessary to accomplish the interceptions unobtrusively and with a minimum of interference with the services that such provider is according the persons whose communications are to be intercepted, with the service provider(s) to be compensated by the Applicant for reasonable expenses incurred in providing such facilities or assistance.

IT IS ORDERED FURTHER that, to avoid prejudice to the government's criminal investigation, the provider(s) of the electronic communications service and its agents and employees are ordered not to disclose or cause a disclosure of the Order or the request for information, facilities and assistance by the (investigative agency), or the existence of the investigation to any person other than those of its agents and employees who require this information to accomplish the services hereby ordered. In particular, said provider(s) and its agents and employees shall not make such disclosure to a lessee, telephone subscriber or any interceptee or participant in the intercepted communications.

IT IS ORDERED FURTHER that this order shall be executed as soon as practicable and that all monitoring of (wire) (oral)communications shall be conducted in such a way as to minimize the interception and disclosure of the communications intercepted to those communications relevant to the pending investigation. The interception of (wire) (oral) communications must terminate upon the attainment of the authorized objectives, not to exceed thirty (30) days measured from the earlier of the day on which investigative or law enforcement officers first begin to conduct an interception of this order or ten (10) days after the order is entered.

Monitoring of conversations must terminate immediately when it is determined that the conversation is unrelated to communications subject to interception under Chapter 119, Title 18, United States Code. Interception must be suspended immediately when it is determined through voice identification, physical surveillance, or otherwise, that none of the named interceptees or any of their confederates, when identified, are participants in the conversation unless it is determined during the portion of the conversation already overheard that the conversation is criminal in nature. If the conversation is minimized, the monitoring agent shall spot check to insure that the conversation has not turned to criminal matters.

IT IS ORDERED FURTHER that Assistant United States Attorney or any other Assistant United States Attorney familiar with the facts of this case shall provide this Court with a report on or about the (tenth), (twentieth), and (thirtieth) days following the date of this Order showing what progress has been made toward achievement of the authorized objectives and the need for continued interception. If any of the above-ordered reports should become due on a weekend or holiday,

IT IS ORDERED FURTHER that such report shall become due on the next business day thereafter.

IT IS ORDERED FURTHER that this Order, the application, affidavit and proposed order(s), and all interim reports filed with this Court with regard to this matter, shall be sealed until further order of this Court, except that copies of the order(s), in full or redacted form, may be served on the (investigative agency/agencies) and the service provider(s) as necessary to effectuate this order.

____

UNITED STATES DISTRICT COURT JUDGE
(District)

Dated this ________________day of___________, 20__

Order to Service Provider

UNITED STATES DISTRICT COURT
_______________DISTRICT OF______________

IN THE MATTER OF THE APPLICATION)
OF THE UNITED STATES OF AMERICA FOR)
AN ORDER AUTHORIZING THE INTERCEPTION)
OF (WIRE) (ELECTRONIC) COMMUNICATIONS)

ORDER TO SERVICE PROVIDER

This matter comes before the Court pursuant to the Application of the United States of America for an order authorizing the interception of (wire) (electronic) communications pursuant to Title 18, United States Code, Section 2518, (if wire: "to and from the telephone(s) bearing the number(s) and located at/billed to") (if electronic: "to and from the pager/facsimile machine/computer bearing the telephone number and located at/billed to").

The Court, having reviewed the Application and found that it conforms in all respects to the requirements of Title 18, United States Code, Sections 2516 and 2518, has this day signed an Order conforming to the provisions of Title 18, United States Code, Section 2518, authorizing the (name the investigative agency/agencies) to accomplish the aforesaid interception.

IT APPEARING FURTHER that the Applicant has requested that the (name the service provider/s) be directed to furnish, and continue to furnish, the Applicant and (name the investigative agency) with all information, facilities and technical assistance necessary to accomplish the interception(s) unobtrusively and with a minimum of interference with the services such provider(s) is according the person(s) whose communications are to be intercepted, and to ensure an effective and secure installation of electronic devices capable of interception of wire communications over the above-described telephone(s) and/or electronic communications over the above-described facsimile machine/pager/computer.

IT IS HEREBY ORDERED that the (name the service provider/s), shall furnish, and continue to furnish, the (name the investigative agency) with all information, facilities and technical assistance necessary to accomplish the interception(s) unobtrusively and with minimum interference with the services that such provider(s) is according the person(s) whose communications are to be intercepted, and to ensure an effective and secure installation of electronic devices capable of interception of wire communications over the above-described telephone(s) and/or facsimile machine/pager/computer.

IT IS ORDERED FURTHER that the service provider(s) is to be compensated by the Applicant for reasonable expenses incurred in providing such facilities or assistance.

IT IS ORDERED FURTHER that the furnishing of said information, facilities, and technical assistance shall terminate thirty (30) days measured from the earlier of the day that assistance is provided under this order or ten (10) days from the date this Order is entered, unless otherwise ordered by this Court; and

IT IS ORDERED FURTHER that this Order is sealed, except that copies of this Order may be served on the (name the investigative agency/agencies) and (name the service provider(s)), and, to avoid prejudice to the criminal investigation, that the (name the service provider(s)) and its agents and employees shall not disclose or cause a disclosure of this Order or the request for assistance or the existence of this investigation to any person other than those of its agents and employees who require this information to accomplish the services hereby ordered, unless and until otherwise ordered by this Court. In particular, no such disclosure may be made to a lessee, telephone subscriber, or any interceptee or participant in the intercepted communications.

DATED this____________ day of__________, 20__

____

UNITED STATES DISTRICT COURT JUDGE
(District)

Sample Minimization Instructions for Oral and Wire Communications

MEMORANDUM
TO: Monitoring Agents
FROM: AUSA ______________
RE: Minimization Instructions
DATE: ____________________

1. All agents must read the affidavit, application, order and these instructions and sign these instructions before monitoring.

2. The Order of______________________ only authorizes the interception of conversations of (name the interceptees listed in the Order) with anyone else occurring (to and from telephone number subscribed to by_______________) (at the premises known as_____________________ and located at___________________), regarding offenses involving (list the offenses).
3. Agents may spot monitor for a reasonable period not to exceed two minutes to determine whether the subject is present and participating in a conversation. This spot monitoring may occur as often as is reasonable, but in any event at least one minute should elapse between interceptions.
4. If, during this spot monitoring, it is determined that additional individuals are engaged in criminal conversation, intercepts may continue despite the fact that the named subject is not engaged in conversations, until the conversation ends or becomes non-pertinent. If individuals other than the subject are participating in criminal conversation, continue to monitor and advise the case agent or supervisor immediately. If these individuals can be identified, provide this information also.
5. If the subject is engaged in conversation, interception may continue for a reasonable time, usually not in excess of two minutes, to determine whether the conversation concerns criminal activities.
   a) If such a conversation is unclear but may be related to (name the offenses), interception may continue until such time as it is determined that the conversation clearly no longer relates to that topic.
   b) If such a conversation is unclear but may relate to other criminal activities, interception should cease after about two minutes unless it can be determined within that time that the conversation does in fact relate to such other criminal activities, in which case interception may continue.
6. The above instructions regarding the number of minutes of permissible interception will vary once experience has been gained. If experience shows that conversations between certain people are invariably innocent, interception of such conversations should be ended sooner. If experience shows that other individuals always discuss criminal activities, a longer interception may be justified. This is especially true for individuals who can be identified as participants with the subjects in possessing and distributing controlled substances. Read all of the logs of interceptions on a continuing basis and notify the case agent if patterns develop.
7. No conversation may be intercepted that would fall under any legal privilege. The four categories of privileged communications are described below:
   a) **Attorney-Client Privilege**: Never knowingly listen to or record a conversation between a subject and his attorney when other parties are not present. Any time that an attorney is a party to a conversation, call the case agent immediately. If it is determined that a conversation involving an attorney constitutes legal consultation of any kind, notify the case agent, shut off the monitor and

stop recording, unless you are able to determine from the interception of any conversation involving an attorney that third parties who are not involved in the legal matters being discussed are present. If such third parties are present, and only if they are present, you may intercept such conversations following the above-described rules of minimization. In any event, notify the case agent immediately.

b) **Parishioner-Clergyman Privilege**: All conversations and conduct between a parishioner and his clergyman are to be considered privileged. An electronic surveillance order could not be obtained to listen to a subject confess his sins to a priest in a confessional booth, similarly, a subject discussing his personal, financial or legal problems with his priest, minister, rabbi, etc. may likewise not be intercepted. Thus, if it is determined that a clergyman is a party to a communication being intercepted and that the communication is penitential in nature turn off the monitor, stop recording, and notify the case agent.

c) **Doctor-Patient Privilege**: Any conversation a patient has with a doctor relating to diagnosis, symptoms, treatment, or any other aspects of physical, mental or emotional health, is privileged. If it is determined that a person is talking to his doctor and that the conversation concerns the person's health (or someone else's health), turn off the machine and notify the case agent.

d) **Husband-Wife Privilege**: As a general rule, there is also a privilege covering communications between lawfully married spouses. Monitoring should be discontinued and the case agent notified if it is determined that a conversation solely between a husband and wife is being intercepted. If a third person is present, however, the communication is not privileged and that conversation may be monitored in accordance with the previously described rules of minimization. If the conversation is between the named subjects and their respective spouses, the conversation may be monitored in accordance with the previously described rules of minimization regarding monitoring these individuals' conversations to determine whether they are discussing crimes. If the nature of the conversation is criminal, monitoring may continue; otherwise, it may not be monitored.

8. Abstracts or summaries of each conversation are to be made at the time of interception and are to be included in the logs and the statistical analysis sheet. If the conversation is not recorded entirely, an appropriate notation should be made indicating the incomplete nature of the conversation and why the conversation was not recorded completely (*e.g., "non-pertinent" or "privileged").*

9. The logs should reflect all activity occurring at the monitoring station concerning both the intercepted conversations as well as the equipment itself (e.g., "replaced tape, "malfunction of tape recorder," "no overheard conversation"). These logs will be used ultimately to explain the monitoring agent's actions when intercepting communications. It is important to describe the parties to each conversation, the nature of each conversation, and the action taken. All monitoring agents will record the times their equipment is turned on and off.

10. All conversations that are monitored must be recorded.

11. The Log

    The monitoring agents should maintain a contemporaneous log, by shifts, of all communications intercepted, indicating the reel and footage locations of each communication; the time and duration of the interception; whether outgoing or incoming in the case of telephone conversations; the number called if the call was outgoing; the participants, if known; and the subjects and a summary of the content of pertinent conversations. Any peculiarities, such as codes, foreign language used, or background sounds, should also be noted. When the interception of a communication is terminated for purposes of minimization, that fact should be noted. This log should record the names of the personnel in each shift and the function performed by each, malfunctions of the equipment or interruptions in the surveillance for any other reason and the time spans thereof, and interceptions of possibly privileged conversations or conversations relating to crimes not specified in the original interception order. Each entry in the log should be initialed by the person making it.

12. Protection of the Recording

The following procedure should be followed during the period of authorized interceptions:

(a) Either during or at the end of each recording period, copies of the recorded conversations should be made for the use of the investigative agencies and the supervising attorney.

(b) The original recording should be placed in a sealed evidence envelope and kept in the custody of the investigative agencies until it is made available to the court at the expiration of the period of the order, and

A chain of custody form should accompany the original recording. On this for should be a brief statement, signed by the agent supervising the interception, which identifies:

i. The order that authorized the recorded interceptions (by number if possible),

ii. the date and time period of the recorded conversations,

iii. the identity (when possible) of the individuals whose conversations were recorded, and

iv. the place (e.g., location of telephone) where intercepted communications took place.

(c) The form should indicate to whom the case agent has transferred the custody of the original recording and the date and time that this occurred. Each subsequent transfer, including that to the court, should be noted on the form.

(e) The case agent should mark a label attached to the original tape reel/cassette in order to identify it as corresponding with accompanying chain of custody forms. The date of the recording should also be marked on the label and this should be initialed by the agent.

(f) Each agent or other person signing the chain of custody form should be prepared to testify in court that the original tape, while in his custody, was kept secure from the access of third parties (*unless noted to the contrary on the form*) and was not altered or edited in any manner. It is the responsibility of the investigative agencies to ensure that original recordings in their custody will be maintained in such a way as to ensure their admissibility in evidence at trial over objections to the integrity of the recording.

13. Procedure When No Recording Can be Made

In those unusual instances when no recording of the intercepted conversations can be made, the following procedure should be used:

(a) The monitoring agent should make a contemporaneous log or memorandum that is as near to a verbatim transcript as is possible;

(b) The log or memorandum should close with a brief statement signed by the agent indicating the date, time, and place of the intercepted conversation. The order authorizing the interception should be identified. The agent should indicate that the log or memorandum contains the contents of the intercepted communication which he overheard. This should be followed by the agent's signature; and

(c) This log should be treated by the investigative agencies as if it were an original recording of the intercepted communication.

14. If the conversation occurs in a language other than English that no one at the monitoring post understands, the entire conversation should be monitored and recorded and then minimized by

a person familiar with the investigation, but who is not actively involved in it, in accordance with the minimization rules set forth above.

15. If anything appears to be breaking suddenly, please call the case agent or the AUSA (Assistant United States Attorney). Several telephone numbers will be posted at the monitoring post.

Suggested language for pagers:

All monitoring of electronic communications to (and from) the (telephone/digital-display paging device/facsimile machine/computer/Internet account) assigned number ______________, will be recorded and examined by monitoring agents and attorneys to determine their relevance to the pending investigation. The disclosure of the contents or nature of the electronic communications intercepted will be limited to those communications relevant to the pending investigation, in accordance with the minimization requirements of Chapter 119 of Title 18, United States Code.

Suggested language for facsimile machines:

All interceptions will be minimized in accordance with Chapter 119 of Title 18, United States Code. Fax transmissions sent or received by will be minimized as follows: each fax transmission will be printed on the machine used to intercept fax communications. The monitoring agent and AUSA will decide, based on the identities of the sender and recipient and the subject matter of the transmission, whether the fax appears to be pertinent to the criminal offenses listed in the Court's order. If the fax does not appear to be pertinent, the intercepted transmission will be placed in an envelope and sealed. It will then be placed in a locked drawer until it is turned over to the court with the other intercepted transmissions after the interception order has expired.

If the facsimile machine is a dedicated to fax transmissions only or, if the facsimile machine is attached to a telephone, but the government has not applied for authorization to intercept wire communications over the telephone, then add: "It is not the intention of the Government to intercept wire communications during this investigation; only electronic communications will be intercepted."

Because of the type of information intercepted, i.e., typewritten fax communications and not verbal communications, the monitors will be unable to minimize any non-pertinent information until after it has been received at the monitoring location. It is anticipated that the monitoring location will not be staffed at all times, but will be activated electronically. The monitoring location will be kept secure and access will be available only to persons authorized to be involved with this investigation.

Conclusion

Your affiant believes that the facts alleged herein establish that the targets of this investigation are engaged in an ongoing criminal enterprise and that the evidence sought will be intercepted on a continuing basis following the first receipt of the particular communications that are the object of this request. Therefore, it is requested that the interception not be required to terminate when the communications described herein are first intercepted, but be allowed to continue until communications are intercepted which fully reveal the scope of the enterprise, including the identities of all participants, their places and methods of operation, and the various criminal activities in which they are engaged which are in furtherance of the enterprise, not to exceed thirty (30) days measured from the earlier of the day on which investigative or law enforcement officers first begin to conduct an interception under this Court's Order or ten (10) days after the Order is entered.

____

(NAME)
Special Agent
(Agency)

Sworn to before me this __________day of____________, 20__

____

UNITED STATES DISTRICT COURT JUDGE
(District)

Sample Title III Roving Affidavit

UNITED STATES DISTRICT COURT
WESTERN DISTRICT OF MICHIGAN
GRAND RAPIDS DIVISION

IN THE MATTER OF THE APPLICATION *
OF THE UNITED STATES OF AMERICA *
FOR AN ORDER AUTHORIZING THE *
INTERCEPTION OF WIRE COMMUNICATIONS *
OCCURRING TO AND FROM THE CELLULAR *

TELEPHONES BEARING THE NUMBERS *
(616) 555-6068, and accessed through *
IMSI 316000115672568 AND (616) 555-6015 * MISC NO.
And assigned ESN 345678000; AND THE * UNDER SEAL
ROVING INTERCEPTION OF WIRE *
COMMUNICATIONS OVER **VARIOUS** *
AND CHANGING CELLULAR TELEPHONES *
USED BY JACOB RIPLEY *

AFFIDAVIT IN SUPPORT OF APPLICATION[17]

I. INTRODUCTION

____

[17] This sample roving affidavit pertains to a fictitious narcotics trafficking investigation and should be consulted when drafting Title III roving pleadings. When using this affidavit as a reference, assume that it was submitted for authorization to the Office of Enforcement Operations ("OEO") in midDecember2004, taking note that the information in support of probable cause is up-to-date. Specific questions regarding all Title III issues should be addressed to OEO at (202) 514-6809.

I, J. Kenneth Smith, a Special Agent with the Drug Enforcement Administration ("DEA")[18], being duly sworn, state as follows:

1. I am a Special Agent with the DEA, duly appointed according to the law and acting as such, and have been employed by the DEA since February 1993. As a DEA Special Agent, I am an "investigative or law enforcement officer" within the meaning of Title 18, United States Code, Section 2510(7), that is, an officer of the United States who is empowered by law to conduct104investigations and to make arrests for offenses enumerated in Title 18, United States Code, Section 2516.[19]
2. I am currently assigned to the DEA Grand Rapids Field Office. In connection with my official duties as a Special Agent of the DEA, I am responsible for conducting investigations into violations of Title 21 of the United States Code and other federal criminal statutes. During my twelve (12) years as a DEA Special Agent, I have participated in numerous narcotics investigations, including more than fifty (50) investigations in which I have been designated as the lead investigative agent. These investigations have resulted in the arrest of more than seventy (70) persons and the seizure of marijuana, methamphetamine, cocaine, MDMA, heroin, and other controlled substances. During the course of these investigations, I have conducted or participated in physical and electronic surveillance; prepared affidavits which have resulted in court ordered wire interceptions; applied for, obtained, and executed more than 30 federal search warrants, conducted numerous debriefings of informants, cooperating defendants, and other individuals cooperating with the United States; seized and evaluated items of evidence; and reviewed taped conversations, seized narcotics records, and financial documents.
3. This affidavit is submitted in support of an application for an order authorizing the interception of wire communications[20] occurring to and from the prepaid cellular telephone bearing the number (616) 555-6068{ subscribed to by Janis Jenkins, 1555 N. Shore Rd., Grand Haven, Michigan,[21] and accessed through international mobile subscriber identification("IMSI") number 316000115672568[22] **("Target Phone 1")** and the cellular telephone bearing the number (616) 555-6015, subscribed to by Steven Hill, 512 S. Division Street, Grand Rapids, Michigan, and assigned

---

[18] Department of Justice ("the Department") policy precludes the use of multiple affiants except when it is indicated clearly which affiant swears to which part of the affidavit, or that each affiant swears to the entire affidavit. For practical purposes, a single affiant should be used.

[19] If a state or local law enforcement officer is the affiant for a federal electronic surveillance affidavit, he/ she must be deputized as a federal officer of the agency with responsibility for the offenses under investigation. See, 18 U.S.C. § 2516(1) (interceptions are to be conducted by the federal agency responsible for the offenses for which the application is made); United States v. Lyons, 695 F.2d 802(4th Cir. 1982).

[20] Cellular telephones often are equipped with features allowing the transmission of both wire (voice over the phone) and electronic (e.g., text-messages and/or email) communications. Undercurrent Department policy, a separate showing of probable cause for each type of communication sought to be intercepted is needed to obtain Department authorization to apply for a court order to intercept each type of communication. This policy is based on the explicit wording of the Title III statute, as well as the legislative history of Title III. See, 18 U.S.C. § 2518(1)(b) (requiring a particular description of the type of communications sought to be intercepted in each application for an order authorizing or approving the interception of wire, oral, or electronic communications), § 2518(3)(b) (requiring facts showing probable cause to believe that particular communications concerning that offense will be intercepted), and § 2518 (requiring that the order specify the particular type of communications to be intercepted and a statement of the particular offense to which it relates).

[21] When identifying the targeted telephone(s), the telephone number(s) and subscriber address(es) (as it appears in service provider records) should be included. In some instances, no subscriber information will be available, as is often the case with prepaid cellular telephones.

[22] An IMSI number is a fifteen (15) digit number assigned to a removable computer chip located inside certain service providers' cellular telephones. IMSI numbers are unique to each individual subscriber, and the chip on which the IMSI number is encoded can be removed and used in other similarly-equipped telephones. Depending on the service provider, these numbers are labeled international mobile subscriber "identification/identifier/identity" numbers.

electronic serial number *("ESN")*[23] 345678000 **("Target Phone 2");** As discussed below, **Target Phones1 and 2** are used by Jacob Ripley *("Ripley")*.[24] Additionally, this affidavit is submitted in support of an application that seeks authorization, pursuant to Title 18, United States Code, Section 2518 (11) (b)[25], to intercept the wire communications6occurring to and from **various and changing cellular telephones** used by Ripley during the authorization period.[26]

4. As set forth in greater detail below, Ripley is the leader of a large-scale, Grand-Rapids-based cocaine and heroin distribution organization ("the Grand Rapids Cell") which is a distribution cell of a larger, Mexico-based narcotics organization ("the Ramirez Organization") headed by Roberto Ramirez ("Ramirez"). Confidential source information, court authorized wire interceptions, physical surveillance, and the analysis of telephone records have established that Ripley directs the distribution of cocaine and heroin in Western Michigan that is transported there from Chicago, Illinois, following its importation from Mexico from the Ramirez Organization, and that Ripley uses the **Target phones 1 and 2** and **various and changing cellular telephones** to facilitate his illegal activities. Specific information related to Ripley's use of **Target Phones 1 and 2 and various and changing cellular telephones** to facilitate his narcotics trafficking activities is set forth below.

---

[23] An ESN is a serial number embedded in a particular telephone instrument. The ESN number is permanently assigned to that particular piece of telephone hardware, is unique to that facility, and cannot be changed without obtaining a new telephone. New telephone numbers can be assigned to an ESN.

[24] When seeking authorization to intercept roving wire communications of a particular target, the current cellular telephone(s) being used by the roving target at the time of the application should be specifically targeted in the application and order. Specifically targeting a phone in the application and order allows law enforcement to continue tapping that telephone should the roving target hand it off to a co-conspirator. Phones that are not specifically targeted in the pleadings cannot be monitored if the roving target hands off the telephone to someone else, even if that person is going to use the telephone to facilitate criminal activities. Remember, roving authority is person-specific. This process is repeated in any subsequent extensions and/or spinoffs of the roving wiretap, with new phones in the hands of the roving target at the time of the extension/spinoff specifically identified and targeted, along with an extension of the roving authority, as the facts warrant. It is important to note that a regular Title **III** authorization is specific to particular telephones (i.e., interceptions can continue no matter who is using that facility as long as the telephone is being used to facilitate predicate Title **III** offenses). Conversely, a roving Title **III** authorization is specific to the particular person (i.e., telephones wiretapped pursuant to the roving authorization can only continue as long as *the roving target is using those telephones).* If the roving target stops using one of the "various and changing" cellular telephones (i.e., those phones intercepted during the roving authorization period that were not specifically identified in the original Title **III** order), interception over those facilities must cease.

[25] The roving provision of Title **III** is codified in 18 U.S.C § 2518(11). The roving interception of oral communications (18 U.S.C § 2518(11)(a)), and wire and electronic communications (18 U.S.C§ 2518(11)(b) are contemplated under the roving statute. Specific citation to 18 U.S.C(II)(a) and/or(b) should be included be included in the Title **III** affidavit, depending of the type of communications sought to be intercepted.

[26] Generally, to justify a roving wiretap the specifically identified roving target must have dropped three or more telephones in a short period of time. The general Department rule has been that if a criminal subject uses a particular phone for longer than 21 days, a roving wiretap is not appropriate. However, there is flexibility with regard to this 21 day rule when, despite the government's best efforts, the roving target is dropping telephones before effective, regular Title **III** surveillance can be accomplished. For example, when a subject utilizes multiple cellular telephones in succession for longer than 21 days, but in a manner that makes it difficult to obtain a traditional interception order, a roving wiretap may be authorized.

When a subject's use of multiple phones does not justify a roving wiretap, the solution is often to seek authorization to wiretap all of the identified phones (sometimes referred to as a "block" of cellular telephones) that the subject is using. While a showing of probable cause must still be made as to the use of each of these facilities, this can be accomplished in two steps. First, a showing of independent probable cause as to the use of at least one of the phones to facilitate criminal conversations (e.g. through comments to an informant or undercover agent, or through intercepted calls over another tapped facility). Second, once one phone is clearly established as "dirty," the government can seek authorization as to other facilities where an analysis of telephone records shows a similar calling pattern with that of the phone that has independent probable cause, or where there is other credible information indicating that the subject uses multiple phones that can ultimately be identified.

5. Ripley has an established pattern of using **various and changing cellular telephones** to accomplish his criminal goals. Ripley changes or "drops" cellular telephones regularly after short periods of time, with the effect of thwarting the ability of law enforcement to conduct electronic surveillance. Investigative facts to date, discussed below, establish that Ripley typically uses a cellular telephone for an average of 18 days before replacing that cellular telephone with another. Based on my training and experience, I know that narcotics traffickers change telephones in this manner in an attempt to avoid detection by law enforcement.

6. I have participated in an ongoing investigation into the Grand Rapids Cell's illegal activities. As a result of my personal participation in this investigation, through interviews with and analysis of reports submitted by other Special Agents of the DEA and other state and local law enforcement personnel, I am familiar with all aspects of this investigation. On the basis of this familiarity, and on the basis of other information which I have reviewed and determined to be reliable, I allege the facts to show that:

   a. there is probable cause to believe that Jacob Ripley a.k.a "Jack"; Steven Hass ("Hass"); LeAndra Langdon ("Langdon") a.k.a "Molly"; Christopher Succrattao ("Succrattao"); Robert Gemink ("Gemink") a.k.a "Big Bobby"; Stanley Paul ("Paul"); "Mr.C."; Roberto Ramirez ("Ramirez"); Raul LNU, Regatto LNU, and others as yet unknown (collectively referred to as the **"Target** Violators")[27], are committing, and will continue to commit offenses enumerated in Section 2516 of Title 18 of the United States Code, namely, the importation of cocaine and heroin, the distribution of and possession with intent to distribute cocaine and heroin, and attempts and conspiracies to do the same, all in violation of Title 21, United States Code, Sections 841, 846, 952, 960, and 963; use of wire facilities to facilitate the commission of the above narcotics offenses, in violation of Title 21, United States Code, Section 843; money laundering and conspiracy, in violation of Title 18, United States Code, Sections 1956 and 1957; and aiding and abetting the offenses described above, in violation of Title 18, United States Code, Section 2 (collectively referred as **"the Target Offenses")**.[28]

   b. there is probable cause to believe that particular wire communications of Ripley, Hass, Langdon, Succrattao, Gemink, Paul, "Mr. C," Raul LNU, Regatto LNU (collectively referred to as108 "the **Target Interceptees")**[29] concerning the above offenses will be obtained through the interception of such communications to and from **Target Phones 1 and 2** and, pursuant to Title 18, United States Code, Section 2516 (11) (b) the interception of wire communications over **various and changing cellular telephones** used by Ripley.[30]

7. In particular, these communications are expected to concern the specifics of the above offenses, including (i) the nature, extent and methods of the Ramirez Organization's (and Grand Rapids Cell's) narcotics trafficking activities; (ii) the identities and roles of accomplices, aiders and

---

[27] If probable cause exists to believe that a person is involved in the criminal offenses under investigation, that person must be named as a target violator (sometimes referred to as "target subject") in the Title III application. Target violators include everyone involved in the criminal conspiracy, even if those individuals are not expected to be intercepted during the Title III authorization period. If the Title III investigation is directed at their activities, they should be named as targets

[28] The offenses for which you can conduct electronic surveillance are listed in 18 U.S.C. § 2516. Probable cause for at least one Title III predicate must be present in the Title III affidavit. Criminal Division policy requires that non-predicate offenses also be alleged in the Title III application, where probable cause exists for those offenses.

[29] Target interceptees are the violators who are expected to be intercepted over the target phone(s)or the roving phone(s). They are simply a subset of the target violators, and often will include all of the target violators.

[30] As discussed above, the roving authority is person-specific. Law enforcement is permitted to intercept wire communications over various and changing telephones used by the roving target but, unlike a traditional Title III, must cease interceptions over those facilities once the roving target stops using them.

abettors, co-conspirators and participants in their illegal activities; (iii) the distribution and transfer of the contraband and money involved in those activities; (iv) the existence and location of records; (v) the location and source of resources used to finance their illegal activities; (vi) the location and disposition of the proceeds from those activities; and (vii) the locations and items used in furtherance of those activities. In addition, these wire communications are expected to constitute admissible evidence of the commission of the above-described offenses.

8. The statements contained in this affidavit are based in part on information provided by Special Agents of the DEA, on conversations held with detectives and officers from the Michigan State Police ("MSP"), the Grand Rapids Police Department ("GRPD"), on information provided by confidential sources, through court-ordered wire interceptions, and on my experience and background as a special agent of the DEA. Since this affidavit is being submitted for the limited purpose of securing authorization for the interception of wire communications, have not included each and every fact known to me concerning this investigation.

I have set forth only the facts that I believe are necessary to establish the necessary foundation for an order authorizing the interception of wire communications.

I have also set forth below my characterization of various coded conversations that have occurred in this investigation among the **Target Violators** and others. My interpretation of these conversations is based on my training and experience as well as my knowledge of the facts of the investigation, including my conversations with confidential sources about the true meaning of the coded conversations

II. PERSONS EXPECTED TO BE INTERCEPTED

9. The following individuals are expected to be intercepted engaging in narcotics-related conversations over **Target Phones 1 and 2 and over various and changing cellular telephones** used by Ripley during the authorization period. The background information on these individuals was obtained from confidential source information (specific information discussed below) / court authorized wire interceptions over cellular telephones in the Southern District of Texas / and a review of court documents:

a. Ripley is a 31-year-old male and the leader of the Grand Rapids Cell. Ripley directs the transportation of multi-kilogram quantities of cocaine and heroin from Chicago, Illinois, to the Grand Rapids, Michigan area, for distribution. Ripley maintains regular contact with narcotics couriers in Chicago, and manages a network of cocaine and heroin distributors in Michigan. Ripley was convicted in the Kent County, Michigan, Circuit Court in 1994 on a charge of possession with intent to deliver cocaine and served two years in prison.

b. Hass, a 21-year-old male, is one of Ripley's narcotics distributors in Grand Rapids. On October 12, 2004, as discussed below, Hass has sold cocaine on several occasions to a confidential source, who was working under the supervision of the DEA. Hass is an 18-year-old with no known prior criminal history.

c. Langdon is a 24-year-old female and Ripley's girlfriend. Langdon regularly attends narcotics-related transactions and meetings with Ripley, and serves as a "lookout" for law enforcement activity for Ripley in these situations. Langdon has no known prior criminal history.

d. Succrattao is a 20-year-old male and one of Hass' cocaine customers, as observed by physical surveillance conducted by the GRPD and information from a confidential source.

Succrattao has four prior felony convictions for possession of narcotics, the most recent occurring in 1999.

e. Gemink is a 26-year-old male and one of Ripley's narcotics distributors. In July 2003, Gemink was interviewed by the GRPD in conjunction with a traffic stop of Gemink's vehicle for failure to stop at a traffic signal. Pursuant to a consent search GRPD officers found sixteen (16) ounces of cocaine in Gemink's vehicle. A felony, possession of narcotics charge was filed against Gemink as a result of this incident, but was dismissed by the Kent County Prosecutor's Office before trial.

f. Paul, a 28-year-old male, is one of Ripley's narcotics couriers. According to multiple confidential sources, Paul travels regularly between Grand Rapids and Chicago to transport narcotics and narcotics proceeds. Paul lives in the same apartment complex as Ripley, and was incarcerated with Ripley from 1995 to 1996. Paul was convicted of a one count of felony possession of a firearm in 1995.

g. "Mr. C." is one of Ripley's Chicago-based narcotics trafficking associates. Based on consensually-recorded calls with a confidential source in which Ripley mentioned the movement of narcotics shipments from Chicago to Grand Rapids, I believe "Mr. C" is responsible for coordinating the shipment of narcotics and narcotics proceeds to and from Grand Rapids for the Ramirez Organization. No further identifying information is available for "Mr. C."

h. Raul LNU is a narcotics courier responsible for transporting narcotics from Texas to Chicago and Grand Rapids. No further identifying information is available for Raul LNU.

i. Regatto LNU is a narcotics courier responsible for transporting narcotics from Texas to Chicago and Grand Rapids. No further identifying information is available for Regatto LNU.

10. Ramirez is believed to be a mid-level cocaine and heroin supplier based in Mexico. A confidential source identified Ramirez as Raul LNU's and Regatto LNU's supervisor who directs the shipment of multi-kilogram quantities of cocaine and heroin from Mexico into Texas and, ultimately, to Chicago and Grand Rapids. I believe Ramirez attempts to insulate himself from detection from law enforcement by avoiding any direct narcotics related discussions with any of the Chicago and Grand Rapids based **Target Violators.** Based on an analysis of telephone records and court-authorized wire interceptions, discussed below, I believe that Ramirez only speaks telephonically with Raul LNU, one of his narcotics couriers, regarding narcotics shipments sent to Chicago and Grand Rapids. Because Ramirez is not expected to be intercepted over **Target Phones 1 and 2** or over any **various and changing cellular telephones** used by Ripley, he has been named as a target violator, but as a target interceptee, in this affidavit.

III. FACTS ESTABLISHING PROBABLE CAUSE

A. SUMMARY OF PROBABLE CAUSE

11. The primary target of this investigation is the narcotics trafficking and money laundering organization led by Ramirez, including the Ramirez Organization' Chicago- and Grand Rapids-based Cells. The Ramirez Organization is involved in the smuggling of cocaine and heroin from Mexico into the United States. The Ramirez Organization, through lieutenants based in the United States, arranges for the transportation of shipments of cocaine and heroin to Chicago and Grand

Rapids for further distribution. The Ramirez Organization also coordinates the laundering of narcotics proceeds and the return of these proceeds111from the United States to Mexico via the Chicago Cell and the Grand Rapids Cell.

12. The investigation has included the interception of wire communications over two cellular telephones ("Raul phones 1 and2") used by Raul LNU pursuant to Title III orders issued by the United States District Court for the Southern District of Texas. The investigation has also included the use of three confidential sources ("CS-I" through "CS-3"). The information provided by CS-1 through CS-3 is believed to be reliable and has been corroborated through other investigative means, including consensually-recorded and/or consensually-monitored conversations, physical surveillance, and other investigative techniques.[31]

13. The investigation, including court-authorized wire interceptions over Raul phones 1 and 2 and information provided by CS-I through CS-3, has revealed that Ripley is the leader of the Grand Rapids Cell, which distributes multi-kilogram quantities of cocaine and heroin to customers in and around the Grand Rapids, Michigan, area, and oversees the collection of narcotics proceeds from these customers; that "Mr. C." is a member of the Chicago Cell and directs the movement of narcotics and narcotics proceeds to and from Grand Rapids; that Ramirez directs the shipment of cocaine and heroin into the United States and the receipt of narcotics proceeds from the United States to Mexico; and that he uses Raul LNU and Regatto LNU as narcotics couriers to facilitate these activities.

14. The investigation has also revealed that since September I, 2004, Ripley has used at least six different cellular telephones (collectively referred to as "Prior Phones 1 through 6"), not including **Target Phones 1 and 2.**

Ripley has been intercepted engaging in narcotics-related conversations while using Prior Phones 3, 4, and 5. Additionally, CS-3 engaged in consensually-recorded, narcotics-related conversations with Ripley over Prior phones I, 2, and 6. On January 3, 2005, DEA Grand Rapids obtained authorization to intercept wire communications over Prior Phone 6. As discussed below, before those interceptions began, Ripley dropped Prior Phone 6. Based on intercepted calls over Raul phones 1 and 2, consensually recorded calls made by CS-1 and CS-3, and the analysis of telephone records, the DEA believes that Ripley uses a cellular telephone for a short period of time and then discards that telephone in favor of a new cellular telephone, all with the effect of thwarting possible electronic surveillance being conducted by law enforcement. Additionally, based on wire interceptions over Raul phones 1 and 2 and consensually-recorded calls made by CS-1 and CS-3, the DEA believes that Ripley has recently begun compartmentalizing his use of cellular telephones, using separate telephones to communicate with his local distributors in Grand Rapids and other cellular telephones to communicate with his Chicago-based suppliers.

---

[31] All confidential sources included in the Title III affidavit must be qualified. Current Department policy requires that all informants used in the affidavit to establish probable cause be qualified according to the "Aguilar-Spinelli" standards (See, Aguilar v. Texas, 378 U.S. 108 (1964) and Spinelli v. United States, 393 U.S. 410 (1969)), rather than those set forth in the more recent Supreme Court decision of Illinois v. Gates, 463 U.S. 1237 (1983). Such qualification should include the statement that the confidential source(s) information is believed to be reliable and a statement regarding the amount of the corroboration of the confidential source(s) information. Additionally, any facts bearing on the credibility of the confidential source (e.g., to the extent promises of leniency, a criminal history involving crimes of dishonest, and any other factors considered pertinent by your circuit) should also be included in the affidavit so that the reviewing court can make an informed determination on the confidential source's credibility.

IV. USE OF PRIOR PHONES 1 THROUGH 6

A. PRIOR PHONE 1

15. The DEA identified Ripley as the user of Prior Phone 1on September 3, 2004, when CS-3 revealed that Ripley used Prior Phone 1 to coordinate the shipment of narcotics from Chicago to Grand Rapids. An analysis of telephone records obtained from the service provider for Prior Phone 1 by subpoena revealed that Prior phone 1 was activated on August 15, 2004. The first call made over Prior Phone 1 occurred on August 15, 2004. An analysis of telephone records revealed that Prior Phone 1 was not used again until September 1, 2004. Between September 1 and September19, 2004, Prior Phone 1 was used to make and receive 547 calls. Based on my training and experience, I believe the initial call made over Prior Phone 1 on August 15, 2004, was completed to verify that Prior Phone 1 was properly activated, but that sustained use of Prior Phone 1 did not begin until on or about September 1, 2004.[32]

16. On September 6, 2004, under the supervision of a DEA agent, CS-3 called Prior Phone 1 and spoke with Ripley. During this consensually recorded conversation[33], CS-3 asked about the timing of "the tractor" (referring to a semi-tractor trailer containing a shipment of narcotics).[34]

Ripley said that he had not spoken to "his boy (Mr. C.) in Chicago," but that he (Ripley) expected "it (the narcotics) to be here by the end of the week." CS-3 asked if "it (the narcotics shipment)

---

[32] In the case of a roving wire or electronic interception, 18 U.S.C. § 2518(b)(ii) requires a showing that the roving target uses various and changing facilities with the effect of thwarting electronic surveillance by law enforcement. This can be shown through informant information concerning the roving target's fear of wiretaps and his intention to use public telephones or cellular telephones to facilitate his criminal activities, combined with physical surveillance and/ or telephone record analysis showing calls by the roving target to known or suspected criminal associates. In establishing this roving pattern, it is inadequate merely to allege that the roving target has been observed using several different pay telephones or cellular telephones and, therefore, must be effectively thwarting electronic surveillance. A sufficient factual basis must be established to permit the court to make the required finding that the roving target has effectively thwarted (optimally through a pattern covering weeks or months) the ability of law enforcement to conduct electronic surveillance by using various and changing facilities. It is not enough to show that the roving target has used a lot of different telephones. It must be established that the roving target has used a lot of different telephones to facilitate criminal activity. See, United States v. Gayton, 74 F.3d 545 (5th Cir.), cert. denied, 117 S. Ct. 77 (1996); United States v. Petti, 973 F.2d 1441 (9th Cir. 1992), cert. denied, 507 U.S. 1035 (1993); United States v. Villegas, 1993WL 535013 (S.D.N.Y. Dec. 22, 1993) (unreported). Although the statute does not distinguish between public, cellular, and landline telephones, it is the Department's policy that, except in rare instances involving rapidly changing use of telephones located in hotel rooms or restaurants, only cellular or public pay phones may be targeted in a roving wiretap.

As a practical matter in establishing the pattern necessary to obtain roving authorization to intercept wire communications over various and changing cellular telephones, the following factual information should be obtained from the service provider(s) of the prior cellular telephones used by the roving target, as well as his/her currently used cellular telephones: (1) date of activation; (2) date of first use of the facility; (3) date the facility was identified by law enforcement as being used by the roving target; and (4) date of last use of the facility (i.e., the date the roving target dropped the phone). Additionally, facts establishing use of the prior phones by the roving target to facilitate his/her criminal activities should also be included in the affidavit (e.g., prior wire interceptions over other tapped phones, consensually-recorded calls made by informants, etc.). Finally, for each prior phone used by the roving target, the affidavit should reference any attempts to obtain regular Title III orders for those phones, and any actual interceptions and the success, if any, at implementing those efforts.

[33] All calls made by confidential sources referenced in the affidavit should be corroborated by noting if the conversation was consensually-recorded or consensually-monitored, and how the call was verified (e.g., toll records and pen register/trap and trace, agent dialing the target phone number).

[34] Coded conversations that the affiant believes are criminal in nature must be characterized in the affidavit with the law enforcement agent's belief (based on training and experience and the information obtained through the course of the investigation) regarding what the conversations actually mean.

was white (cocaine) or dark (heroin)." Ripley responded, "A bit of both (cocaine and heroin)." Later in the conversation, Ripley said that he would call CS-3 back when "it" (the narcotics shipment) arrived.

17. On September 15, 2004, Ripley, using Prior Phone 1 (as revealed by the caller identification feature on CS-3's cellular telephone and an analysis of toll records), called CS-3. During this consensually-recorded conversation, Ripley said, "It (the narcotics shipment) is here tomorrow .... Not as much as I thought (referring to the quantity of narcotics), but it'll be here." CS-3 asked if he should call Ripley the next day about the narcotics shipment. Ripley responded, "I'll call you. I got me a new phone (believed to be referring to Prior Phone 2, as discussed below)." Communications were intercepted would alert the subjects to the existence and extent of the investigation.

WHEREFORE, I respectfully request that the Court issue an Order granting this Application.

____

Assistant United States Attorney

Order for Sealing

UNITED STATES DISTRICT COURT
__________DISTRICT OF___________

IN THE MATTER OF THE APPLICATION)
OF THE UNITED STATES OF AMERICA)
FOR AN ORDER AUTHORIZING (APPROVING))
THE INTERCEPTION OF (WIRE COMMUNICATIONS)
OCCURRING TO AND FROM TELEPHONE NUMBER)
_____________SUBSCRIBED TO BY_____________)
(ORAL COMMUNICATIONS WITHIN THE PREMISES)
KNOWN AS ____________________, LOCATED AT)
______________ .) (ELECTRONIC COMMUNICATIONS)
OVER THE FACSIMILE MACHINE/PAGER BEARING)
NUMBER____________ AND SUBSCRIBED)
TO BY_________________.))

ORDER

Upon consideration of the attached application of the UNITED STATES OF AMERICA, by Assistant United States Attorney____________, and upon finding that disclosure of the subject electronic surveillance would interfere with an ongoing criminal investigation, and also upon finding that the motion of the UNITED STATES OF AMERICA is made in good faith, it is hereby:

ORDERED

1. That (reel-to-reel, cassette, magneto optical disk, computer printouts) recordings of (wire) (oral) (electronic) communications intercepted between and pursuant to the Order of this Court dated (occurring to and from the telephone number__________ subscribed to by__________) (within the premises known as________________, and located at_______________) be sealed;
2. That the aforementioned recordings be held in the custody of the (name the investigative agency) for a period of ten (10) years from the date of this Order in a manner so as to prevent editing, alteration and/or destruction;
3. That the contents of the said recordings be disclosed only upon the order of this Court or any other Court of competent jurisdiction, except as otherwise authorized by Title 18, United States Code, Section 2517;
4. That the notification requirements to Title 18, United States Code, Section 2518(d) be postponed as to all parties intercepted during the subject electronic surveillance until further order of this Court; and
5. That this Order and Application be sealed until further order of this Court.

____

UNITED STATES DISTRICT COURT JUDGE
(District)

Application for 2703(d) Court Order

UNITED STATES DISTRICT COURT
________DISTRICT OF__________

IN THE MATTER OF THE APPLICATION)
OF THE UNITED STATES OF AMERICA)
FOR AN ORDER PURSUANT TO 18 U.S.C.)
2703 (d) ____________________________)

APPLICATION

_________________, an Assistant United States Attorney for the District of______________, hereby applies to the court for an order, pursuant to 18 U.S.C. 2703(d), directing (provider of electronic communication service or remote computing service) to disclose the **(choose as appropriate:** name; address; local and long distance telephone connection records, or records of session times and durations; length of service (including start date) and types of service utilized; telephone or instrument number or other subscriber number or identity, including any temporarily assigned network address; means and source of payment for such service (including any credit card or bank account number); cell site information) of a subscriber to or customer of such service. In support of this application, I state the following:

I am an attorney for the Government as defined in Rule, (b) (1) of the Federal Rules of Criminal Procedure and, therefore, pursuant to Section 2703© of Title 18, United States Code, may apply for an order as requested herein. I certify that the (investigative agency) is conducting a criminal investigation in connection with possible violation(s) of (list principal violations); that it is believed that the subjects of the investigation are using the **(choose as appropriate:** telephone or instrument number; other subscriber number or identity; temporarily assigned network address) in furtherance of the subject offenses; and that the information sought is relevant and material to an ongoing criminal investigation. (*Offer specific and articulable facts showing that there are reasonable grounds for such belief.*)

Wherefore, the applicant requests that the Court issue an order pursuant to 18 U.S.C. 2703(d) directing (provider of electronic communication service or remote computing service) to provide the requested information forthwith. I request further that this Court's order delay notification of this application and this order to the subscriber or customer for a period not to exceed ninety (90) days, and that the Court command the provider of electronic communication service or remote computing service not to notify any other person of the existence of this application and this order (for such period as the court deems appropriate) because such notification would seriously jeopardize the investigation. I declare under penalty of perjury that the foregoing is true and correct.

Executed on____________, 20___

____
Applicant Signature

____
Title

2703(d) Court Order

UNITED STATES DISTRICT COURT

____________DISTRICT OF_____________

IN THE MATTER OF THE APPLICATION)
OF THE UNITED STATES OF AMERICA)
FOR AN ORDER PURSUANT TO 18 U.S.C.)
2703 (d))

ORDER

This matter having come before the court pursuant to an application under Title 18, United States Code, Section 2703(c) by________________, an attorney for the Government, which application requests an order under Title 18, United States Code, Section 2703(d) directing (provider of electronic communication service or remote computing service) to disclose the **(choose as appropriate:** name; address; local and long distance telephone connection records, or records of session times and durations; length of service (including start date) and types of service utilized; telephone or instrument number or other subscriber number or identity, including any temporarily assigned network address; means and source of payment for such service (including any credit card or bank account number); cell site information) of a subscriber to or customer of such service, and the Court finds that the applicant has offered specific and articulable facts showing that there are reasonable grounds to believe that the records or other information sought are relevant and material to an ongoing criminal investigation;

IT APPEARING that the information sought is relevant and material to an ongoing criminal investigation and that disclosure to any person of this investigation or this application and order entered in connection therewith would seriously jeopardize the investigation;

IT IS ORDERED pursuant to Title 18, United States Code, Section 2703(d) that (provider of electronic communication service or remote computing service) will, forthwith, turn over to agents of the (investigative agency) the *(*name, address, local and long distance telephone connection records, or records of session times and durations, length of service [including start date] and types of service utilized, telephone or instrument number or other subscriber number or identity, including any temporarily assigned network address, means and source of payment for such service (including any credit card or bank account number) of (subscriber to or customer of such service).

IT IS FURTHER ORDERED that the application and this order are sealed until otherwise ordered by the court, that the government may delay notice of this order to the subscriber or customer for a period not to exceed ninety days, and that (provider of electronic communication service or remote computing service) is commanded not to notify any other person of the existence of this application and order (for such period as the court deems appropriate), the court having determined that there is reason to believe that such notifications would seriously jeopardize the investigation.

DATED: ______________________

____
UNITED STATES MAGISTRATE (or DISTRICT) JUDGE

Application for Trap and Trace/Pen Register

UNITED STATES DISTRICT COURT

____________DISTRICT OF___________

IN THE MATTER OF THE APPLICATION)
OF THE UNITED STATES OF AMERICA)
FOR AN ORDER AUTHORIZING)
THE INSTALLATION AND USE)
OF A (PEN REGISTER))
(TRAP AND TRACE DEVICE))

APPLICATION

_________________, an Assistant United States Attorney, being duly sworn, hereby applies to the Court for an order authorizing the installation and use of a (pen register) (trap and trace device) on (telephone line or other facility). In support of this application I state the following:

1. Applicant is an "attorney for the Government" as defined in Rule l(b) (1) of the Federal Rules of Criminal Procedure, and, therefore, pursuant to Section 3122 of Title 18, United States Code, may apply for an order authorizing the installation of a (trap and trace device) (pen register).
2. Applicant certifies that the (investigative agency) is conducting a criminal investigation of (name targets) and others as yet unknown, in connection with possible violations of (list violations); it is believed that the subjects of the investigation are using (telephone line___________ or other facility), (listed in the name of___ (if known)___or leased to___ (if known)___ and located at___(if known)___ in furtherance of the subject offenses; and that the information likely to be obtained from the (pen register) (trap and trace device) is relevant to the ongoing criminal investigation in that it is believed that this information will concern the aforementioned offenses.
3. Applicant requests that the Court issue an order authorizing the installation and use of (a pen register to record or decode dialing, routing, addressing, or signaling information transmitted by identify the targeted instrument or facility from which a wire or electronic communication is transmitted), (and) (a trap and trace device to capture the incoming electronic or other impulses which identify the originating number or other dialing, routing, addressing, and signaling information reasonably likely to identify the source of a wire or electronic communication), for a period of (enter time period, not to exceed sixty (60) days), provided, however, that such information shall not include the contents of any communication.
4. The applicant requests further that the order direct the furnishing of information, facilities, and technical assistance necessary to accomplish the installation of (the pen register) (and/or) (trap and trace device) as provided in Section 3124 of Title 18.
5. (*If trap and trace requested*) The applicant requests further that the order direct that the results of the trap and trace device be furnished to the officer of a law enforcement agency, designated in the court order, at reasonable intervals during regular business hours for the duration of the order.
6. With regard to the requirement of Section 3121(c) of Title 18 that the (investigative agency) use technology reasonably available to it that restricts the recording or decoding of electronic or other impulses to the dialing, routing, addressing, and signaling information utilized in the processing and transmitting of wire or electronic communications so as not to include the contents of any

wire or electronic communications, the (investigative agency) is not aware of any such technology.

WHEREFORE, it is respectfully requested that the Court grant an order for (enter time period, not to exceed 60) days authorizing the installation and use of (a pen register) (trap and trace device), and directing the (communications service provider) to forthwith furnish agents of the (investigative agency) with all information, facilities and technical assistance necessary to accomplish the installation of the (trap and trace device) (pen register).

I declare under penalty of perjury that he foregoing is true and correct.

EXECUTED ON ____________________, 20__

____

Applicant

Order for Trap and Trace/Pen Register

UNITED STATES DISTRICT COURT

_______DISTRICT OF_________

IN THE MATTER OF THE)
APPLICATION OF THE)
UNITED STATES OF AMERICA)
FOR AN ORDER AUTHORIZING)
THE INSTALLATION AND USE)
OF A (PEN REGISTER))
(TRAP AND TRACE DEVICE))

ORDER

This matter having come before the Court pursuant to an application under oath pursuant to Title 18, United States Code, Section 3122 by_____________, an attorney for the Government, which requests an order under Title 18, United States Code, Section 3123, authorizing the installation and use of a (pen register) on (telephone line or other facility), the Court finds that the applicant has certified that the information likely to be obtained by such installation and use is relevant to an ongoing criminal investigation into possible violations of (list violations) by (list targets, if known), and others as yet unknown.

IT APPEARING that the information likely to be obtained by a (pen register) (trap and trace device) installed on (telephone line or other facility), (listed in the name of____________ (if known)) (leased to_____________ (if known)), (and located at_____________ (if known)), is relevant to an ongoing criminal investigation of the specified offenses;

IT FURTHER APPEARING that [conform to application statement] with regard to the limitation in Section 3121(c) of Title 18 concerning pen register technology, the (investigative agency) does not have technology reasonably available to it that restricts the recording or decoding of electronic or other impulses to the dialing, routing, addressing, and signaling information utilized in the processing and transmitting of wire or electronic communications so as not to include the contents of any wire or electronic communications.

IT IS ORDERED, pursuant to Title 18, United States Code, Section 3123, that (investigative agency) is authorized to install and use, anywhere within the United States, on (telephone line_______ or other facility) (a pen register to record or decode dialing, routing, addressing, or signaling information) (and) (a trap and trace device to capture the incoming electronic or other impulses which identify the originating number or other dialing, routing, addressing, and signaling information reasonably likely to identify the source of a wire or electronic communication) for a period of (enter time period, not to exceed 60) days; and

IT IS ORDERED FURTHER, pursuant to Section 3123(b) (2) of Title 18, that upon the request of (attorney for the Government or an officer of the law enforcement agency authorized to install and use the pen register), (provider of wire or electronic communication service, landlord, custodian, or other person) shall furnish such (investigative or law enforcement officer) forthwith all information, facilities, and technical assistance necessary to accomplish the installation of the pen register unobtrusively and with a minimum of interference with the services that the person so ordered by the court accords the party with respect to whom the installation and use is to take place, (and) (if trap and trace ordered) that upon the request of (attorney for the Government or officer of the investigative agency authorized to receive the results of the trap and trace device), (provider of a wire or electronic communication service, landlord, custodian, or other person) shall install such device forthwith on the appropriate line or other facility and shall furnish (investigative or law enforcement officer) all additional information, facilities and technical assistance including installation and operation of the device (including the installation of Caller ID service on telephone line or other facility) unobtrusively and with a minimum of interference with the services that the person so ordered by the court accords the party with respect to whom the installation and use is to take place. The results of the trap and trace device shall be furnished to the (officer of a law enforcement agency, designated in the court order), at reasonable intervals during regular business hours for the duration of the order.

IT IS ORDERED FURTHER that the (investigative agency) will reasonably compensate the provider of a wire or electronic communication service, landlord, custodian, or other person who furnishes facilities or technical assistance for such reasonable expenses incurred in providing such facilities and assistance in complying with this order.

IT IS ORDERED FURTHER, pursuant to Section 3123(d) of Title 18, that this order and the application be sealed until otherwise ordered by the Court, and that the person owning or leasing the line or other facility to which the pen register or a trap and trace device is attached or applied, or who is obligated by the order to provide assistance to the applicant, not disclose the existence of the (pen register) (trap and trace device), or the existence of the investigation to the listed subscriber, or to any other person, unless or until otherwise ordered by the Court.

____

UNITED STATES MAGISTRATE (or DISTRICT) JUDGE

______________Date

Application for Order Permitting Government to Use Its Own Pen Register/Trap and Trace Equipment (Triggerfish/Digital Analyzer or Similar Device)

UNITED STATES DISTRICT COURT

______________DISTRICT OF_________________

IN THE MATTER OF THE)
APPLICATION OF THE)
UNITED STATES OF AMERICA)
FOR AN ORDER AUTHORIZING)
THE INSTALLATION AND USE)
OF A PEN REGISTER)

APPLICATION

______________, an Assistant United States Attorney, being duly sworn, hereby applies to the Court for an order authorizing the installation and use of a pen register to identify the Electronic Serial Number (ESN) and Mobile Identification Number (MIN) of a cellular telephone (being used by ___(if known) (within a (color, make, model of vehicle) (bearing state license plate number ______). In support of this application I state the following:

1. Applicant is an "attorney for the Government" as defined in Rule I (b) (1) of the Federal Rules of Criminal Procedure, and therefore, pursuant to Section 3122 of Title 18, United States Code, may apply for an order authorizing the installation of a trap and trace device and pen register.

2. Applicant certifies that the United States Drug Enforcement Administration is conducting a criminal investigation of (name targets (if known) and others as yet unknown), in connection with possible violations of Title __, United States Code, Section(s)_____; it is believed that the subjects of the investigation are using a cellular telephone within a (color; make, model of vehicle) (bearing state license plate number______) in furtherance of the subject offenses; and that the information likely to be obtained from the pen register is relevant to the ongoing criminal investigation.

3. Applicant requests that the Court issue an order authorizing the installation and use of a pen register for a period of (enter time period, not to exceed 60) days.

WHEREFORE, it is respectfully requested that the Court grant an order for (enter time period, not to exceed 60) days authorizing the installation and use of a pen register. I declare under penalty of perjury that he foregoing is true and correct.

EXECUTED ON ____________________, 20__

Applicant____________________________

Order Permitting Government to Use Its Own Pen Register/Trap and Trace Equipment (Triggerfish/Digital Analyzer or Similar Device)

UNITED STATES DISTRICT COURT

_______DISTRICT OF______

IN THE MATTER OF THE)
APPLICATION OF THE)
UNITED STATES OF AMERICA)
FOR AN ORDER AUTHORIZING)
THE INSTALLATION AND USE)
OF A PEN REGISTER)

ORDER

This matter having come before the Court by an application under oath pursuant to Title 18, United States Code, Section 3122 by_____________, an attorney for the Government, which requests an order under Title 18, United States Code, Section 3123, authorizing the installation and use of a pen register to identify the Electronic Serial Number (ESN) and Mobile Identification Number (MIN) assigned to a cellular telephone (being used by (_if known_)) (within a (color, make, model of vehicle), (bearing state license plate number_______), the Court finds that the applicant has certified to the Court that the information likely to be obtained by such installation and use is relevant to an ongoing criminal investigation into possible violations of Title, United States Code, Sections by (list targets (if known) and others as yet unknown).

IT IS ORDERED, pursuant to Title 18, United States Code, Section 3123, that the (investigative agency) is authorized to install and use, anywhere within the United States, a pen register to identify the ESN and MIN of a cellular telephone (being used by (if known)) (within a (color, make, model of vehicle), (bearing state license plate number)), for a period of (enter time period, not to exceed 60) days; and

IT IS ORDERED FURTHER, pursuant to Section 3123(d) of Title18, that this order and the application be sealed until otherwise ordered by the Court.

____

UNITED STATES MAGISTRATE (or DISTRICT) JUDGE

______________ Date

Combined 3123/2703 Application

[NAME]
United States Attorney
[NAME]
Special Assistant United States Attorney
Chief, Criminal Division
[YOUR NAME]
Assistant United States Attorney
[________] Section
State Bar No. [__________]
[ADDRESS]
[CITY STATE ZIP]
Telephone: (XXX)-[______]
Facsimile: (XXX)-[______]
Attorneys for Applicant
United States of America

UNITED STATES DISTRICT COURT

FOR THE [XXXX] DISTRICT OF [STATE]

IN THE MATTER OF THE	)	No.________________
APPLICATION OF THE UNITED	)	**[NOTE: IF CONTINUATION OF**
STATES OF AMERICA FOR AN	)	**EXISTING PEN REGISTER ORDER,**
ORDER: (1) AUTHORIZING THE	)	**INSERT THE ORIGINAL MISC. NO.**
INSTALLATION AND USE	)	**ABOVE, FOLLOWED BY (A), (B)**
OF A PEN REGISTER AND A	)	**ETC. FOR EACH SUCCESSIVE**
TRAP AND TRACE DEVICE AND	)	**CONTINUATION; ALSO INDICATE**
(2) AUTHORIZING RELEASE OF	)	**"FIRST EXTENSION," "SECOND**
SUBSCRIBER INFORMATION, AND	)	**EXTENSION," ETC. UNDER**
CELL SITE INFORMATION	)	**"APPLICATION"; IF AMENDED OR**

SUPPLEMENTAL APPLICATION,
STATE SAME]
APPLICATION
(UNDER SEAL)

I. INTRODUCTION

[Your name], an Assistant United States Attorney for the Central District of California, hereby applies to the court for an order: [Note: for continuation of existing order, replace "installation and use" with "continued use" throughout this application and order]

1. Pursuant to 18 U.S.C. §§ 3122 and 3123, authorizing the [installation and] [continued] use of a pen register and trap and trace device[35] on the following telephone number(s): *(When it is available, have agents show you the fax from telephone company containing subject telephone and subscriber information and make sure information matches; also, try to have agents confirm that telephone information is current within 48 hours because pens are so expensive.)*

    (a) Area code and telephone number; avoid using "UFMI," which relates to Nextel's "direct connect" walkie-talkie feature, as your subject telephone number unless you cannot get telephone number for reasons stated in footnote below;

    If must use UFMI, insert footnote as follows[36], a [type of telephone, e.g., "cellular"; if prepaid, state "prepaid cellular" issued by [Name of carrier, e.g., Verizon Wireless], with Electronic Serial Number ("ESN")[37][insert ESN] [If T-Mobile or Nextel telephones: instead of ESN, insert International Mobile Subscriber Identity ("IMSI"), and/or International Mobile station Equipment Identity (IIMEI")[38]]; [if Cingular Wireless, insert ESN and/or Subscriber Identity Module ("SIM")[39]]subscribed to by [subscriber's name and address]; [note: if subscriber name and address is unknown because subject telephone is prepaid, then insert footnote as follows[40] [if subscriber name and address is unknown because it is a fugitive investigation and there was no time to get subscriber information, then insert footnote as follows [41] and believed to be used by (target's name) (hereinafter the "Subject Telephone Number") (*For other carriers, check with your agent to determine whether it is MIN/ESN, IMSI/IMEI or SIM.)*

---

[35] A "pen register" is a "device or process which records or decodes dialing, routing, addressing, or signaling information transmitted by an instrument or facility from which a wire or electronic communication is transmitted, provided, however, that such information shall not include the contents of any communication." 18 U.S.C. § 3127(3). A "trap and trace device" is "a device or process which captures the incoming electronic or other impulses which identify the originating number" or other identifiers "reasonably likely to identify the source of a wire or electronic communication, provided, however, that such information not include the contents of any communication. 18 U.S.C. § 3127(4).

[36] UFMI is an acronym for "Urban Fleet Mobile Identifier." The UFMI is the unique telephone number associated with Nextel's "Direct Connect"/"Direct Dispatch" walkie-talkie feature. Nextel cellular telephones with the walkie-talkie feature thus have two identifiable telephone numbers: the mobile identification number (MIN, frequently referred to as the public telephone number) and the UFMI. Like a pen register or trap and trace on the public telephone number, a pen register or trap and trace on the UFMI will not disclose content of the call. The (AGENCY/IES) obtained the UFMI from a (confidential source) (criminal associate). Due to the immediate need to locate the fugitive target before he/she stops using the **Subject Telephone Number(s),** there is insufficient time to obtain the corresponding MIN (public telephone number) from the subject telephone company, which could take up to several weeks, without jeopardizing the fugitive investigation.)

[37] ESN is an acronym for "Electronic Serial Number." The ESN uniquely identifies cellular telephone instruments.

[38] IMSI is an acronym for "International Mobile Subscriber Identity." Every mobile phone that uses GSM format has a SIM (Subscriber Identity Module) card that is installed or inserted into the mobile phone handset. The SIM card contains the IMSI, which is a non-dialable number programmed on a microchip on the SIM card. It is the IMSI that is used to uniquely identify a subscriber to the GSM mobile phone network. The IMSI number is unique to that SIM card and is never re-assigned. Thus, if the target exchanges his cell phone for an updated model and/or changes his phone number, but retains his SIM card, the IMSI will remain the same. The IMEI (International Mobile station Equipment Identity) is similar to a serial number and uniquely identifies the telephone handset itself.

[39] SIM is an acronym for "Subscriber Identity Module." The SIM is a card, sometimes called a "smart" card, which can be installed or inserted into certain cellular telephones containing all subscriber-related data. This facilitates a telephone call from any valid cellular telephone since the subscriber data is used to complete the call rather than the telephone's internal serial number.

[40] (Subscriber information for the Subject Telephone Number/s is not known because telephone companies do not require the subscriber to provide identification when purchasing a prepaid cellular telephone because the fees are paid in advance.)

[41] (Agency/ies) obtained the Subject Telephone Number/s from a (confidential source) (criminal associate). Due to the immediate need to locate the fugitive target before he/she stops using the Subject Telephone Number/s, there is insufficient time to obtain subscriber records from the telephone company, which could take up to several weeks, without jeopardizing the fugitive investigation.

(b) Repeat above for each additional subject phone. If requesting pen on multiple phones, or you plan to request pens on future phones in the same case, then number phones as follows: "Subject Telephone Number One," "Subject Telephone Number Two," etc.

(If requesting pen on more than one subject telephone, be sure to use plural "subject telephone number/s." throughout application and order. Just search for brackets and revise as appropriate.)

2. Pursuant to 18 U.S.C. §§ 2703(C) and 2703(d), directing the electronic service providers to disclose or provide upon oral or written request by Special Agents of the (Agency/ies):

a. Records or other information identifying subscribers or customers (but not including the contents of communications or toll records), namely, subscriber name, address, date of birth, social security number, driver's license (state and number), contact names and numbers, employment information, method of payment, length of service, and type of service utilized, for all published, non-published, listed, or unlisted numbers, dialed or otherwise transmitted to and from the **Subject Telephone Number(s)**;

b. All changes (including additions, deletions, and transfers) in service regarding the **Subject Telephone Number(s)** to include telephone numbers and subscriber information (published, non-published, listed, or unlisted) associated with these service changes; and

c. For the **Subject Telephone Number(s)** ,records or other information pertaining to subscriber(s) or customer(s), including historical cell site information[42] and call detail records[43] (including direct connect records[44]) for the following dates: to the present (the last 10 days is recommended) (but not including the contents of communications) .

d. For the **Subject Telephone Number(s)**, all cell site information provided to the government on a continuous basis contemporaneous with call origination (for outbound calling) and call termination (for incoming calls), and at such other time upon the oral or written request of the government, including if reasonably available, during the progress of a call. Specific disclosure of cell site information will assist law enforcement in identifying the approximate physical location of the Subject Telephone and will not disclose content of the calls.

---

[42] A cell site is located in a geographic area within which wireless service is supported through radio signaling to and from antenna tower(s) operated by a service provider. Cell sites are located throughout the United States. Cellular telephones that are powered on will automatically register or re-register with a cellular tower as the phone travels within the provider's service area. The registration process is the technical means by which the network identifies the subscriber, validates the account and determines where to route call traffic. This exchange occurs on a dedicated control channel that is clearly separate from that used for call content (i.e. audio)—which occurs on a separate dedicated channel. As used herein, "Cell site information" refers categorically to any and all data associated with registration of the Subject Telephone with cell sites/network, as well as other data used by the network to establish a connection with the telephone handset and to maintain connectivity to the network. This includes the physical location and/or address of the cellular tower, cell site sector, control channel number, neighbor cell lists, and any identification numbers, processing data, and parameters not pertaining to the contents of a call.

[43] "Call detail records" are similar to toll records (i.e. historical telephone records of telephone activity, usually listing outgoing calls and date, time, and duration of each call), which are made and retained in the ordinary course of business. However, "call detail records" is the term used when referring to toll records of cellular telephones rather than hard-line telephones. Unlike toll records, however, call detail records also include a record of incoming calls and the cell site/sector(s) used by the cellular telephone to obtain service for a call or when in an idle state.

[44] Ask tech agent: define direct connect. Or better yet, is there a generic term, such as a walkie-talkie feature or two-way radio feature?

II. CERTIFICATION FOR A PEN REGISTER AND A TRAP AND TRACE DEVICE PURSUANT TO 18 U.S.C. §§ 3122 AND 3123

In support of this application, I state the following:

1. I am an "attorney for the Government" as defined in Rule l(b) (1) of the Federal Rules of Criminal Procedure, and therefore, pursuant to 18 U.S.C. § 3122, may apply for an order authorizing the installation and use of pen registers and trap and trace devices.

2. I certify that the information likely to be obtained from the pen register and trap and trace devices on the **Subject Telephone Number(s)** is relevant to an ongoing criminal [fugitive] investigation being conducted by the (agency/ies) in connection with possible violations of federal criminal statutes, including (cite violation/s and statute/s, i.e. narcotics distribution in violation of 21 U.S.C. § 841(a) (1)) by (list main target/s or state "unknown individuals").

3. Therefore, based upon the above Certification[45] and pursuant to 18 U.S.C. §§ 3122 and 3123, I request that the court issue an order authorizing:

    a. The [name agency], to install, or cause the provider to install, and use [continued use] a pen register device(s)anywhere in the United States to record or decode dialing, routing, addressing, or signaling information (including "post cut-through dialed digits"[46]) transmitted[47] (*Since Nextel charges extra $1,500 for pen/trap on direct connect (walkie-talkie) communications, include past footnote requesting pen/trap on "direct connect" only if agents decide that investigation warrants request for such data*) from the **Subject Telephone Number(s)**, to record the date and time of such dialing's or transmissions l and to record the length of time the telephone receiver in question is "off the hook" for incoming or outgoing calls, for a period of 60 days from the date the order is filed by the court.

    b. The (name agency) to install, or cause the provider to install, and use [continued use] trap and trace device/s on the **Subject Telephone Number(s)** anywhere in the United States

---

[45] Section 3122 "was not intended to require independent judicial review of relevance; rather, the reviewing court need only verify the completeness of the certification." In re United States, 10F.3d 931, 935 (2d Cir. 1993) (citing S. Rep. No. 541, 99th Cong.,2d Sess. 47 (1986), reprinted in 1986 U.S.C.C.A.N. 3555, 3601); see also United States v. Fregoso, 60 F.3d 1314, 1320 (8th Cir. 1995) (holding that the judicial role under Section 3123(a) is ministerial in nature because a proper application under Section3122 mandates entry of the order); Brown v. Waddell, 50 F.3d 285,290 (4th Cir. 1995) (Section 3122 does not require the government to establish probable cause to obtain a pen register or trap and trace device); United States v. Newman, 733 F.2d 1395, 1398 (10thCir. 1984) ("No showing of probable cause -- or even 'sufficient cause,' as defendant suggests is necessary to justify authorization of a pen register.")

[46] "Post-cut-through dialed digits," also called "dialed digit extraction features," are any digits that are dialed from the Subject Telephone Number/s after the initial call setup is completed. For example, some post-cut-through dialed digits are telephone numbers, such as when a subject places a calling card, credit card or collect call by first dialing a long-distance carrier access number and then, after the initial call is "cut through," dialing the telephone number of the destination party. That final number sequence is necessary to route the call to the intended party and, therefore, identifies the place or party to which the call is being made. Under these circumstances, the "post-cut -through" digits are the type of information (i.e., "dialing, routing, addressing, or signaling" information) specifically authorized by the statute for capture. Post cut- through dialed digits also can represent call content, such as when subjects call automated banking services and enter account numbers, or call voicemail systems and enter passwords, or call pagers and dial call-back telephone numbers (which are considered numeric messages.) To the extent that additional digits that are content are received, the government will not use such information for any investigative purposes.

[47] Including dialing / routing / addressing / or signaling information transmitted over the communication service providers, network by a two-way radio feature (including, but not limited to, Nextel's "Direct Connect/Direct Dispatch, "Verizon Wireless," "Push to Talk," or Sprint's "Ready Link"*)*. The two-way radio feature, like a walkie-talkie, provides communication between similarly equipped cellular phones by pressing a button on the telephone. Like a pen register or trap and trace on a telephone, a pen register or trap and trace for information transmitted by the two-way radio feature will not disclose content of the call.

to capture and record the incoming electronic or other impulses which identify the originating numbers or other dialing, routing, addressing, or signaling information reasonably likely to identify the source of a wire or electronic communication and to record the date, time, and duration of calls created by such incoming impulses, for a period of sixty (60) days from the date the order is filed by the court.

c. That, pursuant to 18 U.S.C. § 3123 (b) (1)(C), the requested [installation and use] [continued use] of a pen register and trap and trace device permit the use of such a pen register and trap and trace device not only on the **Subject Telephone Number(s)**, but also on any changed telephone number/s subsequently assigned to an instrument bearing the same (*insert as appropriate* ESN/IMSI/SIM) as the **Subject Telephone Number(s)**, or any changed (*insert as appropriate* ESN/IMSI/SIM) subsequently assigned to the same telephone number as the **Subject Telephone Number(s)**, or any additional changed telephone number(s) and/or (*insert as appropriate* ESN/IMSI/SIM), whether the changes occur consecutively or simultaneously, listed to the same subscriber and wireless telephone account number as the **Subject Telephone Number(s)**;[48] (*insert as appropriate—Confirm with Tech Agent whether "target filtering" is possible*) and on any cellular phone that is within close proximity to the government device that may autonomously register with the device,[49] within the sixty (60) day period authorized by this order.

4. Pursuant to 18 U.S.C. § 3123 (a) (1) and § 3123 (b) (2), I further request that the court direct that upon service of the order upon it, the local, long distance, and wireless carriers listed in the proposed order, any other communications service provider providing service to the **Subject Telephone Number(s)**,[50] and any other person or entity providing wire communication service in the United States whose assistance may facilitate execution of the order, furnish forthwith all information, facilities, and technical assistance necessary to accomplish unobtrusively the

---

[48] Section 3123 (b) (1) (C) has been amended to require the Court to specify in the order "the attributes of the communications to which the order applies, including the number or other identifier." 18 U.S.C. § 3123 (b) (1) (C). The account number, when combined with the same subscriber name for the Subject Telephone Number/s sufficiently specifies "the attributes of the communications to which the order applies, including the number or other identifier..." as required by § 3123 (b) (1) (C). Cf. *United States v. Duran*, 189 F.3d 1071, 1083-1086 (9th Cir. 1999) (holding interception of wire communications on a cellular telephone with a changed telephone number followed by a changed ESN was covered by the order authorizing the interception of wire communications even though the court order authorizing the wiretap only anticipated a changed telephone number but did not anticipate a changed ESN).

[49] This is necessary in order to identify the Subject Telephone to the exclusion of others also operating within a particular cell site. We respectfully do not concede that a device used to receive radio signals, emitted from a wireless cellular telephone, that merely identify that telephone to the network (i.e., registration data) constitutes a "pen register" or "trap and trace" device. Cf. In the Matter of the Application of the U.S. for an Order Authorizing the Use of a Cellular Telephone Digital Analyzer, 885 F. Supp. 197, 201 (C.D. Cal. 1995) (interpreting prior definition of pen register device and holding that no court order is required to use a digital analyzer to capture cell phone ESN, telephone numbers, and dialed numbers, because the device does not "attach" to a telephone line) We nonetheless submit this request for authorization out of an abundance of caution, on the chance that the device may collect dialed numbers generated by a phone initiating an outgoing call attempt while it is temporarily registered with the device. To the extent such information is incidentally acquired, it is agency policy not to record or retain it or any data associated with non-target telephones. Moreover, the government uses a number of criteria to limit both the collection of data and to minimize any potential temporary disruption of service, most notably by operating the device for limited duration and only when the cell site information acquired from the provider indicates that the Subject Telephone is operating nearby.

[50] The reference to "another communication service provider" is necessary so that the court order is still effective in the event that the Subject Telephone Number(s) is/are transferred to another carrier pursuant to "Local Number Portability" ("LNP"). LNP allows a telephone user to change his/her telephone company but still keep the same telephone number. However, to transfer (i.e. "port") a telephone number pursuant to LNP, the subscriber information must remain the same. Thus, this reference applies if the Subject Telephone Number(s) is/are transferred (i.e. "ported") to another telephone carrier, but the telephone number and subscriber information remain the same.

installation and use of the pen register and trap and trace devices and with minimum interference with the services that are accorded the persons with respect to whom the installation and use is to take place, with compensation to be paid by the investigative agency for reasonable expenses directly incurred in providing such facilities and assistance.

5. I further request that the order direct the local, long distance, and wireless carriers, and any other person or entity providing wire or electronic communication service in the United States whose assistance is used to facilitate execution of the order, to furnish the results of the pen register and trap and trace devices to Special Agents of the (agency/ies) as soon as practicable, on a continuing basis, twenty-four (24) hours a day for the duration of the order.

III. SPECIFIC AND ARTICULABLE FACTS ESTABLISHING REASONABLE GROUNDS TO BELIEVE THAT SUBSCRIBER RECORDS AND CELL SITEINFORMATION ARE RELEVANT AND MATERIAL TO AN ONGOING CRIMINALINVESTIGATION PURSUANT TO 18 U.S.C. § 2703

1. Title 18, United States Code, Section 2703(d) provides that a court may issue an order authorizing disclosure of a record or other information pertaining to a telephone subscriber or customer (not including the contents of communications) when a government agency provides the court with:

> Specific and articulable facts showing that there are reasonable grounds to believe that the contents of a wire or electronic communication, or the records or other information sought are relevant and material to an ongoing criminal investigation.

The statute, by its own language, precludes holding the government to a higher standard of proof, such as probable cause. See Communications Assistance for Law Enforcement Act, Pub. L. No.103-414 § 207(2), reprinted in 1992 U.S. Code Congo & Admin. News4292. The House Report reflected that "this section imposes an intermediate standard to protect online transactional records. It is a standard higher than a subpoena, but not a probable cause warrant." See H.R. Rep. No. 103-827, at 31–32 (1994), reprinted in 1994 U.S. C.A.A.N. 3489, 3511-12.[51]

2. For the purposes of obtaining a court order for disclosure as described in 18 U.S.C. § 2703(c)(1), and in order to satisfy the requirements of 18 U.S.C. § 2703(d), government counsel, based on discussions with SA [agent's name], hereby sets forth the following specific and articulable facts showing that there are reasonable grounds to believe that the records or other information identifying subscribers (but not including the contents of communications) for telephone numbers identified through the pen register and trap and trace device on the **Subject Telephone Number(s)**, cell site information regarding the **Subject Telephone Number(s)**, subscriber information associated with any service changes regarding the **Subject Telephone Number(s)**, and the records or other information pertaining to subscribers (but not including the contents of communications) for the **Subject Telephone Number**(s) will be relevant and material to an ongoing criminal (fugitive) investigation:

    a. Insert summary of facts relating to investigation and relevance of **Subject Telephone Number(s)** to investigation. Please be aware that this section is separate from the certification under section 3122 because it is made pursuant to section 2703(d), which requires a presentation of proof, not merely a certification. In order to obtain a section

---

[51] Persons calling to and from the Subject Telephone Number/s do not have a Fourth Amendment privacy interest regarding their subscriber information. *United States v. Fregoso*, 60 F.3d 1314, 1321 (8th Cir. 1995) (rejecting defendant's challenge to court order permitting phone company to supply subscriber information "for the telephone numbers obtained from the pen register and the caller identification service," holding, "We agree with the magistrate judge's assessment that because this information is listed in phone books and city directories, and at a bare minimum revealed to the phone company to obtain telephone service, there can be no expectation that this information will remain private."). See *Smith v. Maryland*, 442 U.S. 735, 742-44 (1979), ("a person has no legitimate expectation of privacy in information he voluntarily turns over to third parties.")

2703 order, we must present "specific and articulable facts establishing reasonable grounds to believe that subscriber information and cell site information are relevant and material to an ongoing criminal investigation." (*This is a lower standard than probable cause.)* As a result, you need to make sure you set forth specific facts: your investigation, why agent thinks target(s) is/are using the subject telephone(s), and why getting subscriber and cell site information is relevant to your investigation. You can also include any relevant expert opinions of your agents. Try to limit this section to four to five paragraphs, although more may be necessary depending on the case. If quoting wiretapped calls over the subject telephone use no more than two calls per telephone and include agent's interpretation of any coded language. If wiretapped call is lengthy, summarize. *Important: the magistrate judges now require that if this is an extension of a pen/trap on subject telephone(s), you must include a paragraph containing the date, misc. No., and signing judge of any prior pen register orders, on each subject telephone(s) in your case, and a summary of the results of the prior pens during the most recent 60-day period. If summary of past pen(s) does not indicate criminal activity, the judge may not grant requested extension.*

b. Insert explanation as to why records or other information identifying subscribers for telephone numbers obtained through the pen register and trap and trace devices on the **Subject Telephone Number(s)** are relevant and material to your investigation. The following is a sample for narcotics cases, which you may adapt to your case:

Based upon SA (agent's name's) experience, information identifying the subscribers for numbers obtained from numbers captured by the pen register and the trap and trace devices, and subscriber information associated with any service changes, has yielded information that is relevant and material to narcotics trafficking investigations. Such information includes leads relating to: (1) the names of suspected suppliers, customers and other individuals who assist in the distribution of narcotics; (2) the location of "stash" houses where narcotics are stored; (3) the identity of transportation sources used by the drug traffickers; (4) the locations of money transfer businesses used by members of the operation to launder proceeds of drug trafficking activities or through which money is exchanged with co-conspirators; (5) the geographic breadth of the suspected drug trafficking cell; and (6) the identities of potential organizers, leaders, managers, or supervisors of the suspected trafficking cell by examining the calling patterns revealed by the toll data. SA (agent's name) has advised me that, based upon (his) (her) training and experience, one way to identify co-conspirators is to evaluate the pattern of calls and to obtain information identifying subscribers for calls made to and from the **Subject Telephone Number(s)** which could be potential co-conspirators and then to conduct an investigation concerning those individuals. Based upon the subscriber information, SA (agent's name) would also direct other investigators to conduct surveillance at the addresses and determine if criminal activity was occurring there, which in turn could yield potential names of co-conspirators and potential narcotics storage locations the organization used.

Obtaining the subscriber name, address, date of birth, social security number, driver's license information, contact names and numbers, employment information, and method of payment is critical to accurately identifying such subscribers because: (1) if the subscriber name is a common one and/or the subscriber address is not current, it can be difficult to accurately identify the subscriber without a date of birth, driver's license or social security number, especially in an area with a population as the Central District of California; (2) if the subscriber name and address is fictitious, which is the case when criminals purchase telephones, all or part of the remaining identification information may be truthful and help identify the subscriber or lead to identifying other co-conspirators; (3) by accurately identifying subscribers using the above-requested information, agents can eliminate innocent individuals as targets.

[*If fugitive investigation, insert the following*: in (agent's name) experience, information identifying subscribers for numbers obtained from numbers captured by the pen register and the trap and trace devices, and subscriber information associated with any service changes, has yielded information that is relevant and material to a fugitive investigation. Such information includes leads relating to the names of family members, associates, friends and other individuals who may assist in the apprehension of the fugitive or may aid in the harboring of the fugitive. (Agent's name) has advised me that one way to identify associates may be to obtain information identifying subscribers for calls made to and from the Subject Telephone Number and then conduct an investigation concerning those individuals. Based upon the identifying information, (agent's name) would then direct other investigators to monitor those addresses and determine if the fugitive is present or if the associates or family members may lead investigators to him.

c. Insert following explanation as to why cell site information is needed for the **Subject Telephone Number(s)**: The investigating agents have further advised me that the general geographic location of the **Subject Telephone Number(s)** derived from cell site information used by the **Subject Telephone Number(s)** can be used to corroborate the observations of surveillance agents. More specifically, surveillance agents can compare observations of the user of the **Subject Telephone Number(s)** with cell site information in order to verify the identification and proximate location of the user of the **Subject Telephone Number(s).**

Insert if requesting toll/call detail records:

d. Insert explanation as to why you need records or other information pertaining to subscribers of the subject telephone number(s). For fugitive cases: Historical records (i.e. toll information and/or call detail records) for the **Subject Telephone Number(s)** are important in fugitive investigations to establish a past pattern of activity for the fugitive (i.e. where he/she has been, who he/she has been calling) because it helps to determine where the fugitive is at now. The government is requesting historical records for a (number of days, i.e. 30 OR 60 day period) because (explain need for particular period of time).

2. Accordingly, based upon the above proffer, and pursuant to 18 U.S.C. §§ 2703(c)(1)(B) and 2703(d), because there are reasonable grounds to believe that such information is relevant and material to the ongoing investigation, I further request that the court issue an order requiring the providers listed in the proposed order, lodged concurrently herewith, and any other person or entity providing wire or electronic communications service in the United States whose assistance may facilitate execution of the order, to disclose, or provide upon oral or written request by Special Agents of the (agency/ies) the information set forth above in paragraph A2.

IV. REQUEST THAT ORDER PRECLUDE NOTICE AND THAT APPLICATION AND ORDER BE FILED UNDER SEAL

1. Based upon the information provided in this application, I believe disclosure of the requested court order may result in flight from potential prosecution or the destruction of or tampering with evidence, or may otherwise seriously jeopardize the investigation. Moreover, the exact nature of

the government "pen register" device and its configuration is classified as a law enforcement sensitive investigative technique, the disclosure of which would likely jeopardize other ongoing investigations, and/or future use of the technique. Therefore, pursuant to 18 U.S.C. §§ 2705(b) and 3123(d), I request that this application and order be sealed and that the court direct the local, long distance, and wireless carriers listed in the proposed order, any Internet service provider or other electronic communications provider providing voice-over IP telephony,[52] and any other local, long distance, or wireless carrier servicing the **Subject Telephone Number(s)** who is obligated by the order to provide assistance to the Applicant, not to disclose in any manner, directly or indirectly, by any action or inaction, to the listed subscriber(s)for the **Subject Telephone Number(s)**, the occupant of said premises, the subscribers of the incoming calls to or outgoing calls from the Subject Telephone Number(s), or to any other person, the existence of this order, in full or redacted form, of the pen register or trap and trace devices, or of this investigation, unless otherwise ordered by this court.

2. I further request that the identity of any targets of the investigation may be redacted from any copy of the order served on any service provider or other person, and that this order and application be sealed until otherwise ordered by the court.

I declare under penalty of perjury that the foregoing is true and correct to the best of my knowledge and belief, and that this declaration was executed on (date) at (City), California.

(Before filing, search for all parenthesis in application and remove them. Make sure all parenthetical phrases have been filled in or deleted and that you have removed all bold except for "subject telephone number(s)").

(Your name)
Assistant United States Attorney

(Insert section) Section

[52] Voice-over Internet Protocol telephony, also called Voice-over IP or VoIP, is essentially a type of hardware and software that allows people to use the internet as a transmission medium for telephone calls. In general, this means sending voice information in the form of digital packets of information rather than sending it through the traditional public switch telephone network. Like a pen register or trap and trace on traditional telephone service, a pen register or trap and trace for VoIP service will not disclose the contents of the call.

Combined 3123/2703 Order

(Name)
United States Attorney
(Name)
Special Assistant United States Attorney
Chief, Criminal Division
(Your name)
Assistant United States Attorney
(Section number) Section
State Bar No.____________
(ADDRESS)
(CITY, STATE ZIP)
Telephone: (XXX)-[
Facsimile: (XXX) - [
Attorneys for Applicant
United States of America
UNITED STATES DISTRICT COURT
FOR THE (XXXX) DISTRICT OF (STATE)

IN THE MATTER OF THE	)	
APPLICATION OF THE UNITED	)	No.
STATES OF AMERICA FOR AN	)	(NOTE: INSERT SAME AS APPLIC)
ORDER: (1) AUTHORIZING THE	)	
INSTALLATION AND USE OF A	)	(PROPOSED) ORDER
PEN REGISTER AND A TRAP AND)	(UNDER SEAL)	
TRACE DEVICE; AND (2)	)	
AUTHORIZING RELEASE OF	)	
SUBSCRIBER INFORMATION, AND	)	
CELL SITE INFORMATION	)	

This matter having come before the court pursuant to an application under Title 18, United States Code, Sections 2703(c) (d), 3122, and 3123, by Assistant United States Attorney (your name), an attorney for the Government as defined by Fed. R. Crim. P.[53] (b) (1), requesting an order authorizing the (installation and use) (continued use) of a pen register and trap and trace device, on the following telephone number(s):

(a) (Repeat exact same information from application regarding **Subject Telephone Number(s)**, but without footnotes)

Upon review of the application, the court hereby finds that:

[53] "Post cut-through dialed digits," also called "dialed digit extraction features," are any digits that are dialed from the Subject Telephone Number/s after the initial call set-up is completed, subject to the limitations of 18 U.S.C. § 3121(c). To the extent additional digits that are received are content, the government shall not use such information for any investigative purposes or attempt to decode such information.

Pursuant to 18 U.S.C. § 3123, Applicant has certified that the information likely to be obtained by such use is relevant to an ongoing criminal investigation being conducted by the (agency/ies) in connection with possible violations of (describe exactly as in application).

THEREFORE, IT IS HEREBY ORDERED, pursuant to 18 U.S.C.§ 3123, that Special Agents of the [AGENCY/IES] may [install, or cause to be installed, and use] [continue to use] a pen register anywhere in the United States to record or decode dialing, routing, addressing , or signaling information (including "post cut-through dialed digits[1]")[54] (*include footnote 38 only if requested in application*) transmitted from the Subject Telephone Number, to record the date and time of such dialing's or transmissions, and to record the length of time the telephone receiver in question is "off the hook" for incoming or outgoing calls, for a period of sixty days from the date this order is filed by the court.[55]

IT IS FURTHER ORDERED, pursuant to 18 U.S.C. § 3123, that Special Agents of the (agency/ies) may install, or cause to be installed, and use a trap and trace device on the **Subject Telephone Number(s)** anywhere in the United States to capture and record the incoming electronic or other impulses which identify the originating numbers or other dialing, routing, addressing, or signaling information reasonably likely to identify the source of a wire or electronic communication, and to record the date, time, and duration of calls created by such incoming impulses, for a period of sixty (60) days from the date this order is filed by the court;

Pursuant to 18 U.S.C. §§ 2703 (c) (1) (B) and 2703 (d), Applicant has set forth specific and articulable facts showing that there are reasonable grounds to believe that records or other information identifying subscribers or customers (not including the contents of communications) for telephone numbers identified through the pen register and trap and trace devices on the **Subject Telephone Number(s)**, changes in service regarding the **Subject Telephone Number(s)**, cell site information regarding the **Subject Telephone Number(s)**, and records or other information pertaining to subscribers or customers (but not including the contents of communications) for the **Subject Telephone Number(s)** will be relevant and material to an ongoing criminal investigation.

THEREFORE, IT IS FURTHER ORDERED, pursuant to 18 U.S.C. §§2703 (c) (1) (B), 2703 (c)(2) and 2703 (d), that SBC Communications, Inc. or any subsidiary thereof, Ameritech, Southern New England Telephone Company, Verizon California, Inc., XO Communications, Comcast Cable Communications Inc./AT&T Corporation, Verizon New York, Inc., MPower Communications, Verizon New Jersey Inc., BellSouth Telephone Company, Allegiance Telecom, Cox Communications and Qwest Communications (hereinafter the "local carriers"); AT&T U.S., Sprint, and MCI (hereinafter the "long distance carriers"); Cellco Partnership, dba Verizon Wireless, AT&T Wireless Services, U.S. Cellular, Metro PCS, Cingular Wireless, Nextel Partners, Cricket Communications, Sprint Spectrum L.P., T-Mobile USA, Inc., Virgin Mobile USA, Nextel Communications and Western Wireless Corp. (hereinafter "the wireless carriers"); any Internet service provider or other electronic communications provider providing voice-over IP telephony, and any other local, long distance, or wireless carrier servicing the **Subject Telephone Number(s)**, and any other person or entity providing wire communication service in the United States whose assistance may facilitate execution of the order, shall disclose or provide the following upon oral or written request by Special Agents of the(agency/ies):

---

[54] Including dialing, routing, addressing, or signaling information transmitted over the communication service provider's network by a two-way radio feature (including, but not limited to, Nextel's "Direct Connect/Direct Dispatch," Verizon Wireless' "Push to Talk", or Sprint's "ReadyLink")

[55] As used herein, "the date this order is filed by the court" is the date indicated by the clerk's file stamp on the first page of this order.

1. Records or other information identifying subscribers or customers (but not including the contents of communications or toll records), namely, subscriber name, address, date of birth, social security number, driver's license (state and number), contact names and numbers, employment information, method of payment, length of service, and type of service utilized, for all published, non-published, listed, or unlisted numbers, dialed or otherwise transmitted to and from the **Subject Telephone Number(s)**;
2. All changes (including additions, deletions, and transfers) in service regarding the **Subject Telephone Number(s)** to include telephone numbers and subscriber information (published, non-published, listed, or unlisted) associated with these service changes; and
3. For the **Subject Telephone Number(s)** ,records or other information pertaining to subscriber/s or customer/s, including historical cell site information and call detail records (including direct connect records[56]) for the following dates: to the present (the last 10 days is recommended) (but not including the contents of communications).
   d. For the **Subject Telephone Number(s)**, all cell site information[57] provided to the government on a continuous basis contemporaneous with call origination (for outbound calling) and call termination (for incoming calls), or at such other time upon the oral or written request of the government, including if reasonably available, during the progress of a call.

IT IS FURTHER ORDERED that this authorization for the (installation and use) (continued use) of a pen register and trap and trace device applies not only to the **Subject Telephone Number(s)** listed above, but also to any changed telephone number/s subsequently assigned to an instrument bearing the same (*insert as appropriate ESN/IMSI/SIM*) as the **Subject Telephone Number(s)** or any changed (*insert as appropriate ESN/IMSI/SIM*) subsequently assigned to the same telephone number as the **Subject Telephone Numbe(s)**, or any additional changed telephone number/s and/or *[insert as appropriate* ESN/IMSI/SIM], whether the changes occur simultaneously or consecutively, listed to the same subscriber and wireless telephone account as the **Subject Telephone Number(s)**, *[insert only if requested in application—Confirm with Tech Agent]* and on any cellular phone that is within close proximity to the government device that may autonomously register with the device,[58] within the 60 day period authorized by this order;

IT IS FURTHER ORDERED, pursuant to 18 U.S.C. §§ 3123(a) (1)and § 3123 (b) (2), that upon service of this order upon it, the local, long distance, and wireless carriers listed herein, any other communications service provider providing service to the **Subject Telephone Number(s)**, and any other person or entity providing wire communication service in the United States whose assistance may facilitate execution of this order, shall furnish Special Agents of the (agency/ies) forthwith all information, facilities, and technical assistance necessary to accomplish unobtrusively the installation and use of the pen register and trap and trace devices and with minimum interference with the services that are accorded the persons with respect to whom the installation and use is to take place;

IT IS FURTHER ORDERED that the local, long distance, and wireless carriers, and any other person or entity providing wire or electronic communication service in the United States whose assistance is used

---

[56] Ask tech agent: define direct connect. Or better yet, is there a generic term, such as walkie- talkie feature or two-way radio feature?

[57] "Cell site information" refers categorically to any and all data associated with registration of the Subject Telephone with cell sites/network, as well as other data use by the network to establish a connection with the telephone handset and to maintain connectivity to the network. This includes the physical location and/ or address of the cellular tower, cell site sector, control channel number, neighbor cell lists, and any identification numbers, processing data, and parameters not pertaining to the contents of a call.

[58] Once the Subject Telephone is identified and located any data incidentally collected from non-target telephones shall not be recorded or retained.

to facilitate execution of the order, furnish the results of the pen register and trap and trace devices to Special Agents of the (agency/ies) as soon as practicable, on a continuing basis, twenty-four (24) hours a day for the duration of the order.

IT IS FURTHER ORDERED that the local, long distance, and wireless carriers be compensated by the investigative agency for reasonable expenses directly incurred in providing technical assistance; and

Good cause having been shown, IT IS FURTHER ORDERED, pursuant to 18 U.S.C. §§ 2705(b) and 3123(d), that this order and the application be sealed until otherwise ordered by the court, and that the local, long distance, and wireless carriers listed herein, any Internet service provider or other electronic communications provider providing voice-over IP telephony, and any other local, long distance, or wireless carrier servicing the **Subject Telephone Number(s)** who is obligated by the order to provide assistance to the Applicant, shall not disclose in any manner, directly or indirectly, by any action or inaction, to the listed subscriber(s) for the **Subject Telephone Number(s)**, the occupant of said premises, the subscribers of the incoming calls to or outgoing calls from the **Subject Telephone Number(s)**, or to any other person, the existence of this order, in full or redacted form, of the pen register or trap and trace devices, or of this investigation, unless otherwise ordered by this court.

ITIS FURTHER ORDERED that the identity of any targets of the investigation may be redacted from any copy of the order served on any service provider or other person, and that this order and application be SEALED until otherwise ordered by the court.

(Before filing, search for all parenthesis in application to ensure they have been deleted. Check that all parenthetical phrases have been filled in or deleted and that you have removed all bold except for "subject telephone number(s).")

DATED: ___________

(Insert duty magistrate judge's name)
UNITED STATES MAGISTRATE JUDGE Presented by:

(Your name)
Assistant United States Attorney
[Insert section] Section

Appendix

Regional Information Sharing Systems® (RISS) Nodes

Analysis

RISS analysts create analytical products for investigators and prosecutors to help increase their ability to identify, detect, and apprehend suspects as well as enhance prosecutorial success in court. These products include highly complex and specialized flowcharts, link-analysis charts, crime scene diagrams, telephone toll analysis, financial analysis reports, digital forensics analysis, and audio/video enhancement services. Analysts also prepare crime surveys, activity assessment reports such as terrorism assessments, and national-scope products for distribution to member agencies. RISS analysts also provide their expertise during court preparation and serve as expert witnesses at trial, as requested.

Digital forensics analysis assists member agencies in the retrieval of data from computers seized during investigations. Digital forensics services are also available to recover deleted files, locate evidence, evaluate the structure of computerized files, and analyze other computer and cellular phone activity. Computer and digital crime and the use of these items in the planning of other crimes continue to be an area of concern for law enforcement. RISS assists the law enforcement community by providing experts in the field and the technical knowledge needed to gather critical information from these electronic devices that aid in criminal investigations.

Audio/visual services are also available and pertain to the enhancement of video footage or audio bites to assist law enforcement in identifying suspects. Analysts have the ability to enhance video footage by extracting elements to present a clear picture; this is extremely useful during robbery and burglary cases. Audio enhancements are useful for undercover operations because the analyst can remove background and frequency noise, allowing him to clearly identify the suspects.

Confidential Funds

Member agencies can use these funds to purchase information, contraband, stolen property, and other items of an evidentiary nature or to provide for other investigative expenses related to multijurisdictional investigations. The availability and use of confidential funds are strictly controlled by federal guidelines, and internal policies and procedures are developed by each center.

Equipment Loans

Many agencies cannot afford to purchase and maintain specialized investigative equipment. Without access to this equipment, law enforcement efforts would suffer and some investigations would either go unsolved or require a longer time to resolve. The pieces of highly specialized investigative and surveillance equipment loaned to RISS member agencies make a substantial difference in finding and securing information needed to build cases and to investigate and apprehend criminals. RISS Center equipment staffs consult with and train member agencies on the use of equipment and provide expertise in developing specialized equipment.

Technical Assistance

The integration of field services is unique to RISS whereby individuals initiate regular contact with law enforcement and public safety agencies to ensure that RISS is meeting their needs. RISS field staffs conduct on-site visits to provide technical assistance to member agencies as well as potential member agencies, to facilitate delivery of RISS services. This may include an explanation on how to request services and training activities and using RISSNET. In addition, field staffs conduct 28 Code of Federal Regulations (CFR) Part 23 reviews. Field staffs have law enforcement backgrounds and provide tailored training and consultation to agencies.

Training

RISS sponsors or co-sponsors training classes, meetings, and conferences at the local, county, and state levels across the country, providing law enforcement professionals with a forum to exchange information on regional crime problems and criminal activity. Thousands of criminal justice professionals receive RISS training each year and are educated on methods to combat terrorism activity, gangs and violent crimes as well as computer, financial, and economic crimes. Subject areas are also the use of specialized equipment, investigative techniques, analytical techniques, surveillance techniques, and officer safety.

RISS Centers:

MAGLOCLEN

Middle Atlantic-Great Lakes Organized Crime Law Enforcement Network®
Serving: Delaware, Indiana, Maryland, Michigan, New Jersey, New York, Ohio, Pennsylvania, and the District of Columbia as well as Australia, Canada, and England.
Location: Newtown, PA
Phone: 800-345-1322
Email: info@magloclen.riss.net

MOCIC

Mid-States Organized Crime Information Center®
Serving: Illinois, Iowa, Kansas, Minnesota, Missouri, Nebraska, North Dakota, South Dakota, and Wisconsin as well as parts of Canada.
Location: Springfield, MO
Phone: 800-846-6242
Email: info@mocic.riss.net

NESPIN

New England State Police Information Network®
Serving: Connecticut, Maine, Massachusetts, New Hampshire, Rhode Island, and Vermont as well as parts of Canada.
Location: Franklin, MA
Phone: 800-343-5682
Email: info@nespin.riss.net

RMIN
Rocky Mountain Information Network®
Serving: Arizona, Colorado, Idaho, Montana, Nevada, New Mexico, Utah, and Wyoming as well as parts of Canada.
Location: Phoenix, AZ
Phone: 800-821-0640
Email: info@rmin.riss.net

ROCIC
Regional Organized Crime Information Center®
Serving: Alabama, Arkansas, Florida, Georgia, Kentucky, Louisiana, Mississippi, North Carolina, Oklahoma, South Carolina, Tennessee, Texas, Virginia and West Virginia as well as Puerto Rico and the U.S. Virgin Islands.
Location: Nashville, TN
Phone: 800-238-7985
Email: info@rocic.riss.net
Web: http://www.rocic.com

WSIN
Western States Information Network®
Serving: Alaska, California, Hawaii, Oregon, Washington as well as Canada and Guam.
Location: Sacramento, CA
Phone: 800-952-5258
Email: info@wsin.riss.net

CDR Indexes

At the end of a narcotics or other conspiracy investigation you are going to be left with a ton of information that you did not think was useful to your case at the time. Your investigation may have run its course, but the information you obtain might very well be relevant to another law enforcement officer's case. Make it a point to share that information with the rest of us by submitting your call detail records to one of the Regional Information Sharing System nodes such as the Western States Information Network (WSIN). WSIN operates the Master Telephone Index, which stores all phone numbers related to WSIN/RISS entries and also has records related to state run wire taps and pen registers. In addition to storing records from case submissions, several large county level law enforcement agencies are submitting the records from inmate phone calls made from correctional institutions into the Master Telephone Index. Calls identified as being made from a correctional institution can allow an investigator to request the voice recording of those calls from the submitting agency. While there is no standard time frame for correctional institutions to save these recorded calls, agencies may maintain the records for years.

Cell Phone Investigation

Training Programs

I am frequently asked about various training programs and their pros and cons. Most of the time, the actual question behind the question is, where they can get free training? I get it. Even in good times most law enforcement agencies have a hard time forking over money to go to technical training. But in bad financial times, or at the end of the fiscal year, it seems like the first thing to go is the training budget.

Coupled with reduced training funds is the sticker shock most officers get when they see what some of the training companies charge for their services. As with the forensic tools section I will give you my opinion based on either firsthand experience (I actually took the class) or reliable reporting (someone I know and trust took the course and told me about it).

POLICE TECHNICAL (www.policetechnical.com)

POLICE TECHNICAL works with hundreds of agencies throughout the nation, as an instructor for this company, I teach Cell Phone Investigations ™, which covers search and seizure basics specific to mobile devices. It also reveals what information is available from cell phone companies how to obtain it, and how to avoid legal landmines.

This class is designed for law enforcement officers at any stage in their career. Experience or training in cell phone investigations is not required. Students will be receive samples of search warrant templates, comprehensive guides to major cell phone carriers, and model report language for cell phone searches. POLICE TECHNICAL also offers course on Cell Phone Data and Mapping ™ and Cell Phone Tactics ™.

National White Collar Crime Center (www.nw3c.org)

I have taken multiple NW3C classes. For the most part the classes have been effective. However, several of their recent phone classes have been taught by analysts or technicians who don't speak cop very well. I have never experienced it myself, but I have that information on reliable authority from several recent attendees. However, I don't believe this should dissuade an interested investigator from taking one or more of their classes. Particularly because they have several interesting courses including:

Cyber Investigations 105-Basic Cell Phone Investigations

From the NW3C website: "This 2 day course is for officers, investigators, and analysts that encounter cell phone evidence, and includes the acquisition and analysis of information external to the phone. Class concepts include instruction on how to request, read, and analyze Call Detail Records obtained from cellular providers and plot cellular site locations to determine the position of a suspect during a given time frame. Also, this course provides guidance to first responders on proper procedures to seize and preserve evidence on a cell phone until that phone can be fully analyzed. The student is also shown techniques for tying all this information together for case reports and/or court room presentations. No special hardware or software is required, just a new knowledgebase to apply to cell phone cases. This course focuses heavily on analysis and, as such, a strong working knowledge of Microsoft® Office is highly recommended.

Cyber Investigation 205-Cell Phone Interrogation

From the NW3C website: "This two-day course trains officers and investigators in the acquisition and analysis of information internal to cell phones. In addition, attendees will be introduced to best practices for seizing, preserving, and examining evidence on a cell phone through an abundance of hands-on exercises using many of the most popular cell phone investigation tools available on the market today. Students also learn the proper techniques for conducting a manual investigation of a mobile phone and the concerns that accompany this task."

NW3C classes are tuition free for state and local law enforcement agencies, but federal law enforcement may be charged a free. They do not provide any housing or meals or transportation reimbursement. NW3C offers their training at locations around the United States many times a year. The classes fill up fast so you need to check the training schedule frequently and register early. Don't wait for the training officer to complete the paperwork before reserving a spot. You can always cancel it later. An important note: your agency must be a NW3C member organization. You may need to apply for agency membership before registering for one of their courses.

The National Consortium for Justice Information and Statistics (aka search.org)

SEARCH is one of my favorite sources for training. They offer one of the best cell phone investigations classes in their "Core Skills for the Investigation of Cellular Devices." The class is 32 hours long and lets attendees get hands-on with every major forensic tool on the market. The training is vendor neutral so you don't have to worry about only being trained on one forensic tool while being told everything else is worthless. The class is geared toward entry level phone forensic examiners and investigators and covers cell phone systems, tracking cellular phones, how to properly seize the device, how to use forensic tools, and digital evidence recovery.

SEARCH's training is usually free depending on the availability of training funds. Occasionally, they may charge a nominal fee. The "Core Skills for the Investigation of Cellular Devices" course is offered several times a year and available seating is very limited. Their classes fill up fast. SEARCH does not offer any lodging, meals, or transportation. The classes are offered in different locations around the United States and agencies can request to host the training course at their agency.

Lately, is seems their classes have only been open to Internet Crimes Against Children (ICAC) task force members and affiliated agencies so might have to check with them before registering for a training class.

State Training Agencies

Numerous state law enforcement agencies or Attorneys General Offices offer cell phone forensics training classes. California law enforcement officers can apply to attend the Cellular Phone Forensics/Investigations training class offered by the California Criminalistics Institute. This class was formerly taught under the auspices of the Advanced Training Center but quietly moved to the Criminalistics Institute. The class is free and first rate.

From the CCI website: "This 40-hour course is designed to provide law enforcement personnel with foundational skills for forensic examination of cellular phones used as instruments of crime or when cellular phones contain records of criminal activity. These records can include phone book entries, call history, text messages, calendar entries, images, and videos. Classes will include:"

- Basic understanding of cellular phones and cellular networks
- Description of a variety of tools (both hardware and software) available for extracting data from cellular phones
- Demonstrations of how different cellular phone forensic tools work
- Laws related to seizure and examination of cellular phones
- Hands on exercises for extracting data from cellular phones, SIM cards and flash media cards using a variety of different tools
- Basic understanding of SIM cards, Techniques for isolating cellular phones from cellular networks

Lodging, per diem, and transportation are the responsibility of the student's parent agency and the class is only taught twice per year. One of the best things about the training program is attendees have historically been issued forensic software as part of the class. Past classes have received Secure View 3 with a two year license for free. Attendance is extremely limited to this class and is restricted to California law enforcement officers only.

Federal Law Enforcement Training Center (FLETC)

The Department of Homeland Security administers the Federal Law Enforcement Training Center. One of the many courses offered is the Mobile Device Investigations Program (MDIP.)

Prospective students considering the course are usually dissuaded by the hefty price tag for state and local law enforcement officers. The 40-hour course costs $6,000. This causes most law enforcement officers, particularly those in the training department and command staff, to immediately reject this course as a viable training option. However, the $6,000 tuition covers not only the cost of the class, but students leave the course with more than $13,000 worth of equipment. Recent attendees were issued a Cellebrite Universal Forensic Extraction Device (logical version), Susteen Secure View 3 with a two-year license and a standalone tablet computer, a licensed version of Lantern forensic software, and numerous other tools and equipment necessary to immediately begin processing phones. FLETC also offers a housing and meals option for those electing to stay on post. FLETC also arranges transportation from area airports so there is no need to rent a vehicle. The MDIP program is offered multiple times a year and is one of the best options for law enforcement agencies seeking a combination of comprehensive training and equipment.

The United States Secret Service National Computer Forensics Institute (NCFI)

The National Computer Forensics Institute offers several mobile phone forensics training classes for state and local law enforcement including Basic Mobile Device Investigations, Mobile Device Examiner, and Advanced Mobile Device Examiner. The training course varies in length from two to five days. Attendees may also receive forensic software and other tools as part of the course.

The biggest challenges with attending NCFI are the courses are not offered frequently, class size is limited, and state and local attendees must be sponsored by the Secret Service. If your agency participates with a Secret Service task force or has a good working relationship with them your chances of getting one of the limited training spots are better than if you are applying. For further information on attending one of the NCFI training classes, contact your local office of the Secret Ser

Index

41555191R00189

Made in the USA
Middletown, DE
19 March 2017